LASER APPLICATIONS

Volume 5

CONTRIBUTORS

JAMES H. BECHTEL

A. C. BERI

CAMERON J. DASCH

ALAN C. ECKBRETH

THOMAS F. GEORGE

ROBERT J. HALL

KAI-SHUE LAM

JUI-TENG LIN

RICHARD E. TEETS

ROBERT J. VON GUTFELD

LASER APPLICATIONS

Edited by

JOHN F. READY

Honeywell Systems and Research Center
Minneapolis, Minnesota

ROBERT K. ERF

Optics and Acoustics
United Technologies Research Center
East Hartford, Connecticut

VOLUME 5

ACADEMIC PRESS, INC. 1984

(Harcourt Brace Jovanovich, Publishers)

Orlando   San Diego   San Francisco   New York   London
Toronto   Montreal   Sydney   Tokyo   São Paulo

4/1984
Physics Cont

ACADEMIC PRESS, INC.
Orlando, Florida 32887

United Kingdom Edition published by
ACADEMIC PRESS, INC. (LONDON) LTD.
24/28 Oval Road, London NW1 7DX

LIBRARY OF CONGRESS CATALOG CARD NUMBER: 79-154380
ISBN 0-12-431905-X

PRINTED IN THE UNITED STATES OF AMERICA

84 85 86 87   9 8 7 6 5 4 3 2 1

CONTENTS

Laser Processing of Integrated Circuits and Microelectronic Materials

ROBERT J. VON GUTFELD

Laser Photochemistry

THOMAS F. GEORGE, A. C. BERI, KAI-SHUE LAM, and

JUI-TENG LIN

v

Combustion Research with Lasers

JAMES H. BECHTEL, CAMERON J. DASCH, and
RICHARD E. TEETS

Coherent Anti-Stokes Raman Spectroscopy (CARS): Application to Combustion Diagnostics

ROBERT J. HALL and ALAN C. ECKBRETH

CONTRIBUTORS

Numbers in parentheses indicate the pages on which the authors' contributions begin.

JAMES H. BECHTEL, General Motors Research Laboratories, Warren, Michigan 48090 (129)

A. C. BERI, Department of Chemistry, University of Rochester, Rochester, New York 14627 (69)

CAMERON J. DASCH, General Motors Research Laboratories, Warren, Michigan 48090 (129)

ALAN C. ECKBRETH, United Technologies Research Center, East Hartford, Connecticut 06108 (213)

THOMAS F. GEORGE, Department of Chemistry, University of Rochester, Rochester, New York 14627 (69)

ROBERT J. HALL, United Technologies Research Center, East Hartford, Connecticut 06108 (213)

KAI-SHUE LAM, Department of Chemistry, University of Rochester, Rochester, New York 14627 (69)

JUI-TENG LIN,* Department of Chemistry, University of Rochester, Rochester, New York 14627 (69)

RICHARD E. TEETS, General Motors Research Laboratories, Warren, Michigan 48090 (129)

ROBERT J. VON GUTFELD, IBM Thomas J. Watson Research Center, Yorktown Heights, New York 10598 (1)

* Present address: Laser Physics Branch, Naval Research Laboratory, Code 6450, Washington, D.C. 20375.

PREFACE

In the 23 years since the invention of the laser, many laser applications have reached the status of practical utilization in science, engineering, industry, and education. The purpose of this series of books on Laser Applications is to present articles which describe applications which have been well developed and refined. The articles are prepared by knowledgeable individuals who have helped develop the applications and who have great expertise in solving the problems encountered in the applications. The articles are intended to be helpful to engineers and scientists who desire to use lasers for practical solution of their own problems. The experience of research workers and development engineers who have pioneered these applications should benefit the reader in applying laser technology.

The articles in this volume describe applications that are relatively new, but that are rapidly developing in the early 1980s. They demonstrate the versatility of laser technology as applied to a wide variety of technical areas.

In the first article R. J. von Gutfeld describes processing of integrated circuits and microelectronic materials. Lasers have been well established for many material processing applications, such as welding, hole drilling, and resistor trimming. However, lasers have not been widely used for direct fabrication of semiconductor chips on the wafer level. Many new laser processing operations have become possible in recent years. These developments allow one to fabricate integrated circuit elements on an individual basis, and ultimately to fabricate microelectronic circuitry directly without need for mask sets, chemical processing, or photolithography. Such developments can lead to entirely new procedures in microelectronic circuit fabrication.

The second article by Thomas F. George, A. C. Beri, Kai-Shue Lam, and Jui-teng Lin describes the use of lasers for chemical processing. It includes a description of laser spectroscopy as an extremely high-resolution process for selective monitoring of photochemistry. It also describes laser-driven molecular interactions and reaction dynamics. Laser light can be used to selectively excite molecules and change the course of a chemical reaction. The topic of laser photochemistry overlaps the first topic on microelectronic fabrication because some aspects of laser chemistry, including laser chemical vapor deposition, are employed in microelectronic fabrication.

The third article by J. H. Bechtel, C. J. Dash, and R. E. Teets describes research on combustion processes using lasers. In one sense this may be regarded as a photochemical application also, a specialized application in which laser technology is used to investigate the dynamics of combustion processes. Lasers are used to probe the properties of flames and to determine many aspects of flame chemistry, heat transfer, molecular diffusion, and fluid flow.

The fourth article complements the third and becomes more specific in detail in describing the application of Coherent Anti-Stokes Raman Spectroscopy (CARS) to combustion diagnostics. This article by R. J. Hall and A. C. Eckbreth emphasizes this particular technique for nonintrusive diagnostics of flames and combustion. It is of special interest because of its potential for very high resolution in molecular studies. There is current interest in CARS for measurements in internal combustion engines, furnaces, gas turbines, and propellants for propulsion and ballistics.

The Editors hope that this collection of articles will be of benefit to the reader, and will stimulate the development of new practical applications of laser technology.

CONTENTS OF PREVIOUS VOLUMES

LASER PROCESSING OF INTEGRATED CIRCUITS AND MICROELECTRONIC MATERIALS

Robert J. von Gutfeld

IBM Thomas J. Watson Research Center
Yorktown Heights, New York

I. Introduction

The use of lasers for applications such as scribing, resistor trimming, hole drilling, and local welding has been well established for over a decade. Several reviews describe these applications in considerable detail, to which the reader can refer (Cohen, 1967; Cohen *et al.*, 1968; Gagliano

1

et al., 1969). Within the last decade lasers have found new application in the field of microelectronics, the laser beam interacting directly with circuit boards and semiconductor chips containing memory and logic circuits. These laser fabrication processing applications have become possible in part through research that has resulted in lasers with increasing stability, high repetition rates, wavelengths well into the ultraviolet, and short pulse durations. These features make it possible to heat discrete, micrometer-sized regions reliably and repeatably to very high temperatures above ambient. Applied to local regions of a silicon chip, laser heating can occur without producing damage to neighboring material or adjacent circuitry. This has made it possible to laser "wire" or "personalize" high-density chips for engineering design and circuit repair. Additional laser–microelectronic applications that have received considerable attention include mask and circuit-pattern repair using photoresist and lift-off techniques, maskless metal depositions onto circuit boards and chips, and maskless laser-enhanced etching of microelectronic parts using gaseous atmospheres or liquid etchants. Lasers have also been used to produce metal vapor deposition and simultaneous diffusion of the deposition into the semiconductor substrates to provide high-conductivity ohmic regions for thin-film transistors, and p–n junctions for solar cells and light-emitting diodes. A discussion of experiments relating to these new applications is the subject of this article. Laser annealing has received extensive coverage at a number of recent topical conferences as well as in the technical literature and is not included in the present discussion (White and Peercy, 1980; Gibbons *et al.,* 1980); also not included is the closely related subject of laser annealing of polysilicon patterns for present-day circuit applications.

II. Laser Personalization

A. MOTIVATION AND BACKGROUND

A focused pulsed laser can be viewed as a precision heat source which has the capability of readily accessing micrometer-sized regions on a planar or curved surface. This makes its use ideal for discrete wiring functions by providing local heat to produce appropriate disconnections and connections on integrated circuit chips. Discretionary wiring makes it possible to derive a wide variety of outputs from a logic or memory array using chips designed from a single set of masks and processing steps. Design turnaround time is thereby greatly reduced. Similarly, circuit defects can be corrected by including spare or redundant parts on the chip.

These parts can be laser "wired" through an appropriate set of connections or disconnections which also serve to isolate the defective part. Such repair capability has the overall effect of greatly increasing the yield or total number of usable chips from a given production run which can lead to a substantial cost saving in large-scale manufacture.

Successful laser wiring of integrated circuits requires a very intense, localized heat source in order to achieve micrometer-sized disconnections or connections according to a predetermined circuit design. Disconnections are in general simpler to achieve because they normally involve only the ablation or melting of a small portion of a conducting line on the chip to obtain electrical isolation achieved by local optical absorption. The disconnection becomes slightly more complicated when a protective passivating layer is used, typically SiO_2, SiO, or polyimide, with thicknesses ranging from several hundred angstroms to several micrometers. The laser wavelength is then chosen to have minimal optical absorption in the passivation layer but strong absorption in the material line to be disconnected. Usually the choice of wavelength is relatively easy to achieve with SiO or SiO_2 passivation because both are transparent to visible light, whereas the metal or heavily doped polysilicon lines are highly absorbing.

Connections between a metal line and a semiconducting layer or between two metal lines require penetration and removal of an intervening insulating layer (typically SiO_2 on the order of 0.1–1.0 μm). This makes the connection process somewhat more sophisticated than the disconnection process and places fairly stringent requirements on suitable laser parameters.

B. EARLY WORK IN LASER DISCONNECTS

An early report on the use of lasers for micrometer-sized disconnections or circuit-link deletions was published by Sypherd and Salman (1968). Their circuits consisted of epitaxially grown single-crystal silicon on sapphire wafers, the silicon appropriately doped in discrete patterns to form an array of diodes. Circuit interconnects were formed via evaporated aluminum lines. The entire circuit array represented a read-only memory (ROM) with connected diodes representing a logical "1" and a disconnected or deleted diode representing a "0." From this single-circuit package a custom coding scheme was used to achieve the desired memory configuration. For this application an automated laser system was developed to delete portions of preselected aluminum lines by means of a focused pulsed Nd: YAG laser directed either onto the aluminum line or interfacially, i.e., through the transparent sapphire substrate. It was rec-

ognized by these investigators that the proper choice of laser parameters
is critical in order to (1) avoid damage to the silicon and prevent undesirable leakage currents, (2) avoid cracking of the sapphire substrate, and (3)
achieve total removal of the aluminum link for complete isolation of the
diode. Laser parameters for deletion of a typical 12-μm-wide aluminum
line target were 3–6 mJ/pulse with a 0.1- to 1-msec pulsewidth, focused to
an incident power density of up to ~2 × 10^7 W/cm^2. This produced
deletion of a linkage area of ~12 × 25 μm. A single Q-switched 50-nsec
Nd : YAG pulse was also found to be effective for line deletion with only
0.25 mJ of energy. Automated encoding and testing of the memory were
achieved, the former using 30 pulses per second (pps) from the laser
directed to the appropriate positions of the circuit by means of an xy
table. Circuit testing was undertaken both before and after laser encoding,
the latter to assure successful material removal. The investigators estimated that 20 circuit positions per second could be encoded and tested
such that less than 1 min was required for thè 1024-bit memory.

C. LASER CONNECTION AND DISCONNECTION PROCESS: CIRCUITS ON SILICON

1. Background

The use of lasers for circuit connections on silicon wafers has led to
interesting progress in the area of discretionary wiring of very-large-scale
integrated (VLSI) chips. The question facing investigators in the early
1970s was how good an ohmic contact could be formed between two
elements separated by an intervening insulating layer. A partial answer to
this question was provided by a previously issued U.S. Patent (first issued
in 1968 and reissued in 1973) for a thin-film variable resistor (Hanfmann,
1973). This patent teaches the use of a focused laser for incrementally
changing the resistance of a double-layered thin-film resistor, the two
metal layers separated by a thin-film insulator. To increase the resistance,
material removal of the upper layer is accomplished by local laser ablation. To decrease the resistance, local laser heating interconnects the top
thin-film layer to the underlying metal thin film through the 300- to 500-Å-
thick insulating film. Laser pulsewidths were 1.5 msec for the intercon-
nect process and 0.5 msec for top metal removal without interconnection
formation. Power densities on the order of 10^6 W/cm^2 were used, resulting
in connection diameters on the order of 6 mils.

2. Initial Studies on MOS Structures

The increasing circuit density and decreasing linewidths of LSI and
VLSI metal–oxide–silicon (MOS) circuits have placed stringent limita-

tions on both the laser pulsewidth and power density suitable for circuit wiring. The appropriate pulsewidth can be estimated from the thermal diffusion length, λ, which is a measure of the thermal heat spread as a function of time. For a given pulsewidth τ_0 and thermal diffusivity κ

$$\lambda = 2\sqrt{\kappa\tau_0} \tag{1}$$

The absorption in aluminum of a 5-nsec laser pulse, focused to approximate a point heat source, will result in a radial thermal spread of 1.5 μm in the time τ_0. On the other hand, the 50-nsec Q-switched pulses used in the early disconnect work of Sypherd and Salman (1968) would extend radially ~5 μm in aluminum and ~3 μm in silicon. Thus, these relatively long pulses applied to present high-density circuits would surely result in damage to the underlying n^+ diffusion regions and possibly to adjacent circuit elements as well.

The first MOS laser connection experiments were made on structures shown schematically in cross section in Fig. 1 (Kuhn et al., 1974; Cook et al., 1975). A rhodamine 6G dye laser was used, optically excited by a nitrogen laser, resulting in 5-nsec pulsewidths. The experimental arrangement is shown schematically in Fig. 2. The sample was positioned on the xy stage of a microscope, allowing the target region to be accurately aligned with respect to the laser. Maximum power density at the sample surface was ~2 × 10^9 W/cm^2, sufficient to cause both local melting and ablation of the optically absorbing aluminum and Si. Initially, both connection and disconnection experiments were undertaken. Disconnection experiments consisted of opening 1-μm-thick aluminum lines, 6 μm in width being deposited on top of the insulating SiO$_2$ layer. Connection experiments were undertaken to study the joining of an otherwise discon-

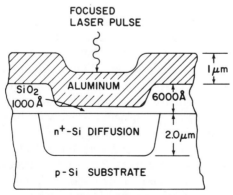

FIG. 1.   Cross section of MOS site. Aluminum is to be connected to the n^+-Si diffusion (Kuhn et al., 1974).

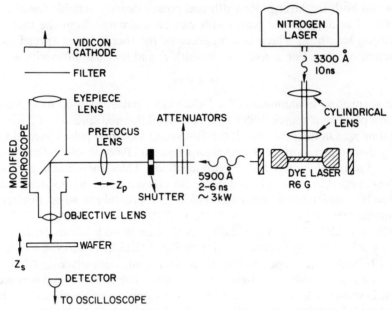

FIG. 2.   Schematic of experimental arrangement for laser connecting MOS sites (Kuhn *et al.*, 1975).

nected Al pad through the SiO_2 layer (thickness ranging from 0.1 to 0.6 μm) to an n^+ diffusion layer on top of a p-type substrate. The I–V characteristics of as-is laser-produced connections were found to be equivalent to those of a degenerate diode. However, annealing for 20 min at 400°C in a nitrogen atmosphere caused the connections to become ohmic.

3. *The Connection Mechanism*

The physical connection process was studied as a function of the number and intensity of laser pulses incident on connection sites. All combinations of these two laser parameters that result in ohmic connections define the coordinates of the process window. The area outside of the process window includes those parameters for which either no connection or a connection with damage giving rise to junction leakage currents results. An example of a typical process window is shown in Fig. 3. For test circuits within the process window requiring four pulses, the following connection process was deduced with the aid of scanning electron micrographs of laser-processed cross-sectioned connection sites: the first two pulses remove the aluminum in the region of laser absorption mainly through ablation, the laser penetrating down to the insulating oxide layer.

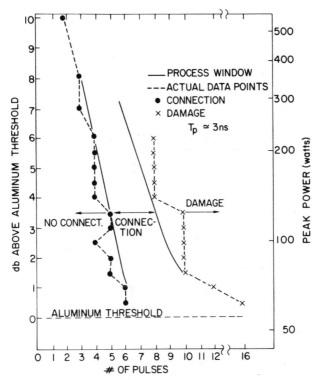

FIG. 3. Process window on MOS structure of site shown in Fig. 1 (Kuhn *et al.*, 1974).

The subsequent pulse removes the oxide layer while also increasing the diameter of the laser-drilled hole in the Al. The final (fourth) pulse produces molten n^+ Si which flows up the sidewalls or periphery of the hole, freezing upon contact with the aluminum, thereby forming a connection (Cook *et al.*, 1975; Kuhn *et al.*, 1974, 1975; Platakis, 1976). The connection process for these four pulses is shown schematically in Fig. 4. In general, for all laser connections examined, the role of the first pulse is mainly that of penetrating the aluminum line, whereas subsequent pulses ablate the intervening dielectric layer, melt the Si underneath, and cause a connecting fillet of Si to form after refreezing onto the Al. A micrograph of the refrozen silicon is shown in Fig. 5 after removal of the Al and SiO_2 by preferential etching (Platakis, 1976).

Effects on the connection process resulting from a passivation layer (SiO_2) deposited over the MOS site have also been investigated (Kuhn *et al.*, 1975). It was determined that connections can be formed for sites containing 2-μm-thick passivations using relatively low laser power lev-

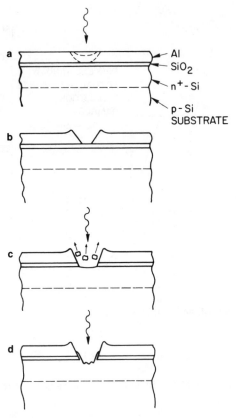

FIG. 4. Schematic representation of connection process with (a) and (b) depicting opening of Al line by laser melting and/or vaporization; (c) removal of SiO_2; (d) melting and refreezing of Si to form the connection.

els, without physically removing any passivation material in the connection process. Electron microscopy studies after contact formation revealed only a slight bump on the upper surface and a small void on the underside. This result suggests that the passivation encapsulation prevents heat losses from the surface. In addition, superheating of the molten Al underneath the passivation layer may also give rise to a somewhat different connection mechanism than that determined for circuits without the passivation (Kuhn *et al.*, 1974). The possibility of achieving a higher temperature at depths below a surface compared to the top absorbing surface has been described for a moving phase boundary and could explain the explosive removal of the passivation layer observed for pulses with higher laser power densities (Dabby and Paek, 1972).

FIG. 5. Scanning electron micrograph of refrozen Si from a laser-connected MOS site after preferential etching of Al and SiO₂ (Platakis, 1976).

In situ experiments to determine the connection formation time have been made with an MOS capacitor to simulate the actual transistor circuit (Kuhn *et al.*, 1974). The capacitor was electrically charged by an external power supply connected in series to a 100-kΩ resistor. Laser connection produces a short in the capacitor, causing it to discharge through a 50-Ω resistor connected in parallel to both the capacitor and a fast oscilloscope. From the time-dependent voltages generated by the discharge it was determined that the initial connection formation occurs in approximately 1 nsec, consistent with the cooling time required for the thin layer of molten Si. However, there is a small resistance change that continues for approximately 100 nsec after application of the laser connection pulse, as indicated by a decay current observed when a small bias voltage is maintained across the circuit.

4. Metal-to-Metal Connections and Circuit Design Using Laser Connections

Early work in metal-to-metal connections, that is, the joining of two metal lines (aluminum) on a silicon chip separated by a thin-film insulator by means of laser pulses, was first described by Kuhn *et al.* (1974). It was observed that the resulting preannealed resistance of such laser-formed contacts was as low as 1 Ω. A further reduction in connection resistance to values as low as 0.3 Ω was achieved by thermal annealing after the laser connection was formed. Connection occurs by the melting and refreezing of Al from the buried line onto the upper metallization in much the same way that the Si-to-metal connection is formed. Connections typically required only two 6-nsec rhodamine 6G-focused laser pulses. These early results gave impetus to the important concept of using a pulsed laser with a reasonable repetition rate (i.e., greater than 10 pps) for the experimental "wiring" of entire portions of chips to model and test new circuit designs. With this technique, optimum designs could be obtained from a single basic chip containing all the required logic elements, while needed changes could be implemented with a minimum turnaround time after circuit testing. Because modifications are made on the same prototype wafer, no new masks or photolithography are required as new designs evolve. Special wafers containing bipolar programmable logic arrays (PLAs) or laser programmable logic arrays (LPLAs) were developed by Logue *et al.* for these VLSI circuit design studies (J. F. Smith *et al.*, 1981; Logue *et al.*, 1981). The LPLA consisted of input latches, bit-partitioning circuits, an AND array, an OR array, output latches, and off-chip devices. Metal-to-metal laser connections (1200–3200) were required to fully personalize or "wire" the logic of each chip.

To facilitate the large number of connections required per wafer and to provide a means for accessing these connection sites, the Logue group designed and built an automated laser tool using a microprocessor and computer to regulate the laser pulse pattern and control an xyz θ-positioning table (Feder *et al.*, 1978). A schematic of the apparatus is shown in Fig. 6. The nitrogen laser was used to pump a stilbene dye laser, operated at 40 pps continuously, with the dye laser pulse appropriately gated onto the connection site by a Pockels cell for a total of two pulses/site. Incident power/pulse ranged from 50 to 200 W. The laser-formed connections were found to withstand on the order of 100 mA of current.

Two methods of utilizing laser-personalized wafers were investigated. In the initial phase, the laser-connected circuits were diced, mounted, and packaged following the laser connection and testing steps on the wafer level. The second method made use of laser processing or personalization following mounting and packaging on pretested modules having good peripheral circuits. In this manner, testing could be accomplished after personalization of each laser-connected crosspoint. The test data were processed on an IBM 370 and served to give instructions to the laser tool for further welds to repair any deficient circuit based on testing results. The

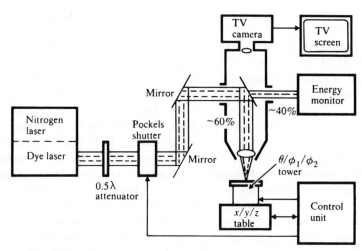

FIG. 6. Schematic of laser tool used for metal-to-metal connections of laser PLAs (Logue *et al.*, 1981).

instruction might be to proceed to a connection site and repeat the laser pulses to wire in a crosspoint not properly wired or to delete the connected circuit by adding connection points that short out that particular portion of the circuit-to-ground. Overall, this module personalization was found to improve both yield and turnaround time. Total time for laser wiring and testing was ~20 min. Although typical connection diameters varied from 5 to 7 μm, the Logue group indicated that minor modifications, mainly in the optics of their automated tool, should bring these dimensions down to 1–2 μm, comparable to line dimensions required for future technologies.

To achieve maximum yields, i.e., good or operable circuits, connection failure mechanisms were investigated. It was found that the aluminum lines and intervening oxide layers required careful thickness control. For oxide layers greater than 2 μm thick, frequent connection failures were observed as a result of an oxide "plug" at the connection site which failed to be removed with subsequent laser pulses. This plug prevents coating of the connection-site sidewalls with molten aluminum and gives rise to a poorly metallized connection. Thin metal lines (<1 μm) also resulted in poor sidewall coating and nonuniform metallization, giving rise to high contact resistance. Electrical tests showed that such contacts eventually lead to circuit opens whereas thick metallizations develop shorts (J. F. Smith *et al.*, 1981).

D. Laser Wiring—Read-Only-Memory Coding and Redundancy

Special metallurgy for current-carrying lines was developed by North and Weick (1976) to achieve laser disconnections for discretionary wiring (coding) of read-only memories with the 1.06-μm YAG laser wavelength (North and Weick, 1976). The circuits on silicon chips used metallic links made of Ti–Pt layers to provide greater absorption at the 1.06-μm wavelength compared to that of Ti–Pt–Au bit lines. With gold as the top layer, only 2% of the YAG light is absorbed, compared to 18% for Pt. Relatively low-energy pulses were used to achieve disconnections, leaving adjacent metallizations unaffected. North and Weick also studied the resistance of a test pattern as a function of the number of laser pulses used for disconnects on each link of the pattern. The results are shown in Fig. 7, indicating that a minimum energy pulse is required with the use of eight pulses. The difference in laser energy required for the desired resistance ($>10^9\ \Omega$) is substantial for one and two pulses, whereas for greater than two pulses this difference becomes smaller. The process reliability was found to increase with more than two laser pulses. Also, cleaner disconnections with more complete metallization removal were found when more than one pulse was used. The optimum laser conditions were found to be 110 μJ/pulse, with a 0.25-μsec pulsewidth and four laser pulses/link. Yields

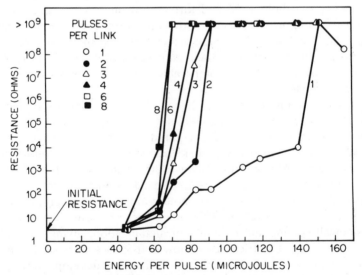

Fig. 7. Test-pattern resistance of 32 parallel links as a function of energy/pulse with the number of pulses/connection as a parameter (North and Weick, 1976).

of 95% were obtained for a 1024-bit read-only memory (ROM). Additional investigations on the effect of the underlying dielectric layers of the circuit on the laser intensity indicated criticality of their thickness in determining the completeness of the disconnect (North, 1977). This effect results from interference of the reflected wave components from these layers, which, as shown by these experiments, must be taken into account in determining the net energy in the region of the Ti–Pt link.

The use of redundancy in combination with laser wiring to bring about large increases in the yields of random-access memories (RAM) has been an active subject for a number of years (Bindels *et al.*, 1981; Cenker *et al.*, 1979; Kuhn *et al.*, 1975; Minato *et al.*, 1981; Schuster, 1978; R. T. Smith *et al.*, 1981). Redundancy utilizes spare parts on the wafer as an integral part of a circuit. In the event of local circuit failure, a spare part can be "wired-in" and the defective part of the circuit made electrically inactive by appropriate disconnection steps. Redundancy techniques have been used with other than laser repair schemes. An example is the fusible link which utilizes a programmable method to permit passage of a large current flow through a circuit element to achieve a disconnect via intense local dissipative heating. The link becomes disconnected in a manner similar to an overdriven fuse.

The importance of redundancy with respect to yield improvement is dramatically illustrated in Fig. 8 (R. T. Smith *et al.*, 1981). Here the result of laser disconnections to implement redundancy is shown in terms of the number of functional chips per wafer versus the yield improvement factor, the latter defined as the ratio of the total number of functional chips divided by the number functional prior to laser processing. This curve represents results obtained over ~2 years of RAM volume production. The highest yield improvement occurs for early circuit models containing large numbers of faults. Later models have fewer faults but an improvement by a factor of 3 is still realized with the laser redundancy technique. The continued increase in the number of functional chips/wafer (going from right to left in the abscissa of the curve) indicates that the laser redundancy technique plays an important role in increasing the yield factor even for wafers with inherently higher yields in their as-processed production-line state. The number of spare parts sufficient for the active design of a chip tends to decrease as overall wafer-manufacturing techniques improve. The use of chip redesign to obtain more area for active circuits by eliminating some of the redundant spare parts constitutes an important trade-off which is still much debated by circuit designers. The overall question reduces to one of cost-effectiveness and has no generic answer.

To implement redundancy, spare columns and rows of decoders and

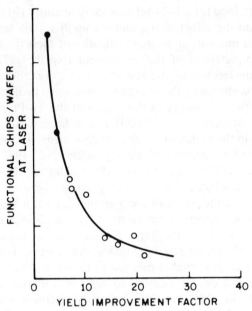

FIG. 8.   Yield improvement resulting from redundancy using laser wiring technique (R. T. Smith *et al.*, 1981).

sense amplifiers are located near an array of memory elements on the wafer to serve as spare parts in case of failure of an element in the array. Details of a typical circuit organization with spare parts are found in R. T. Smith *et al.* (1981). Spare decoders have both address and complement address lines connected to gates of a decode-transistor pair (Fig. 9). Disconnection of a defective memory row is accomplished by deleting a programmable link between the standard row driver and row line. These links, consisting here of heavily doped polysilicon, lie beneath a phosphorus-doped SiO_2 layer through which the laser was directed. The spare decoder is programmed to assume the function of the faulty element by disconnecting either an address or its complement from the six transistor pairs from the spare by laser deletions. In the circuit of Fig. 9, a total of seven laser pulses is required to remove or isolate the faulty element and encode or wire in the spare part. A Q-switched Nd:YAG laser was used, delivering 50-nsec pulses with an energy of $\sim 10 \ \mu J$/pulse. XY translation of the circuit under the incident beam was made using both a step-and-repeat table and a translational lens. By appropriate laser power and beam diameter control, the link, 3 μm in diameter and 14 μm long, was removed by vaporization. The automated system also included a means for both

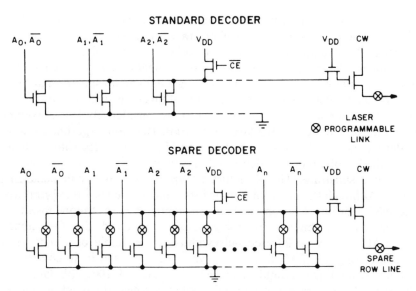

FIG. 9. Schematic of standard and spare decoder with location of laser-programmable links indicated by ⊗ (R. T. Smith *et al.*, 1981).

testing the chips before and after repair and determining the position where additional repair was required. With this scheme it is possible to obtain large-scale volume production of wafers in a fully automated laser system after initial mask design and processing steps are completed.

A group at Hitachi has recently demonstrated an alternate method for connecting spare circuits to implement redundancy using a lateral link (Minato *et al.*, 1981). Here two segments of n^+ diffusion layers are separated by an intrinsic polysilicon region. By applying a suitable laser pulse, the n^+ atoms diffuse into the polysilicon region thereby completing the electrical connection. The pulse does not produce melting, but rather sufficient local heating to cause solid-state diffusion. The impedance through the link decreases from 10^9 to less than 10^3 Ω after laser irradiation and constitutes an operable connection.

E. Laser Personalization with Polyimide Insulators

Polyimide films are becoming increasingly important as insulating layers, replacing SiO_2 and SiO in both wafer and packaging applications. Polyimides are especially desirable because of their durability, thermal stability, dielectric properties, and pinhole-free surfaces. This has made laser deletion cutting of metal lines on polyimide surfaces of interest for circuit personalization. To implement this it is desirable to select laser wavelengths that are not strongly absorbed by the polyimide, thereby avoiding insulator damage. A study of the optical transmission characteristics of a number of polyimide films using a dual-beam spectrometer was undertaken by Perry et al. (1981). Results of this investigation indicate the Nd : YAG laser wavelength (1.06 μm) to be considerably less absorbing than the rhodamine 6G dye laser (0.57 μm). Deletion experiments were undertaken on Cu lines, 4.5 μm thick and 15 μm wide, deposited onto an 8-μm-thick polyimide film deposited in turn onto a 1-μm metallization on a ceramic substrate. This structure is typical of a Si circuit packaging board. Although the nitrogen-pumped rhodamine dye laser (0.57 μm) was successful for line deletions without substantial damage to low-absorbing polyimides, the higher absorbing samples required the Q-switched Nd : YAG laser with 20-nsec pulsewidths. An example of a successful deletion using the YAG laser is shown in Fig. 10a and b. In contrast, a deletion exhibiting polyimide damage is shown in Fig. 10c, with damage in the form of deep grooves cut into the polyimide near the deletion region. Via holes through the polyimide have also been produced by focusing the YAG laser onto the 1-μm underlying metallization. This causes the metallization to become sufficiently hot to produce intense local heating of the otherwise optically transparent polyimide. Such vias have significant importance in advanced packaging of VLSI chips.

Additional work on laser circuit repair by wiring-in redundant parts and/or disconnecting circuits using a pulsed laser is being carried out by a group at MIT Lincoln Laboratory (Raffel et al., 1980). Both a ~100-nsec pulsewidth YAG laser designed for integrated circuit trimming and a millisecond-pulsed argon laser have been used. Connections are made between two metal lines separated by an insulator. Generally, the purpose and techniques are similar to those already described.

F. LASER REPAIR OF VLSI CIRCUITS

The automated laser tool previously described (J. F. Smith et al., 1981; Logue et al., 1981) has been used to repair otherwise fully processed and connected chips. Typical repairs involved the interconnection of broken or missing metal sections such as current pathways, omitted either due to faults in the mask or in the photoresist processing (Feder et al., 1978; Logue et al., 1981). Deletion of unwanted metallization could also be implemented. The repair scheme uses local photoresist processing techniques but without the use of masks. The area to be corrected is first

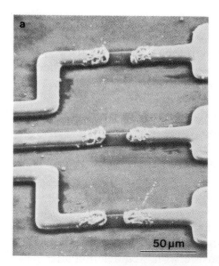
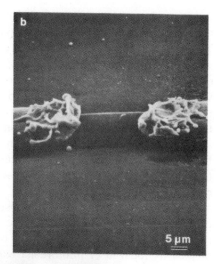

FIG. 10. (a) Deletion of Cu lines on polyimide using a Q-switched YAG laser; (b) detail of a deletion in (a); (c) damage caused to polyimide for a line similar to (a) using a rhodamine 6G dye laser.

covered by a layer of photoresist. This is followed by optical pattern exposure using the pulsed focused dye laser beam as the light source. The computer-controlled stepping table is programmed to trace out the required repair pattern, moving in closely spaced discrete steps to expose, for example, a missing line on the photoresist. The laser is highly attenuated to deliver only on the order of 60 mW/pulse with 7-nsec-wide pulses. The 420-nm stilbene dye laser wavelength falls well within the required wavelength for exposing Shipley AZ-1350 J[1] resist with an energy density of 60 mJ/cm. Following laser exposure, the resist is developed. Metal vapor depositions are then made. Standard photoresist lift-off technique is used to remove the undeveloped resist and unwanted overlay metallization, leaving only the metallization in the region of the developed photoresist. With this technique, corrections to the wafer metallurgy pattern can be made at any stage of wafer processing. Numerous additions and deletions to circuits have been successfully demonstrated and are discussed in reports by Feder *et al.* (1978) and Logue *et al.* (1981). The technique has also been applied to circuit diagnostics. For example, in order to test a portion of a circuit, the removal of certain connections and wiring may be required to isolate a device from other circuitry. This can be accomplished by local laser photoresist exposure with etch deletion after photoresist development. It is estimated that the time saving realized from this technique, in certain cases, can be on the order of 4 weeks because the need for new masks normally required to implement photoresist repair techniques is eliminated. A summary of laser parameters used for the laser personalization work is found in Table I.

III. Laser Deposition and Etching in Gaseous Atmospheres

A. Background

One of the earliest discussions describing the use of lasers for local etching and depositions is found in the U.S. Patent of Solomon and Mueller (1968). To achieve a maskless pattern, a laser beam was directed through a transparent window of a small chamber and focused onto a workpiece such as ceramic, silicon, or a metal. The local heating provided by the laser produced local temperature increases (600–1000°C) sufficient to initiate pyrolytic chemical reactions between a workpiece, for example, tungsten, and low-pressure ambient iodine vapor to produce WI_2, a volatile compound. The formation of WI_2 results in etching of tungsten in those regions absorbing laser energy. The investigators also describe a scheme whereby patterns can be generated on the workpiece with the aid of a movable positioning stage. Vapor depositions are also discussed with the specific example of the decomposition at high temperatures of WI_6 to

[1] Shipley Corp., Newton, Massachusetts.

(1976). A high-energy ($\sim$1 J), pulsed (0.2- to 1-μsec pulsewidth) CO_2 laser was used to obtain nonresonant dielectric breakdown of gases contained in a small cell through which the light was directed. High-intensity electric fields were obtained by focusing the laser through a ZnSe lens with the laser tunable over the P and R branches (9.6- and 10.5-μm bands of the CO_2) by means of a grating. Resulting gaseous products and particulates were examined by mass spectroscopy, infrared spectroscopy, and micro-weighing techniques. The organometallics and their associated products included $Fe(C_5H_5)_2$ giving rise to metallic iron, the hexacarbonyl $Mo(CO)_6$ to metallic Mo and CO, and OCS (carbonyl sulfide) to CO and solid sulfur. The products tend to be selectively stable such that little back reaction or recombination occurs. Scavenging gas is required for those cases where the products of the reaction are less stable. For example, the use of H_2 in combination with SF_6 will scavenge the free fluorine that results from laser-induced dielectric breakdown.

Investigations of multiphoton absorption of SF_6 for sulfur particle formation were described (Lin and Ronn, 1978) in which the dissociation of both SF_6 and the mixture SF_6/H_2 was examined, again using a CO_2 laser. Both off-resonant (dielectric breakdown) and resonant (multiphoton) vibrational excitation effects were investigated. Laser pulsewidths of 500 nsec (FWHH) and repetition rates of 50 pulses per minute (ppm) were used with the light focused to volumes as small as 0.4 cm^3.

With pure SF_6 at pressures between 40 and 100 Torr, strong absorption occurred for the P(20) or 10.6-μm transition, and fluorescence was observed. The fluorescence intensity increased with pressure but decreased with time as the SF_6 content decreased going over to the reaction products SOF_2, SiF_4, CF_4, SOF_4, and SF_4. Additional carbonaceous products at higher temperatures were observed and ascribed to reactions with carbon from the "O" ring and vacuum grease of the reaction chamber. SiF_4 was formed from the reaction

$$4F + SiO_2(glass) \longrightarrow SiF_4 + 2O \qquad (5)$$

with the F atom arising from the reaction

$$SF_6 \longrightarrow SF_5 + F \qquad (6)$$

Some reaction products were also identified during the on-time of the CO_2 pulse, but could not be detected after the pulse was terminated because S and F recombination processes are very efficient. Similar products were found to be formed without vibrational resonance irradiation using the 9.6-μm laser line. Here dielectric breakdown is necessary for product formation to occur. When the SF_6 was mixed with H_2 in a 1 : 1 pressure ratio and irradiated with 10.6-μm resonant radiation, large amounts of S

TABLE II

LASER VAPOR DEPOSITION

Experiment	Laser	Laser parameters	Reference
Photodissociation of organometallic gases to excited states	KrF and ArF (UV excimer)	30 mJ/pulse, 20 nsec; 5 mJ/pulse, 20 nsec	Karny et al. (1978)
Dielectric breakdown; ferrocene and metal carbonyl	CO_2	~1 J/pulse, 0.2–1.0 μsec	Ronn (1976)
Dielectric breakdown; resonant multiphoton photodissociation of SF_6; sulfur particulate formation	CO_2	2 J/pulse, 0.5 μsec	Lin and Ronn (1978)
Dielectric breakdown of metal carbonyl to form Fe, Cr, Mo, and W films	CO_2	~5 J/pulse, 0.1 μsec	Draper (1980a)
Metal carbonyl dissociation; Mo and W films	Cu ion (UV)	18 μJ/pulse, 120 μsec	Solanki et al. (1981)
UV photolysis; photodeposition of Zn via local chemidesorption	KrF and XeCl (UV excimer)	≥20 mJ/pulse, 10 nsec	Coombe and Wodarczyk (1980)
Organometallic photolysis; Cd and Al films	Frequency-doubled argon	~2 × 10⁻⁴ W	Deutsch et al. (1979)
Photodeposition of Cd; doping of InP, laser heated	Argon and frequency-doubled argon	CW, 2.7 W at 514.5 nm; 3 mW at 257.2 nm	Ehrlich et al. (1980d)
Photodeposition and doping to form p–n junctions on Si	ArF (excimer)	7 nsec, 0.1–0.25 J/cm²/pulse	Deutsch et al. (1981a)

Laser microalloying Al with Zn from gaseous photodissociation	Argon and frequency-doubled argon	2 W at 514.5 nm; 3 mW at 257.2 nm	Ehrlich et al. (1981a)
Surface pyrolytic decomposition of SiH_4	Argon	15-W CW	Ehrlich et al. (1981d)
Pyrolytic dissociation of metal alkyls	Krypton	CW, 520.8 and 568.2 nm	Rytz-Froidevaux et al. (1981)
LCVD of Si from silane	CO_2	0.8 J/pulse, 200 nsec	Hanabusa et al. (1979)
LCVD of Si from SiH_4	CO_2	50-W CW, 0.1 ms–10 sec pulses	Christensen and Lakin (1978)
LCVD of Si from $SiCl_4$	CO_2	24-W CW	Baranauskas et al. (1980)
LCVD of Ni, TiO_2, and TiC	CO_2	20-W CW	Allen (1981)
LCVD of carbon from C_2H_2	Argon	5×10^4–3×10^5 W/cm^2, CW	Leyendecker et al. (1981)
Chemidesorption for ZnS deposition	Argon	0.1–1 W/CW, 488.0 nm	Arnone et al. (1980)
Mask repair; Cd photodeposition	Frequency-doubled argon	~1 mW at 257.2 nm, CW	Ehrlich et al. (1980a)

23

were formed and precipitated out uniformly in the experimental cell. Nucleation times were found to be 60–100 μsec with particle size between 1 and 3 μm.

The formation of a number of fine metal particles resulting from the dielectric breakdown of carbonyl vapors can also be made to occur and has been studied by Draper (1980a). A CO_2 pulsed laser was used in the experiments with up to 5 J/pulse, with pulsewidths of $\simeq 0.1$ μsec which could be focused to a spot ~ 0.15 cm in diameter to produce dielectric breakdown of the vapor with an effective power density of 10^{10} W/cm². The solid metal carbonyl, independently heated to $\sim 100°C$ inside the reaction cell, produced an equilibrium carbonyl vapor within the cell volume. Cr, Mo, W, and Fe films were formed using ~ 10 laser pulses at a 0.5-Hz repetition rate. Transmission electron micrographs of the resulting Cr films showed a chain-like agglomeration of fine particles, possibly amorphous or very-fine-grain crystalline. Mean particle size was found to be ~ 75 Å. One possible application suggested for films of this type is for solar cell light collectors. The ability to control particle size by controlling the carbonyl pressure and laser power density also affords a technique for metal alloying in mixed vapor systems when the metal ligands have similar bond strengths. In other deposition experiments with carbonyls, straight-chained hydrocarbons such as 1-hexane and n-hexane were mixed with Mo hexacarbonyl vapor and were found to produce "cobweb-like" structures of Mo using a single pulse from a CO_2 pulsed laser (Draper, 1980b). The films were observed to grow for several seconds even after termination of the laser pulse, a considerably longer time than that observed for CO_2-irradiated $Mo(CO)_6$ growth from pure Mo carbonyl.

Recent work on the photodecomposition of carbonyls to produce metallic films has been more directly oriented toward possible VLSI applications. This has been aided by the availability of UV lasers which, due to their shorter wavelength compared to CO_2 lasers, will yield considerably smaller thin-film structures. A number of UV lasers have been used to achieve the local deposition of refractory metals from both pure metal vapors and from their respective carbonyl molecules. Lasers for those applications include Cu-ion lasers (Solanki $et\ al.$, 1981), the excimer lasers KrF and XeCl, and the frequency-doubled argon-ion laser (Coombe and Wodarczyk, 1980; Ehrlich $et\ al.$, 1981b). Fragments from the metal carbonyl, $M(CO)_n$, following UV multiphoton absorption include $M(CO)_{n-x}$, M, M*, and M^+ (Solanki $et\ al.$, 1981). All of these fragments can contribute directly or indirectly to metal deposition. In the case of laser deposition of pure metal vapors, the effect of the UV laser appears to be one of promoting a surface interaction with the substrate to provide nucleation sites for the colliding metal atoms. Heating or pyrolytic mechanisms appear not to be important factors for these metallic depositions. Ehrlich *et al.* (1981b) have investigated the "writing" of metal lines using a frequency-doubled argon-ion laser (257 nm) aimed perpendicular to a substrate contained in a movable cell with each one of the following metal carbonyls: $Fe(O)_5$, $W(CO)_5$, $W(CO)_6$, and $Cr(CO)_6$. Tungsten lines 2.5 μm wide were produced on quartz substrates by scanning the sample at 0.9 μm/sec with an incident power density of 0.9 kW/cm^2. Resulting linewidths were approximately equal to the focused laser-beam diameter. Some background deposition occurred for Fe deposited from the dissociation of $Fe(CO)_5$, but this could be reduced by lowering the gas pressure or laser intensity, or by the introduction of inert buffer gases such as Ar or He. Deposition rates for Fe and W as a function of laser power density are shown in Fig. 11, obtained with low carbonyl gas pressures. Whereas the W deposition rate is linear over the entire power range, Fe has a sharp discontinuity believed to be due to a gas-phase chain reaction which is followed by homogeneous nucleation of the dissociated gas photoproducts.

Room temperature depositions of Cr, Mo, and W have also been obtained from their respective carbonyls at pressures ranging from 0.1 to 0.3

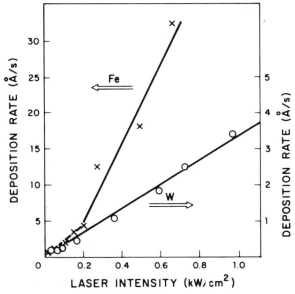

FIG. 11. Deposition rate for Fe and W as a function of laser intensity using a CW 257-nm frequency-doubled argon laser. The Fe deposition resulted from dissociation of 0.5-Torr $Fe(CO)_5$, and W from $W(CO)_6$ at 0.2 Torr (Ehrlich *et al.*, 1981b).

Torr using 150-mW peak power from a pulsed copper-ion laser (Solanki *et al.*, 1981). Experiments were operated with 120-μsec-wide pulses at repetition rates of 40 Hz in the wavelength range 260–270 nm. The cell containing carbonyl (Fig. 12) was pumped out and then backfilled to 1 atm of He to limit the mean free paths of the photofragments to ~1 μm. Identical film deposition results were obtained on several different substrates including Si, a highly optically absorbing material for this wavelength range, and quartz, highly transmitting. If thermal effects were important, the resulting film growth for these two substrates should have shown large differences in thickness, contrary to experimental observation.

Coombe and Wodarczyk (1980) have also reported UV laser-induced vapor depositions on both transparent and absorbing substrates in conjunction with KrF (249 nm) and XeCl (308 nm) excimer lasers. However, metal vapors in these experiments were derived from evaporation of the pure bulk metal in a resistively heated oven contained in an evacuated quartz cell. Such vapors have very high purity and should result in ultraclean depositions. A helium gas stream was used to transport the metal vapor over the substrate, with a resultant cell pressure of approximately 200 mTorr. Laser pulses, typically 10 nsec wide with up to 20 mJ/pulse for the KrF laser and 5 mJ/pulse for the XeCl laser, were focused onto the substrates. Zinc films were deposited on quartz, Pyrex, CaF_2, SnO_2, and aluminum substrates. For those cases in which transparent substrates were used, film deposition resulted on both front and back faces of the substrate. Up to 5-μm-thick zinc films were obtained. Film growth was found to occur even between successive laser pulses. Vapor pressures an order of magnitude lower were required for growth with the laser compared to that with no laser. The explanation for these observations is given in terms of enhanced nucleation sites or increased sticking coeffi-

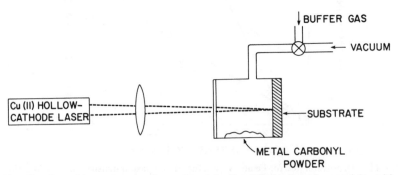

FIG. 12. Deposition geometry of experimental cell and Cu-ion laser for metal depositions from dissociation of carbonyl molecules (Solanki *et al.*, 1981).

cients in those regions of the substrate subjected to the radiation. In particular, very thin hydrocarbon layers appear to be desorbed by the radiation, allowing the metal vapor to have excellent adhesion to the substrate material. Direct heating or pyrolytic effects do not appear to play a role in the deposition mechanism, as both optically absorbing and transparent substrates give similar results.

The effect of increased sticking coefficients on ZnS films deposited on predeposited CdS films has been studied in a separate experiment using 0.1–1 W of argon laser power (488-nm line) focused to a 1.5-cm-diameter spot (Arnone et al., 1980). Clearly, this gives a power density far too low to cause any substantial heating of the CdS or glass substrate. The experimental arrangement is shown in Fig. 13. A wire mesh screen on the glass side was illuminated with the light incident on the glass–CdS interface during deposition of ZnS. The variation in thickness of the ZnS film after deposition corresponded to the light-intensity modulation produced by diffraction from the wire mesh. Depositions of ZnS ceased when a maximum thickness of ~900 Å was reached. The enhancement in ZnS growth occurred only when the CdS film was exposed to air prior to deposition of the ZnS layer and is explained in terms of increased nucleation sites on the CdS surface produced with laser irradiation to produce surface desorption.

Deutsch et al. (1979) and Ehrlich et al. (1980d) used both an argon frequency-doubled CW laser (257.2 nm) and the ArF excimer laser

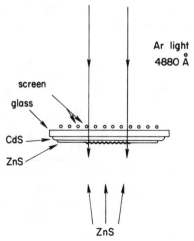

Fig. 13. Apparatus to obtain modulation in ZnS thickness deposited on CdS film (Arnone et al., 1980).

(193 nm) to dissociate the metal alkyl compounds trimethylaluminum [Al(CH$_3$)$_3$] and dimethylcadmium [Cd(CH$_3$)$_2$] for depositions of Al and Cd and to study their application to microelectronics. The experimental setup, Fig. 14, shows the frequency-doubled laser (257.2 nm) focused through a quartz window of a small cell containing the organometallic gas. The transmitted light served as a monitor for the thickness of the deposited film, with the transmitted signal I_{tr} related to the deposition thickness d and the film's absorption constant α by $I_{tr} \propto I_0 e^{-\alpha d}$, with I_0 the incident laser intensity. The incident power of the frequency-doubled argon laser was on the order of 2×10^{-4} W. The transmitted light for various laser power levels as a function of time is shown for Cd(CH$_3$)$_2$ dissociation with both the frequency-doubled argon (Fig. 15a) and the pulsed ArF lasers (Fig. 15b). For the argon frequency-doubled radiation, deposition starts as soon as laser radiation is initiated. For 1 W/cm^2, the deposition rate was found to be 13 Å/sec, with the rate scaling approximately linear for power density $P \leq 1$ W/cm^2 (Fig. 15a). A nonlinear dependence is observed with high laser flux, probably caused by the depletion of available alkyl molecules. Deposition rates up to ~1000 Å/sec were observed for 10^4 W/cm^2 with frequency-doubled UV argon laser flux. With 4 W of focused argon light (514 nm) incident on the surface of the window, no Cd deposition was observed. This strongly suggests that photodissociation rather than heating is the predominant mechanism because the 514-nm intensity is 10^4 times greater than the frequency-doubled light used to obtain micrometer-sized line patterns. Figure 15b shows transmission results obtained with a He–Ne probe beam, collinear with the 1-mJ, 10-nsec ArF (193 nm) photodissociation pulse focused to a power density of 3×10^6 W/cm^2. Here a slight increase in transmission is observed with

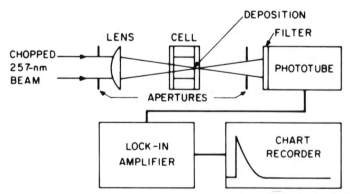

Fig. 14. Experimental setup for metal deposition from metal alkyl compounds (Deutsch *et al.,* 1979).

each laser pulse, indicative of film removal caused by vaporization as a result of intense localized heating. On cooling, however, there is a net increase in the film thickness due to deposition of metallic photofragments. It is believed that both pyrolytic and photolytic processes are involved at these high power densities. In general, the Cd deposition rates were found to be dependent on gas purity, cell wall cleanliness, and the partial pressure of the added buffer gas. Long exposure to $Cd(CH_3)_2$ gas causes a passivation of the walls of the cell and prevents the sticking of Cd atoms. In contrast, freshly deposited Cd films enhance Cd sticking coefficients and the deposition rate, the Cd atoms acting as nucleation centers

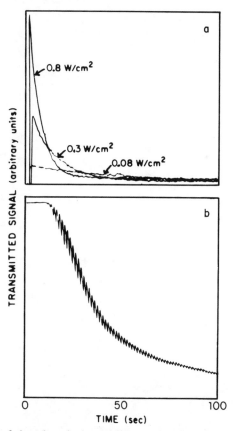

FIG. 15. Growth of photodeposited metal films monitored with transmitting probe beam. (a) 10-Torr dimethylcadmium and 743-Torr He irradiated at three intensities of 257.2-nm light; (b) 1-Torr dimethylcadmium irradiated with 0.03-J/cm² pulsed 193-nm light (Deutsch *et al.*, 1979).

especially during the early stages of deposition. Cd depositions (spots on the order of 105 μm in diameter) containing periodic thickness variations spaced 2 μm apart have also been produced with the 257-nm light (Fig. 16). These variations correspond to a Fresnel diffraction pattern in depo-

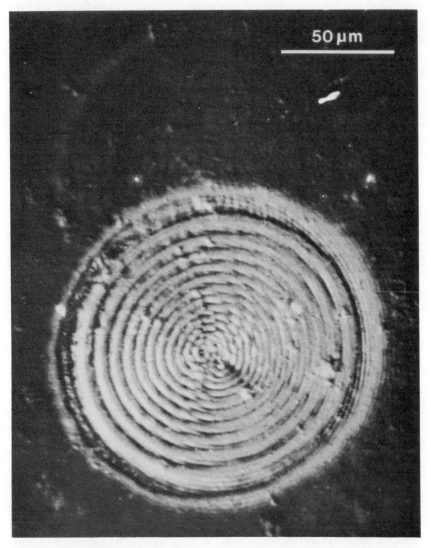

FIG. 16.   Cadmium film Fresnel diffraction deposition pattern from photodissociation of 3-Torr dimethylcadmium and 700-Torr He using 0.1 mW of 257.2-nm light for 60 sec (Deutsch et al., 1979).

sition thickness replicating the laser interference pattern resulting from passing the light through a small aperture prior to entering the reaction cell. This extraordinary resolution is indicative of the technological promise of this deposition technique for VLSI.

Silicon films can also be deposited by photolytic decomposition of silane, as shown by Hanabusa *et al.* (1979). The P(20) CO_2 line (10.59 μm) produced efficient silane decomposition and deposition of silicon whereas the P(16) line (10.55 μm) resulted in a factor-30 thinner film under otherwise similar experimental conditions. The observed nonlinear deposition rate with laser power is ascribed to multiphoton absorption giving rise to silane photodissociation.

2. *Laser Chemical Vapor Deposition*

Chemical vapor deposition (CVD) is a well-known technique for growing thin films from a gaseous atmosphere on solid surfaces. Film deposition generally requires elevated temperatures for the gaseous compound to either decompose or react with other elements to form the desired deposition. In conventional CVD the rise in temperature is obtained with a dc heater or inductive heating of the sample stage. However, maskless film patterning is not possible for that type of deposition. On the other hand, a laser directed at the substrate can provide local heating and cause local laser chemical vapor deposition (LCVD) to occur. Replacement of the conventional CVD heat source with a focused laser makes it possible to achieve maskless patterning, limited thermal distortion of the substrate, and a generally cleaner environment which can result in considerably purer film depositions. The localization of the heat can also be an important factor in inhibiting the gas reaction from occurring spontaneously in regions remote from the desired deposition area.

Chemical vapor deposition of polycrystalline silicon has been demonstrated using a CW and pulsed CO_2 laser to provide the necessary local substrate heating for the pyrolytic decomposition of silane at a pressure of ~10 Torr (Christensen and Lakin, 1978). Substrates (3-mm-thick fused quartz) reached temperatures on the order of 1×10^3–$1.2 \times 10^{3°}$C. Growth rates of several micrometers/minute were observed with undiluted SiH_4, whereas considerably slower rates were observed for gas mixtures consisting of SiH_4/Ar. Film quality was improved when the CO_2 laser was tuned to an absorption minimum for the silane gas, i.e., 10.33 rather than 10.6 μm. CO_2 laser pulses resulted in depositions as small as 50 μm in diameter.

Silicon has also been grown on quartz substrates by laser chemical vapor deposition from the decomposition of $SiCl_4$ using a 50-W CO_2 laser

(Baranauskas *et al.*, 1980). H_2 gas was passed over liquid $SiCl_4$ to provide
the vapor mix of $H_2/SiCl_4$ for the pyrolytic dissociation and deposition.
From a plot of deposition rate versus laser power and from a fit of these
data to calculated temperatures, the investigators determined the activation energy for film growth to be 1.6 eV. A study of the resulting film
thickness as a function of power revealed nucleation patterns corresponding to temperature variations in the plane of the substrate consistent with
the incident Gaussian laser-beam profile. At intermediate powers, ~19 W,
mesa structures were found as a result of Si etching that occurs in the
cooler peripheral regions while simultaneously hotter regions undergo
vapor deposition. The temperature at which etching rather than deposition occurs is a function of the $H_2/SiCl_4$ concentration ratio, and for these
experiments occurred at substrate temperatures below ~900°C.

Polysilicon lines as small as 1 μm in width were obtained from LCVD of
SiH_4 in conjunction with an argon laser (Ehrlich *et al.*, 1981d). Lines on
the order of 2–20 μm wide and 1 μm thick were fabricated by scanning the
sample at speeds of ~100 μm/sec. Si films were grown on both Si and
glass substrates. Fastest growth rates were obtained for small film dimensions, i.e., dimensions comparable to mean free paths of the gas molecules. In contrast to standard CVD, the small dimensions obtainable with
the focused laser permit the product gas, in this case H_2 produced in the
deposition reaction $SiH_4 \rightarrow Si + 2H_2$, to diffuse away more rapidly from
the surface, allowing the reactant gas (SiH_4) to enter the pyrolytic region.
LCVD Si film growth rates of 15 μm/sec have been observed, 30 times
the maximum rate typical for large-area CVD.

LCVD experiments using from 0.5 to 10 W of CO_2 laser power have
been reported by Allen (1981) for depositions described by the reactions

$$Ni(CO)_4 \xrightarrow{140°C} Ni + 4CO \tag{7}$$

$$2TiCl_4 + 4H_2 + 4CO_2 \xrightarrow{900°C} 2TiO_2 + 8HCl + 4CO \tag{8}$$

$$TiCl_4 + CH_4 \xrightarrow{1200°C} TiC + 4HCl \tag{9}$$

Depositions were studied as a function of laser intensity and irradiation
time. The Ni and TiO_2 were deposited on quartz substrates whereas TiC
was deposited on both quartz and stainless-steel samples. For the Ni
depositions it was found that the resulting spot diameter is approximately
proportional to $\sqrt{\tau_0}$ (irradiation time), consistent with thermal diffusion in
the substrate. The spot diameters are generally smaller than the laser-
beam diameter because a temperature threshold is required for the deposition reaction. This result is also consistent with the nonuniform temperature expected from absorption of a Gaussian beam. In the region where
the spot diameter is linear with temperature, the deposition thickness as a
function of distance from the spot's center was double valued, with mini-

mum thickness at the spot's center. This effect is likely due to convection of the gas favoring transport of molecules to the spot's periphery as well as overheating at the center, the latter tending to decrease the sticking coefficient of the vapor. Some hardness, adhesion, electrical conductivity, and grain-size measurements of the Ni-LCVD films were made, though a comparison to films grown by other techniques is not described here. Film thickness ranged from 0.01 to 1 μm with maximum deposition rates as high as ~16 μm/sec. However, deposition rates are found to vary with the length of irradiation time, generally decreasing at longer times.

For TiO_2 deposition the range of irradiation conditions is limited because the reaction temperature is close to the SiO_2 melt temperature. Film thicknesses were found to be approximately linear with the irradiation time, with a single-thickness peak at the center and the deposition. Deposition rates varied from 2 to 20 μm/min and films were found to be clear and adherent.

LCVD films of TiC were made with a partial pressure of 220 Torr of CH_4 in combination with liquid $TiCl_4$. For depositions on steel, a 1.4-kW CO_2 laser was used to compensate for the relatively low optical absorption of the film and the high thermal conductivity of the substrate. Coatings up to 1 μm in thickness were obtained for both stainless-steel and quartz substrates without the use of H_2 buffer gas, normally used in CVD depositions. TiC film thickness on SiO_2 substrates was found to depend linearly on laser irradiation time.

Laser chemical vapor deposition of carbon from reacting C_2H_2 gas on graphite, alumina (ceramic), and tungsten wire has been demonstrated by Leyendecker et al. (1981). An argon-ion laser rather than a CO_2 laser was used mainly for its inherently smaller, diffraction-limited spot size, which results in smaller maskless film patterning as previously noted. A further advantage of the argon laser for LCVD is its transparency to most gases, reducing the probability for spontaneous reactions in undesired regions of the cell. Motion of the substrate in front of the incident laser beam resulted in stripes of carbon as narrow as 81 μm. Figure 17 shows examples of carbon stripes made with a constant substrate-scanning velocity for three different power densities. Figure 18 is an SEM micrograph of a carbon needle grown on a tungsten wire substrate. Temperatures up to 2300 K are estimated for the tip of the needle during formation. A volume deposition rate as high as 2×10^5 μm^3/sec for an incident power density of 3×10^5 W/cm^2 has been observed.

3. *Laser Depositions and Surface Doping*

The interest in metal deposits on semiconductor substrates lies in the possibility of fabricating local diffusion or alloyed regions for a variety of

FIG. 17. Polycrystalline carbon stripes grown on polished alumina at $\sim 3.4 \times 10^3$, $\sim 2.9 \times 10^3$, and $\sim 2.5 \times 10^3$ W/cm^2 of argon laser light. Resulting stripe widths are 26, 23, and 18 μm. Scanning velocity was 42 μm/sec (Leyendecker *et al.*, 1981).

FIG. 18. Carbon needle grown on W wire with 1.37×10^3 W/mm^2 (Leyendecker *et al.*, 1981).

applications. Here the laser can cause both the local deposition and incorporation of the metal atoms into a thin subsurface layer. Deposition of metal films from thermal decomposition of trimethylaluminum, dimethylzinc, and dimethylcadmium for the purpose of alloying on GaAs surfaces using a CW krypton laser as the local heat source has been reported (Rytz-Froidevaux et al., 1981). Spikes of amorphous Al containing some crystalline Al_3C_4 were grown to thicknesses up to 500 μm, and growth rates of 1–5 μm/sec were observed. Growth rates as a function of time were also investigated for Zn depositions. Deutsch et al. (1981b) have used BCl_3 and PCl_3 as parent gases in conjunction with UV excimer lasers to obtain doping of amorphous and polycrystalline Si. ArF and XeF lasers with 7-nsec pulses and ~0.5 J/cm^2/pulse were focused to a line pattern near the surface of the samples such that metal atoms were released onto the surface with simultaneous surface melting to incorporate them in a thin layer near the substrate surface. Translation of the samples inside the gas cell resulted in narrow strips of doped regions whose resistance was subsequently measured with a four-point probe.

For experiments with BCl_3, it was found that the boron was released via photodissociation when the 193-nm laser was used. However, BCl_3 has a small absorption coefficient at 351 nm and the deposition with the XeCl laser is believed to occur via gas pyrolysis. Evidence that surface-adsorbed atoms of BCl_3 are important for the doping process was verified by the decrease in sheet resistivity of a Si sample, laser irradiated in a vacuum subsequent to exposure to 100 Torr of BCl_3. The observed factor-6 decrease in resistivity of these Si samples is believed to be due to photodissociation of adsorbed BCl_3 at the surface followed by diffusion into the Si. Similar adsorption effects were observed with PCl_3. The incorporation of the dopant atoms is consistent with diffusion constants of molten Si, resulting in dopant thickness layers of several thousand angstroms. A plot of sheet resistance as a function of laser fluence for amorphous and single-crystal silicon is shown in Fig. 19. Similar values were obtained for both excimer lasers even though the BCl_3 optical absorption for the two laser wavelengths differs by orders of magnitude, indicative of pyrolytic effects with XeCl irradiation.

4. Applications

The possibility of creating maskless patterns on a local scale using the laser to produce gaseous dissociation has resulted in work on a number of microelectronic applications. An example is found in the repair of a faulty simulated-metal mask consisting of thin-film chromium patterns, ~1000 Å thick, deposited on a quartz substrate (Ehrlich et al., 1980a). For the

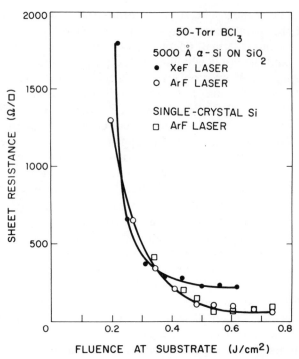

FIG. 19. Sheet resistance of single-crystal and CVD Si as a function of ArF or XeF laser fluence at the substrate sample. Scan rate, 1 mm/min; laser pulse rate, 2 Hz; BCl_3 pressure, 50 Torr (Deutsch *et al.*, 1981b).

repair step, the mask was mounted in a glass cell containing several Torr of dimethylcadmium. A focused 1-mW UV laser beam (frequency-doubled argon) was used to photodissociate and deposit Cd in those regions containing missing metallization. Typically, depositions required from ~2 to 120 sec to complete, depending on the area of additive material required for the repair. Deposition rates were on the order of at least 100 Å/ sec. A mask, before and after repair, is shown in Fig. 20. The mask is shown with a break prior to repair (a), after repair (b), and after laser trimming (c) with the aid of a separate laser trimmer to vaporize regions of unwanted deposition.

Ohmic contacts and the doping of InP by simultaneous heating and photodecomposition of dimethylcadmium have been achieved (Deutsch *et al.*, 1980; Ehrlich *et al.*, 1980d). In another application, p–n junctions were fabricated for use as solar cells by the simultaneous deposition and doping of silicon (Deutsch *et al.*, 1981a). An ArF laser (7-nsec pulsewidth) was used both to produce aluminum or boron atoms from the photolytic

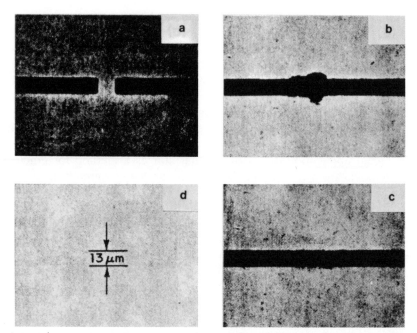

Fɪɢ. 20. Mask repair using a 1-mW UV laser to make Cd deposition from the photodissociation of dimethylcadmium: (a) before repair; (b) after deposition; (c) after laser trimming
with a Nd : YAG laser; (d) width marker (Ehrlich *et al.*, 1980a).

dissociation of $Al(CH_3)_3$ or $\cdot B(CH_3)_3$, respectively, and to bring about
liquid-phase diffusion of these atoms via intense local heating. Typically,
25 laser pulses with 0.1 J/cm^2 per incident pulse were required. Depositions were made on n-type silicon, phosphorus doped to resistivity values of 2–5 Ω-cm. Active regions ~250 μm on a side were produced.
Standard photolithographic techniques were used to define the mesa regions on the silicon while plasma etching was used to remove the n-Si
material surrounding the mesa area. The best solar cells, based on open-
circuit voltage measurements, were achieved with photodissociation of
$B(CH_3)_3$. Solar conversion efficiencies as high as 9.6% were realized
without the use of antireflection coatings.

C. Lᴀsᴇʀ Eᴛᴄʜɪɴɢ—Gᴀsᴇᴏᴜs Aᴛᴍᴏsᴘʜᴇʀᴇs

1. *Introduction*

Interest in laser etching has accelerated during the last few years in part
as a result of the numerous applications related to VLSI and other preci-

sion microelectronic parts. Semiconductor etching is of interest for p–n-junction mesa formation, the fabrication of ink-jet nozzles, and for a number of packaging applications where blind via holes and through holes are required. The direct modification of VLSI circuits for redundancy or personalization can also be achieved with etching as an alternative to the laser ablation techniques discussed earlier. The advantage of laser etching using photodissociation of a gaseous atmosphere is the very small temperature change that usually occurs, thereby lessening the possibility for thermal damage to the wafer and adjacent circuitry.

A recent application of silicon etching utilizes a method for efficient heat removal from the circuit chip which can lead to greatly increased circuit density and speed (Tuckerman and Pease, 1981). Heat removal is accomplished by etching a series of parallel mil-sized grooves on the underside of a chip, covering the grooves with a cemented glass platelet, and forcing water under high pressure through these open-ended channels. Experiments with this configuration have shown that a power density of 800 W/cm^2 applied to the top surface of the Si can be removed, maintaining a steady-state temperature of $\sim 70°C$ above ambient, an enormous improvement over present-day heat-sinking techniques. The increasing use of ceramics in applications such as magnetic disk heads and circuit boards has also made etching an important processing step. Generally, etching is one of the few means of shaping and contouring ceramics because conventional cutting and grinding techniques are good only for planar material removal.

2. Fundamental Experiments

Photolytic and pyrolytic gaseous dissociations to produce etching have been investigated with CO_2, argon, frequency-doubled argon, and excimer lasers. Gases include alkyl halides such as CH_3Br, CH_3I, and CH_3Cl, as well as XeCl and SF_6. Simple inorganic gases such as Br_2, Cl_2, and HCl have also been used. Major interest has centered on silicon as a substrate material. A summary of experiments to be described and the lasers associated with each is given in Table III. Laser etching in a gaseous atmosphere via pyrolytic dissociation is described in a U.S. Patent by Solomon and Mueller (1968), which serves as one of the earliest references for this technique. Slots, apertures, and grooves were formed in a variety of substrates. The sample undergoing etching was housed in a gas cell containing a transparent window to allow the laser beam to be scanned across the workpiece and form the desired pattern. Etching experiments utilizing photodissociation of gases were subsequently described by Sullivan and Kolb (1968) and Baklanov et al. (1974). Sullivan

TABLE III

LASER ETCHING IN GASEOUS ATMOSPHERES

Experiment	Laser	Laser parameters	Reference
Pyrolytic etching and deposition	Unspecified		Solomon and Mueller (1968)
Etching of single-crystal Ge with Br_2	Argon	488.0 nm, CW, up to 40 W	Sullivan and Kolb (1968)
Photoetching of Ge films for mask fabrication	Argon	1-W CW (also 200-W mercury lamp)	Baklanov et al. (1974)
Etching of Si with bromine gas	Argon	488.0 nm, CW, up to 40 W/cm^2	Sveshnikova et al. (1977)
Etching of Si with SF_6—photodissociation	CO_2	Pulsed, 1.4 J/pulse, 50 nsec	Chuang (1980)
Etching of Si with XeF_2—laser-enhanced chemisorption	CO_2	Pulsed, 1.4 J/pulse, 50 nsec	Chuang (1981a)
CF_3Br to etch silicon oxide and nitride	CO_2	Pulsed, 0.2–1.5 J/pulse, 45 nsec	Steinfeld et al. (1980)
Etching of n-InP and n-GaAs with methyl halides—photodissociation	Argon	Frequency doubled, 257.2 nm, up to 100 W/cm^2, 4-mW CW	Ehrlich et al. (1980c)
Etching of Si, photolytic and pyrolytic, using Cl_2 or HCl vapor	Argon	CW multiline, up to 7 W	Ehrlich et al. (1981c)

and Kolb used both a 1-W argon laser and a 200-W mercury lamp with several halogen gases to produce etch patterns through a mask on germanium films. Photoexcited Br_2 gave rise to patterns with 1-μm resolution.

In the work of Baklanov et al. (1974), molecular bromine was also used to etch patterns in Ge. To prove photodissociation rather than pyrolytic dissociation of the gas as the mechanism, the experimenters directed the argon light nearly parallel to the substrate surface, thereby producing minimal substrate heating above ambient. Upon exposure to the 488-nm argon line it was found that Br_2 dissociation occurs, the reaction described by

$$Br_2 + h\nu \longrightarrow Br + Br^* \tag{10}$$

with both the ground state and excited products extremely reactive. It was determined that a 40-W argon laser yields a degree of dissociation of 10–20% near the Ge specimen immediately after the laser is switched on, but that this yield decreases with time and is functionally dependent on the gas pressure. Photodissociation is estimated to be three orders of magnitude greater than thermal dissociation in the temperature range 200–500°C. The etching reaction follows dissociation in several steps starting with

$$Br + Ge \longrightarrow [GeBr]ads \tag{11}$$

where the right side of Eq. (11) is a chemisorbed state on the germanium surface and accounts for the main decrease in the number of bromine molecules in the reaction chamber. The energy of this adsorbed complex can give rise to bond breaking between surface Ge atoms and interior atoms according to

$$[GeBr]ads + Br \longrightarrow [GeBr_2]ads \tag{12}$$

and a final step which results in material removal

$$[GeBr_2]ads + Br_2 \longrightarrow [GeBr_4]ads \longrightarrow [GeBr_4]gas \tag{13}$$

Etching of single-crystal silicon with 488-nm laser radiation and Br_2 has also been studied by Sveshnikova et al. (1977). Etching rates were determined by accurate measurements of sample-weight changes. The rates obtained with Br_2 and a sample temperature of $\sim$400°C with no laser light are $\sim$10^3 times slower than those obtained with $\sim$40 W/cm^2 of argon laser light and a temperature of $\sim$250°C, indicative of a predominantly photolytic process.

Laser isotope separation of SF_6 (Ambartsumyan et al., 1975) has served as fundamental background for the more recent etching experiments that utilize vibrational resonance as a means for achieving molecular dissociation. In the early experiments, a pulsed CO_2 laser irradiated the SF_6 gas in a reaction cell from which the infrared and mass spectra could be monitored both before and after irradiation. When the CO_2 laser was tuned to 947 cm^{-1}, the $^{32}SF_6$ molecules disappeared from the mixture due to collisionless selective dissociation. With 2×10^3 pulses, each 90 nsec wide, and a power density of 1–2 GW/cm^2, the ratio of the two isotopes, $^{34}SF_6/$ $^{32}SF_6$, increased by a factor of 3×10^3. It was also possible to tune the laser to the selective $^{34}SF_6$ absorption band, thereby obtaining an increase in the ratio of $^{32}SF_6/^{34}SF_6$. Similar experiments were performed by Lyman et al. (1975), who describe the dissociation reaction in terms of multiphoton absorption by SF_6 to produce SF_5 + F by either sequential photon absorption or simultaneous absorption of several photons.

Experiments on Si etching using CO_2 laser-excited SF_6 gas were first reported by Chuang (1980, 1981b). It was shown that vibrationally excited fragments, SF_5^* and F*, form SiF_4, which results in Si etching with high-intensity laser fluxes. At low fluxes, vibrationally excited SF_6, i.e., SF_6^*, is formed and is also reactive with silicon. The degree of etching was determined by the change in weight measured in terms of a frequency shift of a quartz crystal microbalance containing the silicon film undergoing etching. With CO_2 irradiation ($\sim$50-nsec pulses) incident on the film, but in the absence of the SF_6, no effect on the microbalance was observed. In the presence of SF_6, gaseous dissociation occurs, resulting also in a reaction on the microbalance, as shown in Fig. 21. The etch rate is

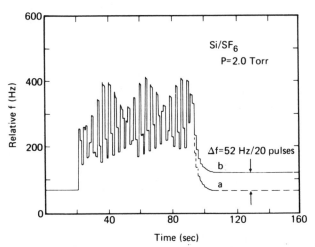

FIG. 21. Frequency response of Si microbalance for 1 J/cm^2 of CO_2 radiation, normally incident, tuned to 942.4 cm^{-1}, shown for (a) no gas and (b) SF_6 at 2.0 Torr. Δf is the change in microbalance frequency occurring for 20 pulses of radiation (Chuang, 1981a).

proportional to the number of laser pulses and is a function of both SF_6 pressure and laser wavelength. At 2.0 Torr of SF_6 and with 1 J/cm² of laser energy/pulse tuned to ~942.4 cm⁻¹, 4.4 × 10¹⁴ Si atoms/pulse are removed. The observed wavelength dependence for the Si etching indicates that photodissociation rather than pyrolytic dissociation must occur for significant etching to take place. It was also found that etching of the Si surface can occur when the laser beam is directed parallel to the sample. The etch rate is then dependent on the distance of the laser beam from the sample because the dissociated SF_6 fragments must diffuse to the sample surface in order to react with the silicon.

Etching experiments similar to those just described for SF_6 were also undertaken in a XeF_2 atmosphere, again with the CO_2 laser (Chuang, 1981a). XeF_2 molecules, however, do not have infrared absorption bands in the CO_2 tunable-laser frequency range. In these experiments, XeF_2 gas was admitted to the reaction chamber from the vapor pressure generated by solid XeF_2 at room temperature. Laser radiation incident on the Si film–quartz microbalance sample in vacuum prior to admission of $XeFe_2$ gas produced no change in the microbalance frequency, i.e., no Si removal. Upon exposure of the Si to XeF_2 gas, but without laser irradiation, a decrease in the frequency of the microbalance was observed, indicative of chemisorption on the Si surface followed by an increase in frequency due to etching. These reactions are believed to proceed via the steps

$$XeF_2(g) + Si(s) \longrightarrow SiF_2(ads) + Xe(g) \uparrow \qquad (14)$$

$$XeF_2(g) + SiF_2(ads) \longrightarrow SiF_4(ads) + Xe(g) \uparrow \qquad (15)$$

With the CO_2 laser incident on the surface in the presence of XeF_2, enhancement in etching is observed. It is believed that the laser promotes an increase in the sticking coefficient for XeF_2 on the Si surface which leads to an increase in the net surface reaction rate. The laser radiation also produces surface excitation and a rearrangement of the atoms to enhance the formation of the volatile adsorbate, SiF_4. This adsorbate leaves the surface as a gas via the intermediate step

$$SiF_2(ads) + SiF_2(ads) \xrightarrow{hv} SiF_4(ads) + Si \qquad (16)$$

$$SiF_4(ads) \longrightarrow SiF_4(g) \uparrow \qquad (17)$$

The effects are independent of laser frequency over the range 920–980 cm⁻¹ and thus not due to direct photolysis of the XeF_2 gas. Similar results have been observed for tantalum surfaces in the presence of XeF_2 with CO_2 laser irradiation, which again results in the formation of volatile adsorbates on the surface to produce etching (Chuang, 1981a).

Dissociation of CF_3Br with a CO_2 laser can also lead to localized etching of SiO_2 (Steinfeld et al., 1980). The laser in these experiments was focused on a spot 1 mm distant from the surface to be etched. Twelve thousand pulses of 0.4 J/pulse of 9.23-μm irradiation were required to achieve an etch depth of 350 Å. It is estimated from the experimental conditions that the lifetime of the radical CF_3 is ~1.5×10^{-6} sec, making the effective etch rate 200,000 Å/sec based on the actual on-time or duty cycle of the laser. This rate is more than three orders of magnitude higher than typical plasma etch rates.

3. High-Resolution Etching of Semiconductors

Several halide gases, CH_3Br, CF_3I, and CH_3Cl, have been used in conjunction with a 4-mW, 257.2-nm UV laser to produce photoetching of both n-type InP and GaAs (Ehrlich et al., 1980b,c). Exposure of either GaAs or InP causes rapid etching which is almost linear over a wide range of intensities. An enhancement in etch rate of only a factor of 6 was observed for GaAs with CH_3Br when the surface temperature was raised to 600°C with the argon laser fundamental at 514.5 nm. The mechanism for CH_3Br etching of GaAs is described by the reactions

$$CH_3Br \xrightarrow{\ h\nu\ } CH_3 + Br \qquad (18)$$

$$Br + GaAs \longrightarrow \begin{Bmatrix} GaBr \\ AsBr \end{Bmatrix} \begin{matrix} {}^{gas} \\ \uparrow \\ {}_{ads} \end{matrix} \qquad (19)$$

The step following Br photodissociation is one of surface adsorption followed by the formation of volatile adsorbates. A summary of etch rates for these semiconductors is given in Table IV. Localization of the excited gas for producing small patterns is aided by the addition of a buffer gas to promote recombination of the reactive atoms and prevent atomic migration to regions where etching is not desired. Such localization has made it possible to obtain microscopic etching in the form of Fresnel diffraction

TABLE IV

LASER-INDUCED ETCH RATES FOR InP AND GaAs USING A
CW 257.2-nm LASER (SPOT SIZE 19 μm FWHM)[a]

Substrate	Etchant	Rate at 100 W/cm² (Å/sec)
(100) n-GaAs	750-Torr CH_3Br	5.2
(100) n-InP	750-Torr CH_3Br	9.4
GaAs, amorphous	750-Torr CH_3Br	9.7
GaAs, amorphous	25-Torr CF_3I	9.9×10^{-2}
(100) n-InP	1000-Torr CH_3Cl	—

[a] From Ehrlich et al. (1980c).

rings. These etch patterns result from the light-intensity variation caused by the optical diffraction pattern of the UV laser directed onto the sample in much the same way as the Fresnel deposition patterns previously discussed. Variations in etch height of 0.5 μm and Fresnel ring spacings on the order of 1.5 μm were obtained, indicating the potential for high-resolution laser etching with this technique.

The argon laser has also been used in the presence of reactive gases to produce etching in both single-crystal and polycrystalline Si (Ehrlich *et al.*, 1981c). Both Cl_2 and HCl gases were used as etchants. Cl_2 photodissociates with 514.5-nm light, but HCl requires wavelengths shorter than 280 nm. Silicon etch rates were found to be a function of crystal orientation, gas pressure, laser intensity, and wavelength. Etch rates as a function of laser intensity are shown in Fig. 22. The nonlinear etch-rate dependence with laser power for (100) silicon suggests that photolysis and thermally activated etching occur up to the silicon melting point. For (111) Si, a greatly enhanced etch rate occurs near the melt temperature, with the rate much smaller than the rate for (100) Si at lower laser powers. The slow etch rate for the (111) direction is consistent with the known anisotropy for Si etching. However, near the melt temperature the crystal symmetry is destroyed and etching proceeds relatively much more rapidly, i.e., very much like that found for (100) Si.

At low power levels, the etch rate for (100) Si was found to increase with decreasing laser wavelength by as much as a factor of 20 in going from the 514.5- to the 457.9-nm argon line due to enhanced Cl_2 photodissociation. Etch rates were also found to increase linearly with Cl_2 pressure between 1 and 400 Torr. The much slower etch rate found for HCl below the Si melt temperature is explained by the fact that HCl does not absorb argon laser light, thus photodissociation cannot occur.

Through holes in 250-μm-thick Si wafers have been etched at an average rate of ~5–7 μm/sec using 7 W of argon laser light with 200 Torr of Cl_2. Thin films of polysilicon deposited onto a Si_3N_4 layer have been etched by scanning the argon laser in the presence of Cl_2 gas to form high-resolution grooves on the order of 1 μm in width and spaced ~5 μm apart. A depth of 4000 Å was obtained with a single scan, the etch rate greatly decreasing once the Si_3N_4 layer was reached.

IV. Laser Etching in Solution: Semiconductors and Ceramics

A. Low-Intensity Photolytic Processes

A number of schemes to obtain etching in solution by photolytic processes have been proposed using either an intense, incoherent light

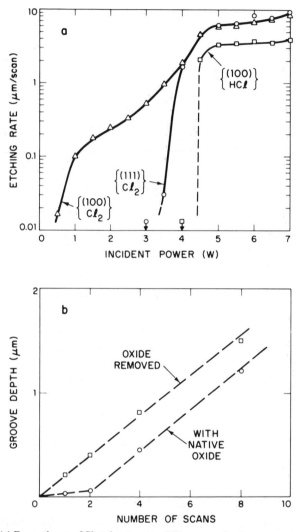

FIG. 22. (a) Dependence of Si etch rate on multiline argon-ion laser power with scan rate of 27 μm/sec; native oxide removed using Cl_2 or HCl gaseous atmosphere. (b) Depth of Cl_2-etched groove vs number of superimposed scans for (100) Si with and without native oxide using 2-W, 488-nm argon light, scanning at 0.9 μm/sec. Both (a) and (b) are focused by an $f/20$ system (Ehrlich *et al.,* 1981c).

source or a low-intensity laser. For example, a U.S. Patent for the photo-etching of gold was issued in 1969 to Schaefer. Both iodoform in benzene and an aqueous solution of potassium ferrocyanide are suggested for use with the gold etching. Etching occurs only upon exposure of the solution to light, typically for periods up to an hour with a 200-W xenon lamp to obtain ~3–4 μm of etching. Patterns are etched by shining the light through a suitable mask. Variable spatial etch rates are available by using a semitransparent mask with varying opacity, as in photographic film.

A second patent by Schaefer describes the use of a photosensitive solution (1-fluorodecane) plus concentrated HCl to etch SiO_2 deposited on a silicon wafer (Schaefer, 1970). The liquid was allowed to evaporate, producing etching of 6000 Å in 4 hr of exposure to a 1000-W mercury lamp. The light was also shone through a mask to achieve silicon etch patterns. A number of other low-optical-intensity etching experiments have been carried out using photodecomposable etchants or electron–hole pair generation. Generally the resulting etch rates are on the order of 2–100 Å/sec.

A more recent experiment in the photoetching of n-type GaAs uses a solution of KOH and a nonlaser light source with a power density on the order of 3 mW/cm^2 to give an etch rate of 2 Å/sec (Hoffman $et\ al.$, 1981). Maximum etch rates occurred when the GaAs/electrolyte contact was maintained at short-circuit conditions, thereby confirming that a net flow of minority carriers at the interface is required for etching. Forward biasing resulted in a greatly reduced etch rate. An example of laser-induced photoelectrochemical etching of p-type GaAs has been described using alternating cathodic and anodic pulses to dissolve the GaAs in two steps (Ostermayer and Kohl, 1981). For a negative pulse applied to the GaAs/electrolyte interface relative to a reference electrode, the reaction

$$GaAs + 3e^- \longrightarrow Ga + As^{3-} \tag{20}$$

occurs, which should lead to surface dissolution was it not for the fact that Ga remains on the surface of the sample and acts to stabilize the surface, preventing further dissolution. Oxidation of the free Ga surface occurs upon making the GaAs surface positive such that

$$Ga + 3h^+ \longrightarrow Ga^{3+} \tag{21}$$

with the Ga^{3+} going into solution. A helium–neon laser with a peak intensity of 0.75 W/cm^2 incident on the GaAs was used to generate electron–hole pairs at the surface to enhance the reactions. Electrolytes for these experiments included a number of solutions of both acids and bases, including H_2SO_4 and NaOH. Etch rates varied from 3 to 22 Å/sec. The dark etch rate, i.e., the etching that occurs without the laser radiation, was found to be approximately 70 times slower compared to etching in the presence of the laser radiation.

The etching of metals and semiconductors with laser-excited aqueous solutions containing dissolved bromine and iodine salts has resulted in patterns with micrometer- and even submicrometer-sized dimensions (Haynes *et al.*, 1980). Both visible (632.8 nm from a He–Ne laser) and UV (413.1 nm from a krypton laser) radiation have been used as excitation sources with power varying between ~0.1 and 2.1 mW. Patterns in the form of spots were etched with a simple positive lens to focus the laser whereas more elaborate patterns were obtained by irradiating the samples through a mask. The observed linear etch-rate dependence with power is indicative of a nonpyrolytic mechanism. The reverse current is supplied here by the semiconductor hole current and the electrolytic electron current and obviates the need for an external bias voltage. The electrolytic electron current is provided by the optical dissociation of Br_2 followed by a recombination of a Br atom with a Br ion. The ion is obtained from dissociated KBr salt added to the solution. With this etching technique it was possible to produce grating patterns with as many as 200 lines/mm. Etch rates are on the order of 300 Å/sec for a helium–neon power density of $\sim 4.0 \times 10^{-2}$ W/cm^2.

Etching of both undoped and semiinsulating compound semiconductors for micromachining purposes has also been observed using liquid etchants that include H_2SO_4 and H_2O_2 mixtures, as well as solutions of HCl, HNO_3, and KOH (Osgood *et al.*, 1982). Materials successfully processed include semiinsulating GaAs, n-type GaAs, CdS, and InP, into which blind via holes, through holes (in ~125-μm-thick samples), and gratings were etched using any one of several visible and near-UV lasers. Gratings with better than 0.5-μm resolution were obtained. At low laser power densities, the mechanism for etching is described in terms of electron–hole pair production. Etch rates are dependent on the pH of the solution and crystal orientation. The latter relates to the differences in band energy with orientation and hence to the band bending of the semiconductor solution interface or Schottky barrier. Etch rates up to 2 μm/min for 100 W/cm^2 of laser power density were observed.

B. Etching in Liquids—High-Temperature Effects

A recent scheme for the etching of ceramics, semiconductors, and certain glasses has been described in which localized melting and etching occur simultaneously (von Gutfeld and Hodgson, 1982). A focused argon laser beam shines into a transparent quartz cell containing the sample and a concentrated solution of potassium hydroxide (2–18 M). The incident laser-beam intensity is on the order of 10^6–10^7 W/cm^2. This technique has been used to fabricate slots, blind holes, and through holes in three ceramic materials, i.e., alumina/TiC, MnZn ferrite, and alumina, and in

⟨111⟩ and ⟨100⟩ Si, WC (tungsten carbide), and several optically absorbing glasses. Results for volume removal of alumina/TiC and ⟨111⟩ Si are shown in Fig. 23. Each point is an average volume measured from several holes made with a fixed laser exposure time. For Si there is no measurable removal below 3.3 W, whereas for Al_2O_3/TiC, removal stops below approximately 1 W. For the material removal process it is postulated that melting and/or vaporization contribute to (1) direct removal of material, (2) increasing the surface area in contact with the etchant, and (3) local temperature increase of both sample and etchant to promote thermally activated kinetics.

In general, silicon etches anisotropically in KOH, the ⟨111⟩ direction

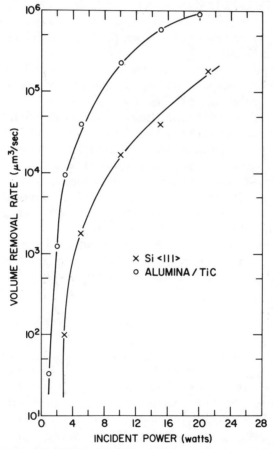

FIG. 23.   Volume removal from Al_2O_3/TiC ceramic and (111) Si as a function of laser power (curves added to aid the eye) (von Gutfeld and Hodgson, 1982).

etching only a negligibly small amount with KOH and the $\langle 100 \rangle$ direction etching rapidly. Laser etching of $\langle 111 \rangle$ Si in the present scheme becomes possible in KOH by the local melting of the sample. Melting destroys the crystal symmetry such that etching results are similar for both $\langle 111 \rangle$ and $\langle 100 \rangle$ Si samples.

An example of a laser-etched hole in alumina/TiC is shown in Fig. 24a (Al–TiC). The quality of the etched hole stands in sharp contrast with holes etched in air (Fig. 24b) or water (Fig. 24c). For samples etched either in air or water there is evidence of melting and refreezing without complete material removal. Voids and cracks are also generated. Holes made in water show little or no material removal whereas holes made in air have a refrozen peripheral rim.

A groove in a $\langle 111 \rangle$ Si sample made by scanning the sample in KOH through a focused argon laser beam is shown in Fig. 24d. Material removal is uniform, and smooth wall surfaces result. From this sample as well as from the volume removal rates (Fig. 23) it is clear that etch rates using the present method can be extremely rapid. The technique may become useful for the micromachining of Si and ceramic parts used in the microelectronics industry, especially where variable depths of a single part are required. Such contouring is not possible with present reactive ion-etching fabrication techniques.

An experiment in which dopant atoms of Zn were deposited onto Al films has been reported by Ehrlich et al. (1981a). The 200- to 500-Å-thick deposition by laser photodissociation with simultaneous local laser heating of the Al caused diffusion of the Zn at temperatures well below the Al melt temperature. This local change in composition makes possible subsequent selective etching of the alloyed region. The photodissociation of $Zn(CH_3)_2$ was obtained with 3 mW of frequency-doubled argon light while 2 W of the fundamental argon light was used to cause the Zn diffusion. An aqueous acetic acid solution was used to etch the alloyed region for 1 or 2 min. In contrast, the untreated Al film etched at a rate of only 0.1 Å/min at room temperature. Etched lines have been fabricated by this method in conjunction with a stepping motor to translate the sample relative to the laser beam.

In a somewhat related technique, Salathè et al. have used high-intensity laser power to produce local melting and regrowth of semiconductors in order to obtain anisotropic etching (Salathè et al., 1979). A CW argon-ion laser was scanned across an AlGaAs structure, forming regrown regions that were found to etch more rapidly than the unexposed AlGaAs. After regrowth, the sample was etched in aqueous NH_3 and H_2O_2 which removed the regrown material, leaving AlGaAs mesa structures with widths as small as 2.5 μm.

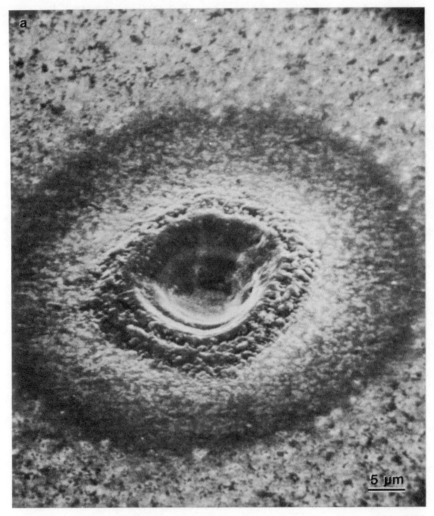

FIG. 24.   Effect of laser-etching Al_2O_3/TiC ceramic in (a) air, (b) water, and (c) KOH. Note the rim formation around holes as well as voids and refrozen material in (a) and (b) not present in (c). Holes were made with 1.5 W of argon laser light applied for 1 sec. (d) Slot scanned at 16 μ/sec with 1.5 W of argon laser light (von Gutfeld and Hodgson, 1982).

V. Laser-Enhanced Plating and Etching: Pyrolytic Effects

A. INTRODUCTION

Laser-enhanced electroplating (and electroetching) using simple non-photosensitive salt solutions such as $CuSO_4$ and $NiCl_2$ as electrolytes

was investigated over the entire polarization curve, the curve of plating current versus applied overpotential between reference electrode and cathode. A special cathode was designed with a very small electroactive area using photoresist to cover the entire cathode surface except for a small circular aperture. Focused laser light with a spot of approximately the aperture diameter was directed into the pinhole opening. A polarization curve for this cathode is shown in Fig. 26a with the laser turned on at 50-mV intervals for several seconds. At those intervals, large current spikes are observed. It is especially interesting that current enhancement (hence plating enhancement) occurs over the entire curve. This same curve on a magnified current scale is shown in Fig. 26b, without the laser incident on the pinhole. An analysis of these two curves led to the conclusion that the laser enhancement in the low-overpotential region (called the kinetic or charge-transfer region) is due to increased kinetics or charge-transfer rates brought about by locally higher temperatures produced by the laser absorption on the cathode. This result is readily understood in terms of the Butler–Vollmer charge-transfer equation which contains thermally activated terms (Bockris and Reddy, 1974). The charge-transfer region for this case spans the overpotential η from 0 to ~ -300 mV. At higher overpotentials ($|300|$ mV $< \eta < |700|$ mV), the plating current (plating rate) without the laser becomes limited by the available ions in solution and the current remains nearly constant with increasing voltage. In this plateau region of the overpotential curve the plating is mass-transport or diffusion limited. In commercial plating, agitation of the plating bath by mechanical or ultrasonic stirring is used to bring a fresh supply of ions into the otherwise ion-depleted cathode region. Alternatively, the cathode workpiece may be mechanically oscillated in the bath. With the laser, large current enhancements are observed in this plateau region (Fig. 26a) due to local hydrodynamic stirring brought about by large local temperature gradients in the electrolyte near the region of laser absorption.

An additional factor related to the laser plating enhancement is a shift that occurs in the rest potential with temperature variation. In the case of Cu/Cu^{2+} this shift is toward a more positive direction with increasing temperature. A positive shift in the rest potential displaces the overpotential curve in a direction to give effectively greater current at a given overpotential. For a Cu/Cu^{2+} electrode maintained at a fixed overpotential, the local rest potential of the region receiving laser radiation will be shifted and result in a higher plating current in that region. This permits laser plating at zero-applied overpotential. Under these conditions, laser plating occurs without any background plating because the nonilluminated regions are maintained at a voltage that results in zero net current flow.

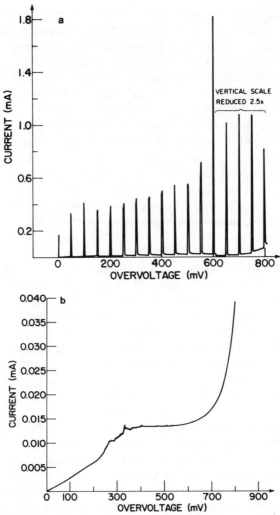

FIG. 26. (a) Polarization curve for copper electrodeposition with periodic laser illumination (represented by current spikes). Electrode consisted of glass slide with thin-metal predeposition and photoresist covering all but a small circular aperture. (b) Polarization curve for same electrode as (a) but with no incident laser light (Puippe *et al.*, 1981).

C. LASER PLATING WITHOUT EXTERNAL EMF

The positive shift in local rest potential with laser radiation (heating) creates a local thermobattery with a potential difference arising between the heated and unheated region of the electrode. Plating will occur with laser radiation even without the use of any external electromotive source (von Gutfeld *et al.*, 1979b, 1982; Puippe *et al.*, 1981). Laser radiation incident on such a Cu/Cu^{2+} electrode results in plating in the region absorbing radiation, with simultaneous etching in areas peripheral to the incident laser beam. The etching must occur in order to maintain a net charge conservation. The depth of etching, however, is generally small because a relatively large surface area is available to contribute the ions necessary to compensate for the relatively heavy plating occurring in the much smaller area irradiated. Extensive work on gold plating using the thermobattery effect has been undertaken on both thin-film substrates on glass and bulk copper alloys. The latter has particular interest for local gold plating of contacts on electronic materials used for microcircuitry connectors (von Gutfeld *et al.*, 1981, 1982).

A second type of laser-enhanced plating using no external power source is electroless plating (von Gutfeld *et al.*, 1979b, 1982). This is a process in which charge conservation is maintained via a catalyst in solution rather than by simultaneous etching. Experiments on laser-enhanced electroless nickel plating have been reported using nickel plating solutions containing sodium hypophosphate as the catalytic reducing agent. Large enhancements are possible here because plating speeds of electroless solutions are very temperature dependent. Thus background plating can be reduced to almost zero by designing a solution that requires temperatures higher than ambient or room temperature to produce a substantial plating rate, the higher local temperature being supplied by laser absorption.

D. DEPOSITIONS ON InP

Metal depositions on undoped n- and p-type InP samples immersed in various aqueous metal salt solutions have been made without an external EMF using 10-nsec-wide dye laser pulses. The purpose of these deposits is to form ohmic contacts (Karlicek *et al.*, 1982). The maximum laser energy used was 5 mJ/pulse which, when directed onto the sample with a 1-mm beam diameter, gave a power density of 5×10^7 W/cm^2. This flux is sufficient to cause thermal decomposition of the InP, but without surface damage. Platinum, gold, and nickel from solutions of chloroplatinic, chloroauric, and nickel sulfate (H_2PtCl_6, $NiSO_4$), respectively, have been deposited with from 1 to 5000 pulses. The platinum deposits were examined by Auger analysis and were found to be free of solution contaminants such as oxygen or chlorine. The investigators suggest a thermal deposition mechanism because the solutions cannot absorb at the laser wavelengths used, 580–720 nm. However, details of the mechanism are not described further in their report.

E. APPLICATIONS

An important example of an application of laser-enhanced electroplating is the bridging of two copper lines to demonstrate circuit repair (Puippe *et al.*, 1981). The copper lines, 12 μm in width and 5 μm high, were fabricated on a glass substrate coated with predeposited 200-Å Nb/ 1500-Å gold layers using standard masking and plating techniques. This simulated circuit board was immersed in a $CuSO_4$ solution, subjected to focused argon laser light, and scanned between the two copper lines. The circuit board was held at zero overpotential such that virtually no background plating occurred. A scanning electron micrograph of the laser-plated bridge is shown in Fig. 27. Electrical conductivity of the bridge was measured and found equal to that of the original copper lines. In other experiments, copper lines as narrow as 2 μm have been produced on copper-coated glass substrates (Fig. 28) using the thermobattery effect. A number of gold-plating experiments on bulk samples of nickel-plated copper and beryllium–copper have been performed (von Gutfeld *et al.*, 1981). In spite of the high thermal conductivity of the substrates, spot diameters as small as 50 μm were obtained, indicative of a high degree of plating localization. Some cross-sectioning of laser-electroplated and thermobattery-deposited gold has been completed, indicating dense deposits, as required for electrical contacts. Work in this area is continuing and initial experiments for this application have been reported.

VI. Concluding Remarks

Based on recent and continuing research, it appears evident that lasers will continue to play an increasingly important role in the area of VLSI and microelectronic processing. A more detailed understanding of the many photolytic and pyrolytic chemical processes is still needed. In the case of film depositions, grain structure and wear characteristics must be ascertained. A number of ongoing experiments are proceeding in these directions. New understanding will surely enlarge the increasing field of possible applications. At present, laser personalization of VLSI for circuit repair is already in place in the manufacturing environment. The other topics of this article, i.e., laser depositions for maskless patterning and alloying, as well as laser etching, are still in a more exploratory stage. Their acceptance as routine processes in the manufacture of microelectronic parts would appear to be at hand in the not too distant future.

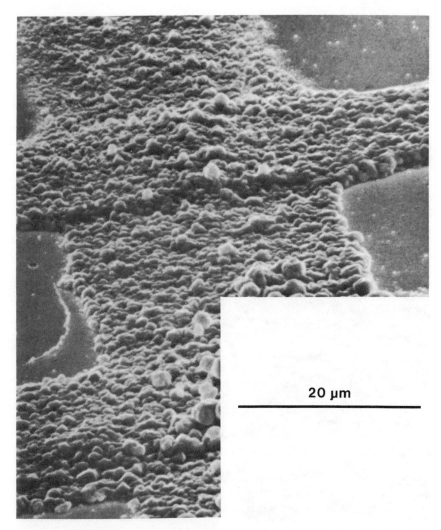

FIG. 27. SEM micrograph of laser-enhanced electroplated bridge on a glass substrate containing predeposited Nb/Au, 200/500 Å, with 5-μm-high electroplated copper lines. The bridge was formed by scanning the laser between the Cu lines. The sample was maintained at zero overpotential using a potentiostat (Puippe *et al.*, 1981).

VII. Addendum

Additional papers of interest have appeared relating to laser processing of circuits and microelectric materials since the completion of this article.

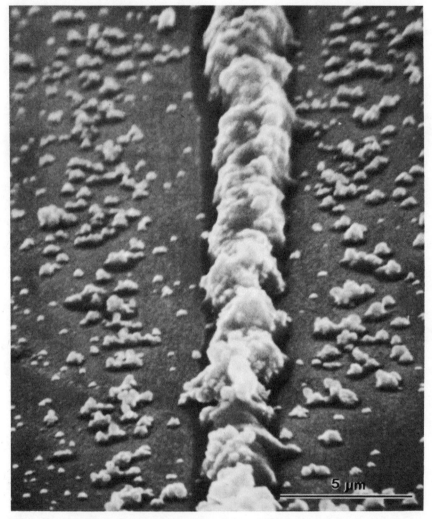

FIG. 28. Laser-plated copper using thermobattery effect to produce a 2-μm-wide line (Puippe *et al.*, 1981).

Some of these publications are listed in Table V. In addition, three conferences devoted in part to this topic were held over the past 14 months; these are given in the footnote to Table V and the interested reader is encouraged to refer to the published abstracts and papers of these talks for detailed updates. One subject not described earlier in this text, the ablative photodecomposition of organic polymer films by far-UV excimer laser irradiation, has developed rapidly with important potential applications and is summarized herein.

Radiation shorter than 200 nm is intensely absorbed by nearly all organic polymers, with more than 95% of the photons absorbed in a depth of ~3000 Å. Such radiation has an efficiency of >50% for breaking bonds in organic material. When a laser pulse of 193 nm impinges on an organic film, numerous small fragments are created in a small volume. These fragments ablate explosively above a threshold value of ~10 mJ/cm^2 and burst out of the irradiated volume. The energy of the photons exceeds the dissociation energy of any of the bonds of the organic molecule such that excess energy is carried away by the fragments. This process is termed "ablative photodecomposition" (Srinivasan and Mayne-Banton, 1982). Examination of the substrate by optical and scanning electron microscopy has shown that the area of removal of the material by this photodecomposition process is completely defined by the light. The substrate shows no

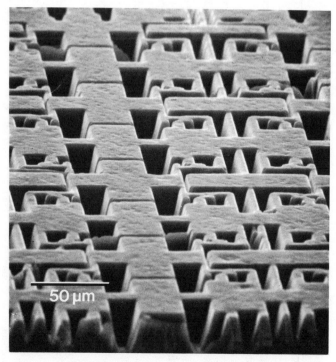

FIG. 29. Scanning electron micrograph of commercial photoresist patterned by ablative photodecomposition. Thickness of polymer, 50 μm; etch depth, 160 nm/pulse at 245 nJ/pulse. (Photo courtesy R. Srinivasan.)

TABLE V

RECENT DEVELOPMENTS IN LASER APPLICATIONS[a]

Subject	Reference	Description
Ablative photodecomposition using far UV	Srinivasan, R., and Mayne-Banton, V. (1982). *Appl. Phys. Lett.* **41**, 576.	Experimental results and theoretical description of far-UV (193-nm) ablation of organic polymers to create etch patterns
	Srinivasan, R., and Leigh, W. J. (1982). *J. Am. Chem. Soc.* **104**, 6784.	Analysis of ablative products from polyethylene terephthalate using 193-nm laser
Thermal ablation for patterning	von Gutfeld, R. J., Gelchinski, M. H., and Romankiw, L. T. (1983). *J. Electrochem. Soc.* **130**, 1840.	Method for using a Q-switched Nd:YAG laser for creating patterns defined by ablating photoresist or other organic on Ni-plated Be–Cu: the ablated region is then gold plated
LCVD	Bauerle, D., Leyendecker, G., and Wagner, D. (1983). *Appl. Phys.* **A30**, 147.	Growth of single-crystal Si from silane using a krypton laser
	Krauter, W., Baurle, D., and Fimberger, F. (1983). *Appl. Phys.* **A31**, 13	Growth of Ni needles and thin-film stripes from Ni(CO)$_4$ using a krypton laser
	Leyendeker, G., Noll, H., Bauerle, D., Geittner, P., and Leydtin, H. (1983). *J. Electrochem. Soc.* **130**, 157.	Determination of activation energies from carbon deposition rates by LCVD
Photochemical deposition	Tsao, J. Y., Becker, R. A., Ehrlich, D. J., and Leonberger, F. J. (1983). *Appl. Phys. Lett.* **42**, 559.	Photodeposition of Ti from TiCl$_4$ using excimer lasers; importance of surface catalyzation. Used to dope LiNbO$_3$ optical waveguides with Ti

62

	Description	Reference
	A method of separating photochemical deposition effects of organometallics into effects occurring in the gas volume and in the adsorbed layer on the substrate surface	Wood, T. H., White, J. C., and Thacker, B. A. (1983). *Appl. Phys. Lett.* **42**, 408.
	Use of argon frequency-doubled laser to deposit methyl methacrylate. Importance of surface catalysis in the photopolymerization, and use of process to pattern Al	Tsao, J. Y., and Ehrlich, D. J. (1983). *Appl. Phys. Lett.* **42**, 997.
	Large area depositions and characterization of ZnO films photodeposited with an excimer laser from dimethylzinc and nitrogen oxide	Solanki, R., and Collins, G. J. (1983). *Appl. Phys. Lett.* **42**, 662.
Laser-formed connections	Laser electrical link structures of ~1-kΩ resistance are formed between conducting lines using 1-msec argon pulses	Raffel, J. K., Freidin, J. F., and Chapman, G. H. (1983). *Appl. Phys. Lett.* **42**, 705.
Laser-enhanced electroplating	Laser plating on copper–zinc alloys. Argon laser with thin layers of low-thermal-conductivity material to increase temperature rise produced by laser	Kuiken, H. K., Mikkers, F. E. P., and Wierenga, P. E. (1983). *Electrochem. Soc.* **130**, 554.
	Study of laser plating on both Ni and Ni-plated Be–Cu. Effects of plating current, laser intensity, and laser-beam scanning on the morphology of laser-plated gold	Gelchinski, M. H., Ramankiw, L. T., and von Gutfeld, R. J. (1982). *Ext. Abstr., Meet.—Electrochem. Soc.* (Abstr. No. 131).
Laser-enhanced etching	Difference between nonthermal and thermal laser-enhanced Si-etching effects in terms of etch products using XeF_2 gas	Houle, F. A. (1983). *Chem. Phys. Lett.* **95**, 5.

[a] Conferences: Materials Research Society, November 1–4, 1982, Boston Massachusetts: symposium on ''Laser Diagnostics and Photochemical Processing for Semiconductor Devices'' (proceedings published by Elsevier, Amsterdam); Conference on Lasers and Electrooptics (CLEO 1982), April 13–15, 1982, Phoenix, Arizona; Conference on Lasers and Electrooptics (CLEO 1983), May 17–20, 1983, Baltimore, Maryland.

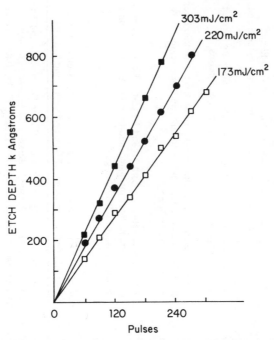

FIG. 30. Plot of etch depth as a function of the number of pulses at constant fluence per pulse. (Graph courtesy of R. Srinivasan.)

sign of heating to its softening point for a softening temperature as low as 100°C. Analytical data on the products of ablative photodecomposition of polyethylene terephthalate films substantiate the view that the photodecomposition is a nonlinear process (Srinivasan and Leigh, 1982). The ablation process shows considerable promise for the patterning of polymer-based circuits. An etch pattern prepared with a mask with 193-nm laser light is shown in Fig. 29. The thickness of the polymer (photoresist) is 50 μm, with the etch depth 160 nm/pulse using 245 mJ/pulse. The dependence of etch depth on the number of pulses, with fluence/pulse as a parameter, is shown in Fig. 30.

Ablation of organic films at longer excimer laser wavelengths (248 or 308 nm) has also been observed, but is believed to depend more strongly on a combination of both thermal and photochemical decomposition processes.

ACKNOWLEDGMENTS

The author is grateful to the IBM management for their encouragement and help in the preparation of this material. In particular, S. E. Schuster and J. M. Woodall contributed valuable suggestions and discussions. In addition, the assistance of J. P. Moruzzi in preparing the manuscript is gratefully acknowledged.

References

Allen, S. D. (1981). *J. Appl. Phys.* **52**, 6501–6505.

Ambartsumyan, R. V., Gorokhov, Y. A., Letokhov, V. S., and Makarov, G. N. (1975). *JETP Lett. (Engl. Transl.)* **21**, 171–172.

Arnone, C., Daneu, V., and Riva-Sanseverino, S. (1980). *Appl. Phys. Lett.* **37**, 1012–1013.

Baklanov, M. R., Beterov, I. M., Repinski, S. M., Rzhanov, A. V., Chebotaev, V. D., and Yurshina, V. I. (1974). *Sov. Phys.—Dokl. (Engl. Transl.)* **19**, 312–314.

Baranauskas, V., Mammana, C. T. Z., Klinger, R. E., and Greene, J. E. (1980). *Appl. Phys. Lett.* **36**, 930–932.

Bindels, J. F. M., Chlipala, J. D., Fischer, F. H., Mantz, T. F., Nelson, R. G., and Smith, R. T. (1981). *Dig. Tech. Pap., Int. Solid-State Circuit Conf., 1981,* pp. 82–83.

Bockris, J. O'M., and Reddy, A. K. N. (1974). *In* "Modern Electrochemistry," Vol. 2, Chapter 8, p. 845. Plenum, New York.

Cenker, R. P., Clemons, D. G., Huber, W. R., Petrizzi, J. B., Procyk, F. J., and Trout, G. M. (1979). *Dig. Tech. Pap., Int. Solid-State Circuit Conf., 1979,* pp. 150–151.

Christensen, C. P., and Lakin, K. M. (1978). *Appl. Phys. Lett.* **32**, 254–256.

Chuang, T. J. (1980). *J. Chem. Phys.* **72**, 6303–6304.

Chuang, T. J. (1981a). *J. Chem. Phys.* **74**, 1461–1466.

Chuang, T. J. (1981b). *J. Vac. Sci. Technol.* **18**, 638–642.

Cohen, M. I. (1967). *Bell Lab. Rec.* **45**, 247.

Cohen, M. I., Unger, B. A., and Milkosky, J. S. (1968). *Syst. Tech. J.* **47**, 385–485.

Cook, P. W., Schuster, S. E., and von Gutfeld, R. J. (1975). *Appl. Phys. Lett.* **26**, 124–126.

Coombe, R. D., and Wodarczyk, F. J. (1980). *Appl. Phys. Lett.* **37**, 846–848.

Dabby, F. W., and Paek, U.-C. (1972). *IEEE J. Quantum Electron.* **QE-8**, 106–111.

Deutsch, T. F., Ehrlich, D. J., and Osgood, R. M., Jr. (1979). *Appl. Phys. Lett.* **35**, 175–177.

Deutsch, T. F., Ehrlich, D. J., Osgood, R. M., Jr., and Liau, Z. L. (1980). *Appl. Phys. Lett.* **36**, 847–848.

Deutsch, T. F., Fan, J. C. C., Turner, G. W., Chapman, R. L., Ehrlich, D. J., and Osgood, R. M., Jr. (1981a). *Appl. Phys. Lett.* **38**, 144–146.

Deutsch, T. F., Ehrlich, D. J., Rathman, D. D., Silversmith, D. J., and Osgood, R. M., Jr. (1981b). *Appl. Phys. Lett.* **39**, 825–827.

Draper, C. W. (1980a). *Metall. Trans., A* **11A,**349–351.

Draper, C. W. (1980b). *J. Phys. Chem.* **84**, 2089–2090.

Ehrlich, D. J., Osgood, R. M., Jr., Silversmith, D. J., and Deutsch, T. F. (1980a). *IEEE Electron Device Lett.* **EDL-1**, 101–103.

Ehrlich, D. J., Osgood, R. M., Jr., and Deutsch, T. F. (1980b). *IEEE J. Quantum Electron.* **QE-16**, 1233–1243.

Ehrlich, D. J., Osgood, R. M., Jr., and Deutsch, T. F. (1980c). *Appl. Phys. Lett.* **36**, 698–703.

Ehrlich, D. J., Osgood, R. M., Jr., and Deutsch, T. F. (1980d). *Appl. Phys. Lett.* **36**, 916–918.

Ehrlich, D. J., Osgood, R. M., Jr., and Deutsch, T. F. (1981a). *Appl. Phys. Lett.* **38**, 399–401.

Ehrlich, D. J., Osgood, R. M., Jr., and Deutsch, T. F. (1981b). *J. Electrochem. Soc.* **128**, 2039–2041.

Ehrlich, D. J., Osgood, R. M., Jr., and Deutsch, T. F. (1981c). *Appl. Phys. Lett.* **38**, 1018–1020.

Ehrlich, D. J., Osgood, R. M., Jr., and Deutsch, T. F. (1981d). *Appl. Phys. Lett.* **39**, 957–959.

Feder, M. P., Smith, J. F., and Liberman, H. E. (1978). *Dig. Tech. Pap., Conf. Laser Electropt. Syst., 2nd, 1978*, pp. 92–93.

Gagliano, F. D., Lumley, R. M., and Watkins, L. S. (1969). *Proc. IEEE* **57**, 114.

Gibbons, T. F., Hess, L. D., and Sigmon, T. W., eds. (1980). "Proceedings of Material Research Symposia." North-Holland Publ., Amsterdam.

Hanabusa, M., Namiki, A., and Yoshihara, K. (1979). *Appl. Phys. Lett.* **35**, 626–627.

Hanfmann, A. M. (1973). U.S. Patent 3,400,456 issued Sept. 10, 1968; U.S. Patent Re. 27,772 issued Oct. 2, 1973.

Haynes, R. W., Metze, G. M., Kreismanis, V. G., and Eastman, D. E. (1980). *Appl. Phys. Lett.* **37**, 344–346.

Hoffman, H. S., Woodall, J. M., and Chappell, T. I. (1981). *Appl. Phys. Lett.* **38**, 564–566.

Karlicek, R. F., Donnelly, V. M., and Collins, G. J. (1982). *J. Appl. Phys.* **53**, 1084–1090.

Karny, Z., Naaman, R., and Zare, R. N. (1978). *Chem. Phys. Lett.* **59**, 33–37.

Kuhn, L., Schuster, S. E., Zory, P. S., Jr., Cook, P. W., and von Gutfeld, R. J. (1974). *Tech. Dig.—Int. Electron Devices Meet.*, pp. 557–560.

Kuhn, L., Schuster, S. E., Zory, P. S., Jr., Lynch, G. W., and Parrish, J. T. (1975). *IEEE J. Solid-State Circuits* **SC-10**, 219–228.

Leyendecker, G., Bäuerle, D., Geittner, P., and Lydtin, H. (1981). *Appl. Phys. Lett.* **39**, 921–923.

Lin, S. T., and Ronn, A. M. (1978). *Chem. Phys. Lett.* **56**, 414–418.

Logue, J. C., Kleinfelder, W. J., Lowry, P., Moulic, J. R., and Wu, W. W. (1981). *IBM J. Res. Dev.* **25**, 107–115.

Lyman, J. L., Jensen, R. J., Rink, J., Robinson, C. T., and Rockwood, S. D. (1975). *Appl. Phys. Lett.* **27**, 87–89.

Minato, O., Masuhara, T., Sasaki, T., Yoshizaki, K., and Sakai, Y. (1981). *Dig. Tech. Pap., Int. Solid-State Circuit Conf., 1981*, pp. 14–15.

North, J. C. (1977). *J. Appl. Phys.* **48**, 2419–2423.

North, J. C., and Weick, W. W. (1976). *IEEE J. Solid-State Circuits* **SC-11**, 500–505.

Osgood, R. M., Jr., Sanchez-Rubio, A., Ehrlich, D. J., and Daneu, V. (1982). *Appl. Phys. Lett.* **40**, 391–393.

Ostermayer, F. W., Jr., and Kohl, P. A. (1981). *Appl. Phys. Lett.* **39**, 76–78.

Perry, P. B., Ray, S. K., and Hodgson, R. T. (1981). *Thin Solid Films* **85**, 111–117.

Platakis, N. S. (1976). *J. Appl. Phys.* **47**, 2120–2128.

Puippe, J.-C., Acosta, R. E., and von Gutfeld, R. J. (1981). *J. Electrochem. Soc.* **128**, 2539–2545.

Raffel, J. I., Naiman, M. L., Burke, R. L., Chapman, G. H., and Gottschalk, P. G. (1980). *Tech. Dig.—Int. Electron Devices Meet.*, pp. 132–135.

Ronn, A. M. (1976). *Chem. Phys. Lett.* **42**, 202–204.

Rytz-Froideveaux, Y., Salathè, R. P., and Gilgen, H. H. (1981). *Phys. Lett. A* **84A**, 216–218.

Salathè, R. P., Gilgen, H. H., Rytz-Froideveaux, Y., Lüthy, W., and Weber, H. P. (1979). *Appl. Phys. Lett.* **35**, 543–545.

Schaefer, D. L. (1969). U.S. Patent 3,482,975 issued Dec. 9, 1969.

Schaefer, D. L. (1970). U.S. Patent 3,520,687 issued July 14, 1970.

Schuster, S. E. (1978). *IEEE J. Solid-State Circuits* **SC-13**, 698–703.

Smith, J. F., Yanavage, K. T., and Moulic, J. R. (1981). *1st Int. Laser Proc. Conf., 1981* (Paper No. 17).
Smith. R. T., Chlipala, J. D., Bindels, J. F. M., Nelson, R. G., Fischer, F. H., and Mantz, T. F. (1981). *IEEE J. Solid-State Circuits* **SC-16,** 506–514.
Solanki, R., Boyer, P. K., Mahan, J. E., and Collins, G. J. (1981). *Appl. Phys. Lett.* **38,** 572–574.
Solomon, R., and Mueller, L. F. (1968). U.S. Patent 3,364,087 issued January, 1968.
Srinivasan, R., and Leigh, W. J. (1982). *J. Am. Chem. Soc.* **104,** 6784.
Srinivasan, R., and Mayne-Banton, V. (1982). *Appl. Phys. Lett.* **41,** 576.
Steinfeld, J. I., Anderson, T. G., Reiser, C., Denison, D. R., Hartsough, L. D., and Hallahan, J. R. (1980). *J. Electrochem. Soc.* **127,** 514–515.
Sullivan, M. V., and Kolb, G. A. (1968). *Electrochem. Technol.* **6,** 430–434.
Sveshnikova, L. L., Donin, V. I., and Repinski, S. M. (1977). *Sov. Tech. Phys. Lett. (Engl. Transl.)* **3,** 223–224.
Sypherd, A. D., and Salman, N. D. (1968). *Proc. Natl. Electron. Conf.* **24,** 206–208.
Tuckerman, D. B., and Pease, R. F. W. (1981). *IEEE Electron Device Lett.* **EDL-2,** 126–129.
von Gutfeld, R. J., and Hodgson, R. T. (1982). *Appl. Phys. Lett.* **40,** 352–354.
von Gutfeld, R. J., and Puippe, J.-C. (1981). *Oberfläeche—Surf.* **11,** 294–297.
von Gutfeld, R. J., Tynan, E. E., Melcher, R. L., and Blum, S. E. (1979a). *Appl. Phys. Lett.* **35,** 651–653.
von Gutfeld, R. J., Tynan, E. E., and Romankiw, L. T. (1979b). *Ext. Abstr., Meet.—Electrochem. Soc.* (Abstr. No. 472), 1185.
von Gutfeld, R. J., Gelchinski, M., and Romankiw, L. T. (1981). *Proc.—Electrochem. Soc.* (Abstr. 663 RNP).
von Gutfeld, R. J., Acosta, R. E., and Romankiw, L. T. (1982). *IBM J. Res. Dev.* **26,** 136–144.
White, C. W., and Peercy, P. S., eds. (1980). "Laser and Electron Beam Processing of Materials." Academic Press, New York.

LASER PHOTOCHEMISTRY

Thomas F. George, A. C. Beri, Kai-Shue Lam, and Jui-teng Lin[1]
Department of Chemistry
University of Rochester
Rochester, New York

I. Introduction

The phenomenon of "light amplification by stimulated emission of radiation" (the laser), discovered about two decades ago, has had a significant influence on the research programs of the physical, engineering, and biological communities. Although the major effort has been in laser development, a variety of laser applications have been explored, such as the use of lasers as scalpels in surgery, as means of communication by fiber optics, and as components in programs of nuclear energy development. The application of lasers in chemistry is still in the early stages, and whereas this has been generally confined to basic research projects, it appears that such application is beginning to see practical advantages in government and industry.

The organization of this article presents two sections apart from this introductory section. Section II deals with spectroscopy, and Section III deals with molecular interactions and reaction dynamics. The distinction between the content of these two sections parallels the manner in which chemical research has historically been carried out. Spectroscopists tend to be concerned with the interaction of light with matter (i.e., molecular

---

[1] Present address: Laser Physics Branch, Naval Research Laboratory, Washington, D.C.

69

systems) as a topic of interest in itself, where the concern for dynamical processes exists only in how they modify such interaction. On the other hand, researchers in molecular interactions and reaction dynamics have either ignored problems in which a molecular system interacts with an external radiation source, or have viewed such a source as just a modifier or disturbance of the dynamics of interest. Exceptions to this distinction can certainly be found, particularly in the area of photochemistry. In fact, the availability of lasers has begun to erase this distinction in many laboratories, i.e., many chemists are now interested and active in both spectroscopy and molecular dynamics. Nevertheless, this distinction still serves as a useful guideline for presenting a review of photochemistry.

II. Spectroscopy

A variety of laser-induced processes serve to illuminate the *structure* of atomic and molecular species. By structure is meant chiefly the distribution of quantized energy levels (or states) in a system—electronic, vibrational, and rotational—and the extent to which these can be made to interact with each other via the radiation field. It is hoped to demonstrate here that an external radiation field such as a laser, in addition to being useful as a *probe* of structure, may also be essential in *altering* structure to suit our needs. This latter aspect of the radiation field will be revealed later in a discussion of chemical dynamics; the focus will first be on the laser as a probe as well as on the versatility of the laser as an agent to selectively prepare state-specific chemical species. This last function is of enormous importance from the viewpoint of chemistry because certain reactions will occur only when the reagents are in specific states. Herein lies the claim (as yet unrealized) that laser photochemistry may fulfill the age-old dream of alchemy.

It is useful to distinguish between two types of laser-induced processes: single photon and multiphoton. A single-photon process can be either a single-step or multistep process involving a single photon in each step, and the photons in the different steps may come from different lasers. A multiphoton process, on the other hand, involves a number of photons (maybe as many as 30) in a single step. This distinction will be classified by examples in the following sections.

A. Single-Photon Processes

Functioning as a probe or preparation agent, the laser is essential in the following processes: photoionization, photodissociation, photoisomerization, and photodeflection, among others. The first two processes have been the most widely studied, and will be addressed chiefly.

The difference between atomic and molecular systems in laser photo-processes is that atoms in general present much more stringent resonance requirements. In other words, the energies of the laser photons have to match quite well with those of the atoms. This is due to the fact that there is only one kind of excitation in atoms—discrete electronic excitations. Molecules, on the other hand, are much less selective because, in addition to electronic excitations, they exhibit excitations due to nuclear motion, namely, vibrational and rotational excitations. This leads to much more complex spectra which can "accommodate" photons of widely varying energies. In general, there is correspondence between electronic transitions and photons in the visible and ultraviolet (UV) range, vibrational transitions and the infrared (IR) range, and rotational transitions and the microwave range. Laser photochemistry is thus mainly concerned with electronic and vibrational transitions. (The effects of the rotational states on these transitions will be ignored here.) There is another very important feature in molecular systems which helps to relax the resonance requirements on laser photons. This is the Born–Oppenheimer adiabatic (slow) variation of electronic energy levels with nuclear separation. In situations where electronic states do not support bound levels, the electronic energies vary *continuously* as the nuclei in a molecular system move with respect to each other, and resonance requirements (at least over a certain range of frequencies) can be relaxed accordingly. This relaxation of resonance requirements due to nuclear dynamics forms the cornerstone of laser-induced dynamical processes (both reactive and nonreactive).

With respect to atoms, selective multistep photoionization is the most interesting and potentially useful photophysical process made possible by lasers, the most attractive application being laser isotope separation (LIS). All selective ionization schemes for atoms involve the following sequence of processes: (1) selective excitation and (2) ionization of the excited atoms. Figure 1 illustrates two typical schemes which are frequently employed in LIS: Fig. 1a represents a scheme in which the second laser (with photon energy $\hbar\omega_2$) brings the atom to an autoionizing state. (An autoionizing state is a configuration of discrete electronic energy where two electrons are excited, with the discrete energy level higher than the ionization limit of the valence electron.)

An example of the process in Fig. 1a is the ionization of Rb by a dye laser tuned to $\omega_1 = 7947.6$ Å (Ambartzumian and Letokhov, 1972). This photoexcites the 5p $^2P_{3/2}$ state of Rb. A second photon of $\omega_2 = 3471$ Å,

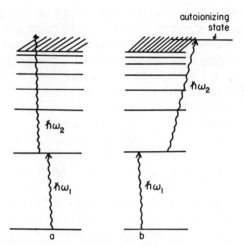

FIG. 1. Two schemes of selective multistep photoionization of atoms by laser radiation
(a) Two-step photoionization; (b) two-step ionization via an autoionizing state.

which is generated by the second harmonic of the same laser, then brings
the excited state to the continuum

$$\text{Rb(vapor)} + \hbar\omega_1(7947.6 \text{ Å}) \longrightarrow \text{Rb}(5p\ ^2P_{3/2}) \xrightarrow{\hbar\omega_2(3471\ \text{Å})} \text{Rb}^+ + e^-$$

A second example of this process (Tuccio *et al.*, 1975) is the LIS of ^{235}U

$$^{235}\text{U} + \hbar\omega_1(3781 \text{ Å}) \longrightarrow\ ^{235}\text{U}^* \text{ (low-lying metastable state at 620 cm}^{-1})$$
$$\Big\downarrow \hbar\omega_2 \binom{3075\ \text{Å}}{3564\ \text{Å}}$$
$$\text{U}^+ + e^-$$

An example of the process in Fig. 1b is the two-step selective autoioni-
zation of Ca (Brinkmann *et al.*, 1974). Ca atoms are prepared in the
metastable states $4s4p\ ^3P_{2,1,0}$, which are then excited by a first laser ($\omega_1 =$
6162 Å) to the excited state $4s5s\ ^3S_1$. A second photon ($\omega_2 = 4880$ Å)
subsequently further excites the atom to the autoionizing state $3d5p\ ^3P_1$

$$\text{Ca}(4s4p\ ^3P_{2,1,0}) + \hbar\omega_1(6162 \text{ Å}) \longrightarrow \text{Ca}(4s5s\ ^3S_1)$$
$$\xrightarrow{\hbar\omega_2(4880\ \text{Å})} \text{Ca}(3d5p\ ^3P_1) \longrightarrow \text{Ca}^+ + e^-$$

For molecular photoionization there is the single-step ionization of O_2^-
(Cosby *et al.*, 1975) and the two-step ionization of formadehyde (Andreyev *et al.*, 1977)

$$O_2^- + \hbar\omega \longrightarrow O_2 + e^-$$

$$\text{H}_2\text{CO}(X^1A_1) + \hbar\omega_1(3371 \text{ Å}) \longrightarrow \text{H}_2\text{CO}^*(^1A_2) \xrightarrow{\hbar\omega_2(1600\ \text{Å})} \text{H}_2\text{CO}^+ + e^-$$

It is now appropriate to consider the topic of selective multistep photo-dissociation of molecules. As mentioned before, the spectroscopy of molecules is much more complex and interesting than that of atoms due to the presence of the nuclear degrees of freedom. This extra complexity provides the opportunity for a judicious combination of IR and UV (or visible) lasers for effecting the dissociation process. Hence selectivity can be considerably enhanced. Figure 2 illustrates two important schemes of selective multistep photodissociation which are also of great potential value for LIS. Figure 2a represents the process of two-step photodissociation: the first laser provides an IR photon exciting a high vibrational level of the ground electronic state, and a second laser with a UV photon then accesses the repulsive excited electronic state, causing the molecule to dissociate. This process can be considered the molecular analog of atomic selective two-step photoionization (Fig. 1a). Figure 2b represents the process of single-step selective laser-induced predissociation: a UV laser excites the molecule to an excited vibrational level in an excited electronic state which crosses a repulsive excited electronic state. By virtue of the crossing, vibrational motion in the bound electronic state is coupled to continuum motion in the repulsive state, thus allowing dissociation to take place. (A dissociation of this type as a result of curve crossing, whether induced by laser radiation or not, is known as predissociation.)

The following presents several examples. A prototype of a two-step photodissociation (Letokhov, 1973) is the process

$$HCl(X^1\Sigma^+, \nu = 0) + \hbar\omega_1(1.19 \ \mu m) \longrightarrow HCl(X^1\Sigma^+, \nu = 3)$$
$$\xrightarrow{\hbar\omega_2(2650 \ \text{Å})} HCl(A^1\Pi) \longrightarrow H + Cl$$

In this example the ground electronic state of HCl, $X^1\Sigma^+$, supports a number of vibrational levels. The first laser, supplying an IR photon of wavelength $1.19 \ \mu m$, excites the system from the ground vibrational level ($\nu = 0$) to the third excited vibrational level ($\nu = 3$). The second laser, supplying a UV photon of 2650 Å, then excites the system to the repulsive electronic state $A^1\Pi$, on which the molecule can dissociate. [In the experiment on HCl reported by Letokhov (1973), the second photon is actually generated by the fourth harmonic of the first laser, which is a dye laser pumped by a Nd : glass Q-switched laser.] The photodissociation cross section is observed to be $\sim 10^{-19} \ cm^2$. Other examples of photodissociation based on this process include the isotope separation of ^{15}N from $^{15}NH_3/^{14}NH_3$, B from BCl_3/O_2, ^{50}Ti from $TiCl_4$, and D, T, ^{16}O, and ^{18}O from either H_2O/CO or $H_2O/CO + C_2H_4$.

The following process (Yeung and Moore, 1972) provides a good example of laser-induced predissociation:

$$H_2CO + \hbar\omega(3472 \ \text{Å}) \longrightarrow H_2CO^* \longrightarrow H_2 + CO$$

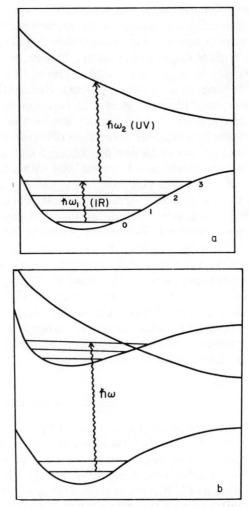

FIG. 2. (a) Two-step (IR + UV) photodissociation. The IR photon accesses an excited vibrational level of the ground electronic state and the UV photon then accesses the excited electronic state on which the molecule dissociates. (b) One-step photopredissociation. A UV photon accesses an excited vibrational level in an excited electronic state which crosses an excited repulsive electronic state. Dissociation occurs through the coupling near the crossing.

In this example only one laser is required. The 3472-Å photon excites the system from the ground vibrational level of the ground singlet electronic state to an excited vibrational level of the excited singlet electronic state. Dissociation then occurs through crossing of the latter electronic state with a repulsive electronic state. By selectively choosing laser frequencies to excite isotopically distinct vibrational levels, the isotopic species ^{13}CO, ^{14}CO, and $C^{18}O$ can be separated. Also, H_2CO can be separated from H_2CO/D_2CO mixtures.

In many instances LIS can be effected by employing a scavenger which reacts preferentially with a species in its excited state. An intense laser can then be used to selectively excite an isotopic species. The unwanted isotopic species would simply absorb no photons and hence would not react with the scavenger. A prototype of this process (Zare, 1977) is

$$I^{37}Cl + \hbar\omega(6050 \text{ Å}) \longrightarrow I^{37}Cl^*$$

$$I^{37}Cl^* + C_6H_5Br \longrightarrow I + Br + {}^{37}ClC_6H_5$$

The bromobenzene (C_6H_5Br) is acting as the scavenger and reacts only with ICl^* (excited electronic state). The $I^{35}Cl$ in the sample will not absorb any 6050-Å photons and remains in the ground electronic state. Hence it is incapable of reacting with the C_6H_5Br.

B. MULTIPHOTON PROCESSES

1. Atomic Systems

In multiphoton processes in atomic systems, electronic transitions in atoms involve the simultaneous absorption or emission of more than one photon. By using sufficiently high-energy photon fluxes, a variety of non-linear electronic transitions may be observed (Eberly and Lambropoulos, 1978). Some of the important examples are (1) multiphoton ionization (simultaneous absorption of more than one photon leading to the emission of an electron); (2) electronic Raman scattering, elastic/inelastic photon scattering, and resonance scattering (absorption of photons with frequency ω followed by the emission of photons with frequency ω'); (3) third-harmonic generation (absorption of three photons of equal frequency ω with the emission of one photon of frequency 3ω); and (4) infrared up-conversion (absorption of three photons of different frequencies with the emission of one photon of the sum frequency 3ω).

A number of theoretical and experimental investigations of multiphoton processes in atomic systems have been reported in the literature (Eberly et al., 1979). Here the focus is only on the multiphoton transition probability and the ionization probability of a multilevel atomic system.

It is known that the multiphoton transition probability of an atomic system can be greatly increased via intermediate states with frequency nearly resonant with the driving radiation field. As long as the pumping rate does not exceed the binding energy of the electron, or for laser powers not higher than 10^{16} W/cm^2, perturbation theory is usually adequate for a theoretical description. Time-dependent first-order perturbation theory gives the probability of finding an electron in the upper state of a two-level atom (Sargent *et al.*, 1974) as

$$P_{0\rightarrow1}(t) = \{|V|^2/[\Delta^2 + (\gamma/2)^2]\}[1 + e^{-\gamma t} - 2\cos(\Delta t)e^{-\gamma t/2}] \qquad (1)$$

where V is the pumping rate (proportional to the electric field of the radiation), and Δ and γ are the detuning and the width of the upper level, respectively. The above functional form for a single-photon transition can also be used to describe a two-photon transition by defining the two-photon detuning $\Delta = \omega_f - 2\omega$ and $\gamma = \gamma_f$, where ω_f and γ_f are the frequency and width of the final state and ω is the field frequency. In this two-photon process, the pumping rate V is now replaced by

$$|V^{(2)}| = \left|\sum_j \frac{V_{fj}V_{jg}}{i\Delta_j + \gamma_j/2}\right| \qquad (2)$$

where j is the intermediate state with frequency ω_j and width γ_j, and $\Delta_j = \omega_j - \omega$ is the detuning.

The above approach is satisfactory if no intermediate state is situated at resonance. This process, without an intermediate resonance, is called "coherent," and, in general, for a system consisting of a set of intermediate states, there can be both coherent and incoherent excitations.

The transient behavior of a two-photon process has been experimentally observed using Doppler-free excitation of a cell of sodium atoms (Bassini *et al.*, 1977). By using Eq. (1) it is possible to generate the absorption profiles shown in Fig. 3: these curves qualitatively show the experimentally observed absorption transients in sodium for various two-photon detunings (Lin and George, 1981a).

A generalization of the two-photon excitation gives the ionization probability (Stenholm, 1979) for an N-photon process with nearly resonant intermediate states

$$P_{\text{ion}} = \beta I^N \qquad (3)$$

where β is a constant of proportionality and I is the intensity of the laser field. The type of behavior characterized by the order N has been experimentally observed for many atoms with N ranging from 5 to about 20 (Delone, 1975). However, when an intermediate resonance occurs, there are deviations from the simple law of Eq. (3). Figure 4 shows the intensity dependence of a three-photon ionization in cesium at the ruby laser frequency (Georges and Lambropoulos, 1977). It is seen that the plot of log P_{ion} versus log I departs from the straight line of slope 3 found in nonresonant processes [Eq. (3)]. This behavior has been observed in a number of

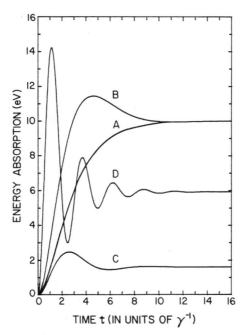

FIG. 3. The universal energy absorption profiles of the pumped mode for different sets of the optimum detuning Δ_{opt} and damping factor γ (in units of cm^{-1}). Curve A, $\Delta_{opt} = 0$, $\gamma = 2 \times 10^{-3}$; curve B, $\Delta_{opt} = 0.5\gamma = 5 \times 10^{-4}$; curve C, $\Delta_{opt} = \gamma = 10^{-3}$; curve D, $\Delta_{opt} = 2.5\gamma = 2.5 \times 10^{-4}$; for low-power laser, $I = 100$ W/cm². Note that the time scales are shown in units of γ^{-1}. (From Lin and George, 1981a.)

experiments and can be understood in terms of the strong radiative interaction with nearly resonant intermediate states which are coupled to the continuum (ionization) states and shifted and broadened due to multilevel effects. The field-induced Stark shift and level broadening for an atomic system are similar to the frequency shift and damping effect in a gas/surface system where many-body effects can be treated by the Wigner–Weisskopf approximation (Louisell, 1973).

2. Molecular Systems

In molecular systems, although visible and ultraviolet photons have the most pronounced effects on the chemical properties of molecules, much attention has been paid to infrared photochemistry since the first report of laser isotope separation (Ambartzumian et al., 1974). A considerable amount of theoretical and experimental research has resulted from the suggestion that multiphoton excitation (MPE) could be a novel method for

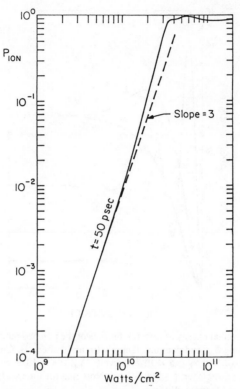

FIG. 4.   Log–log plot of the intensity dependence of three-photon ionization in cesium at the ruby laser frequency. (From Georges and Lambropoulos, 1977.)

vibrational mode control of molecular decomposition (Schulz *et al.*, 1979; Letokhov, 1980; Lee and Shen, 1980). Sulfur hexafluoride (SF_6) was one of the first molecules to be dissociated by a high-power CO_2 laser. The pressure dependence of isotopic selectivity in the multiphoton decomposition (MPD) of SF_6 (Ambartzumian *et al.*, 1974, 1978; Lyman *et al.*, 1977) suggests that the dissociation takes place under collisionless conditions.

A well-accepted qualitative model for MPD of SF_6 has provided a general picture where the molecular energy levels are divided into three regions: (1) the discrete region characterized by coherent excitation, where the ν_3 (active) mode absorbs three to six photons and the anharmonicity of the vibrational potential is nearly compensated for by allowed rotational transitions; (2) the quasi-continuum region where the level density is very high and incoherent excitation is essential; and (3) the true continuum region. Region 1 processes are responsible for isotopic selectivity, coherent effects (multiphoton resonance, photon echoes, coherent wave propagation, etc.), and the intensity dependence of MPE with high selectivity. For the excitation processes in regions 2 and 3, it has been shown experimentally that high laser fluence (energy), not high laser power (intensity), is necessary for driving the molecule through the quasi-continuum and is thus the important parameter for determining the dissociation yield (Grant *et al.*, 1978).

The phenomenon of infrared MPE and MPD seems well understood qualitatively. However, quantitative understanding hinges on the following fundamental questions:

1. What are the roles of coherent and incoherent absorption processes in MPE and MPD? What is the magnitude of the absorption cross section and how does it change with molecular parameters (dipole moment, level width, anharmonic potential, etc.) and intensity, fluence, frequency, and degree of coherence of the laser radiation?

2. How selective is the particular MPD? How fast does the energy in the pumped active mode randomize with the other degrees of freedom via inter- and intramolecular energy relaxations?

3. What is the energy distribution pattern of the absorbed photons among the vibration–rotation states of the molecule?

4. What are the dynamics of the dissociation event and what are the parameters that determine the rate of unimolecular decomposition?

5. How can selectivity in laser chemistry be achieved by novel methods such as multistep excitations?

In order to answer these questions either qualitatively or semiquantitatively, the following first discusses some recent theoretical approaches using classical and quantum models and subsequently presents the selectivity aspects of laser chemistry.

a. Classical Models. Due to the high density of states at most energies in a typical molecule, rigorous quantum mechanical treatments of multiphoton processes are presently impractical. Classical studies, however, do not suffer from such a difficulty, although many quantum effects cannot always be successfully incorporated (Bloembergen, 1975; Cotter *et al.*, 1976; Walker and Preston, 1977; Steverding *et al.*, 1977, 1978; Lin, 1979).

Consider a polyatomic molecule subject to an intense infrared radiation field. By singling out the active vibrational mode (e.g., the ν_3 mode for the SF_6 molecule) as an anharmonic classical oscillator and the remaining

modes as the heat bath, the dynamics of the system may be described by
the following nonlinear equation of motion (Lin, 1979)

$$\ddot{Q} + 2\gamma\dot{Q} + \omega_0^2 Q + \alpha Q^2 + \beta Q^3 = f(t)/m \tag{4}$$

where Q is the normal coordinate of the active mode (with fundamental
frequency ω_0), α and β are the anharmonic coefficients, m is the reduced
mass, and $f(t)$ is the driving force. This single-body equation may be
rigorously derived from a many-body system (Lin and George, 1981a). By
the harmonic balance method (Minorsky, 1962), Eq. (4) can be linearized
by introducing an amplitude-dependent effective frequency (Lin, 1979).
The solution of the linearized equation gives the power absorption of the
anharmonic oscillator. This simple system with a constant damping factor
γ obtains the power absorption and also the absorption cross section as a
Lorentzian. However, for a more realistic system, the damping factor
could be time dependent and the phase of the oscillator could also relax
due to incoherent processes. For this purpose, instead of Eq. (4), consider
the generalized Langevin equation (GLE) (Lin and George, 1980a)

$$\ddot{Q} + \int_0^t \beta(t - t')\dot{Q}(t')dt' + \omega_{\text{eff}}^2 \int_0^t M(t - t')\dot{Q}(t')dt'$$
$$= [f(t) + R(t)]/m \tag{5}$$

Equation (5) introduces the damping kernel β and the dephasing kernel M
to describe the interaction dynamics between the active mode and the
bath modes; $f(t)$ and $R(t)$ are the laser driving force and the bath-induced
random force, respectively. ω_{eff} is the effective frequency of the active
mode related to the fundamental frequency ω_0, the anharmonicity K^*,
and the bath-induced frequency shift $\delta\omega$ by

$$\omega_{\text{eff}} = \omega_0 - K^* A^2 - \delta\omega \tag{6}$$

where A is the steady-state amplitude of the oscillator and is proportional
to the applied laser intensity. The above GLE enables the study of the
dynamics of classical multiphoton absorption via T_1 (energy) and T_2
(phase) relaxations, which are governed by $\beta(t)$ and $M(t)$, respectively.
For given forms of the memory functions $\beta(t)$ and $M(t)$, calculations can
be made of the velocity autocorrelation function, which in turn gives the
time- and ensemble-averaged energy absorption rate. The case of Markovian processes and an exponentially decaying phase obtains

$$\langle dE/dt \rangle = (qE_0)^2 P(T_0)[\gamma_2 A + \omega B]/[A^2 B] \tag{7}$$

$$P(T_0) = (2\hbar\omega)^{-1}(kT_0/m)[1 - \exp(-\hbar\omega/kT_0)] \tag{8}$$

$$A = \omega_{\text{eff}}^2 - \omega^2 + \gamma_1\gamma_2 \tag{9}$$

$$B = \omega(\gamma_1 + \gamma_2) \tag{10}$$

where q and T_0 are the classical charge and the initial temperature of the molecule, respectively, E_0 is the electric field of the laser radiation, and γ_1 and γ_2 are the energy-damping factor and the dephasing factor, respectively. Note that the absorption cross section (or the lineshape), which is proportional to the energy absorption rate, is characterized by the overall broadening $(\gamma_1 + \gamma_2)$ and detuning $(\omega_{eff} - \omega)$ and is, in general, asymmetric due to the dephasing term $\gamma_2 A$. For the case of incoherent processes, $\gamma_2 \gg \gamma_1$, and the line broadening is dominated by the T_2 dephasing. On the other hand, for the coherent processes, $\gamma_1 \gg \gamma_2$, and the lineshape reduces to the symmetric Lorentzian governed by the simple equation of motion, Eq. (4).

In addition to the above phenomenological models, a classical trajectory calculation of MPE has been made for SF_6 (Poppe, 1980). The results show that the energy transfer and dissociation rate depend on laser fluence (intensity $\times$ time), and the intramolecular energy relaxation rate is estimated to be on the order of a picosecond.

Another classical treatment of MPE and MPD for a system of two nonlinearly coupled oscillators was studied by means of the Krylov–Bogoliubov–Mitropolsky theory (Ramaswamy et al., 1980). The results show two regions of behavior for the exchange of energy between the system and the laser field: (1) the regular region with well-defined frequencies, and (2) the erratic region. Motion in the latter region leads to dissociation of the pumped molecule.

b. Quantum Models. Two distinctly different approaches have been used for building simplified quantum mechanical models of MPE and MPD. In the first approach (referred to as the heat-bath model), attention is focused on the active vibrational mode. The second approach treats all vibrational modes on an equal footing and deals with transitions between true molecular eigenstates induced by the laser field. First to be discussed is the heat-bath model. The total system may be described by the Hamiltonian

$$H(t) = H_A + H_B + H_{AB} + H_{AF} \qquad (11)$$

where H_A and H_B represent the unperturbed Hamiltonians of the active and bath modes, respectively, H_{AB} represents the anharmonic coupling between the active and the bath modes, and H_{AF} represents the interaction between the active mode and the laser field.

In the Heisenberg–Markovian picture (Louisell, 1973), the above Hamiltonian, expressed in a second-quantization form, has been used to calculate the average excitation of the active mode for the case of linear coupling in H_{AB} (Narducci et al., 1977) and high-order coupling (Gan et al.,

1978; Lin and George, 1981b). The key feature of multiphoton absorption in the heat-bath approach may be clearly described by the equation

$$dn_A/dt = \sigma I - \gamma(n_A - \bar{n}) \tag{12}$$

where n_A and $\bar{n}$ are the average excitations of the active mode and the heat-bath modes, respectively. I is the laser intensity, γ is the damping factor describing the T_1 (energy) relaxation of the pumped mode, and σ is the absorption cross section. Equation (12) is based on energy conservation and can be derived rigorously from a microscopic Hamiltonian (Lin, 1980). The steady-state excitation may be easily found by $n_A^{ss} = \sigma I/\gamma + \bar{n}$, which gives $n_A^{ss} \propto I$ for harmonic or low excitations (σ independent of n_A^{ss}) and $n_A^{ss} \propto I^\alpha$, with $\alpha < 1$ for anharmonic or high excitations (σ is excitation dependent). Furthermore, the average number of photons absorbed by the molecule (active plus bath modes) is given by the time integral of the pumping rate σI, which yields the fluence (ϕ)-dependent form $\langle n_m \rangle \propto \phi^\beta$, with $\beta = 1$ and $\frac{1}{3}$ for the harmonic and highly anharmonic cases, respectively.

The second quantum approach treats all vibrational modes on an equal footing, with the eigenstates of the system being mixtures of all the normal modes. Due to the high density of vibrational states at high excitation energy, the molecule can easily absorb more photons through resonant incoherent transitions between so-called quasi-continuum levels, where rate equations have been widely used for collisionless MPD (Quack, 1978; Thiele et al., 1980; Grant et al., 1978; Fuss, 1979; Barker, 1980; Baldwin and Barker, 1981) and collisional MPD (Troe, 1977; Tardy and Rabinovitch, 1977; Stone et al., 1980).

For multiphoton excitation in the quasi-continuum region, the Fermi Golden Rule is valid and the full Schrödinger equation reduces to a set of incoherent rate (master) equations. For the collision-free condition (low pressure), the rate equations for the energy population are given as

$$dP_n/dt = W_{n-1}^a P_{n-1} + W_n^e P_{n+1} - (W_n^a + W_{n-1}^e)P_n - k_n P_n \tag{13}$$

where P_n is the population in the nth level (n photons absorbed). W_n^a (W_n^e) is the transition rate constant for absorption (emission) from level n to $n + 1$ ($n + 1$ to n), and is related to the absorption cross section (σ_n) and frequency of the field (with density I) by $W_n^a = \sigma_n I/\hbar\omega$ and $W_n^e/W_n^a = g_n/g_{n+1}$, where g_n is the molecular density of states at energy $n\hbar\omega$. The unimolecular decomposition rate constant, k_n, can be calculated by Rice–Ramsperger–Kassel–Marcus (RRKM) theory (Forst, 1973) or quantum RRK theory (Shultz and Yablonovitch, 1978). From Eq. (13) it is realized that collisionless MPE and MPD are characterized by the laser intensity and frequency, the absorption cross section, the density (or degeneracy) of states, and the unimolecular reaction constant. During the past few years, the rate equations describing MPD have been studied by different approaches, such as a thermal model for Boltzmann-type energy populations (Black *et al.*, 1979), a diffusion model for continuum populations (Fuss, 1979), an exact stochastic model (Baldwin and Barker, 1981), the model of restricted intramolecular relaxation (Stone and Goodman, 1979), and the random-coupling model (Carmeli and Jortner, 1980).

In the thermal model, the population P_n is given by a Boltzmann function

$$P_n = A \, \exp(n\hbar\omega/kT_{\text{eff}}) \tag{14}$$

A is the preexponential factor and T_{eff} is the effective vibrational temperature given by the energy conservation equation

$$S\bar{n} = \langle n \rangle \hbar\omega \tag{15}$$

$$\bar{n} = [\exp(\hbar\omega_0/kT_{\text{eff}}) - 1]^{-1} \tag{16}$$

Here ω_0 is the mean frequency of the molecule with S vibrational modes and average number $\langle n \rangle$ of absorbed photons. It is noted that for multiphoton processes with $kT_{\text{eff}} >> \hbar\omega_0$, Eq. (15) reduces to the simple form $kT_{\text{eff}} \approx \langle n \rangle \hbar\omega/S$. The corresponding dissociation probability P_d can be easily found by Eq. (14) based on the threshold number of photons for dissociation n^*

$$P_d = \sum_{n=n^*}^{\infty} P_n \propto \exp(-Sn^*/\langle n \rangle) \tag{17}$$

which is the usual Arrhenius form.

Depending on the forms of the transition rate constants W_n^a and W_n^e, the solution of the rate equation gives different populations and the corresponding dissociation probabilities. Some important populations and their corresponding dissociation probabilities are shown in Fig. 5 (Lin and George, 1979b; George *et al.*, 1980).

c. Discussion. i. Selectivity in laser photochemistry. Selectivity is characterized not only by the coherent properties of the laser field but also by the molecular properties of the excited system. Therefore, the types of selectivity in infrared multiphoton excitations can be classified according to the relation between the various relaxation times of the excited system and the energy pumping rate of the laser field (Letokhov, 1980). R^{intra} and R^{inter} are defined as the intra- and intermolecular vibrational energy transfer rates, respectively, R^{VT} as the relaxation rate for molecular vibration–translation coupling, i.e., $(R^{\text{VT}})^{-1}$ is the time for

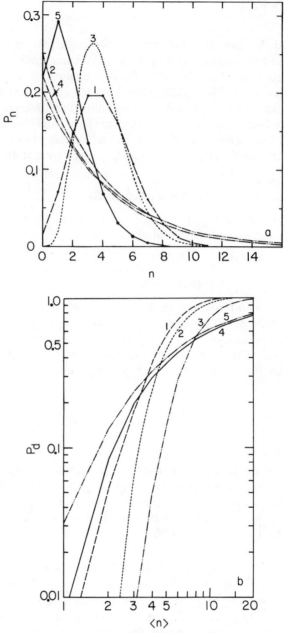

FIG. 5. (a) The distribution functions of four-photon excitations, $\langle n \rangle = 4$, for Poisson population (1), diffusion-model population with $S = 1$ (2) and $S = 6$ (3), Boltzmann population with $S = 1$ (4) and $S = 6$ (5), and quantal population with $S = \alpha = 1$ (6). (b) The desorption probabilities of $n^* = 5$ for Poisson distribution (1), diffusion model with $\beta = 1$, $S = 6$ (2), Boltzmann distributions with $S = 6$ (3) and $S = 1$ (4), and quantal model with $S = \alpha = 1$ (5).

complete thermal equilibrium to be reached in the molecular mixture, and W_{exc} as the rate of vibrational multiphoton excitation of the molecule. Four different types of selectivity can be distinguished depending on the relative magnitudes of the relaxation rates and the laser excitation rate (note that the relaxation rates R^{inter} and R^{VT} are pressure dependent, although for low pressures it is expected that $R^{VT} \ll R^{inter} \ll R^{intra}$):

1. Mode (bond)-selective excitation ($W_{exc} \gg R^{intra}$). A certain mode or a functional group of a polyatomic molecule is in a nonequilibrium state which has a higher vibrational temperature T_{eff} [defined by Eq. (15)] as compared with the remaining modes or functional groups. This is the situation of a long lifetime or high pumping rate.

2. Molecular-selective excitations ($R^{intra} \gg W_{exc} \gg R^{inter}$). In this case the absorbed photon energy is rapidly randomized within the excited molecule in which the local vibrational temperature is higher than the overall translational temperature of the mixture of different molecules.

3. Vibrational-selective excitation ($R^{inter} \gg W_{exc} \gg R^{VT}$). In this more moderate condition, vibrational equilibrium among all the mixed molecules is reached, but there is still no overall thermal relaxation. This situation prevails for low pressures where vibration–translation relaxation rates are lower than vibration–vibration relaxation rates.

4. Nonselective thermal excitation ($R^{VT} \gg W_{exc}$). This is the situation of thermal excitation of all molecules in a mixture by a low-power CW laser. This field of infrared thermal chemistry is of interest for a heterogeneous system (e.g., species adsorbed on a solid surface) where the laser radiation is used to excite the adsorbed gas molecules without significantly heating the solid surface (assuming the phonon coupling to be small). This type of excitation will be discussed in Section III,B,2.

The attraction of infrared laser chemistry is that if the photon energy can be deposited and maintained in a specific vibrational mode (or functional group), a selective reaction involving that mode (functional group) may be induced. Therefore, one of the critical questions concerning MPD processes is whether the photon energy remains localized in the pumped mode (or molecule) long enough to result in a mode-selective (or molecule-selective) reaction. Most theoretical models of MPD assume that intramolecular vibrational relaxation (IVR) is very fast (on the order of a picosecond), and statistical approaches such as RRKM theory have been applied successfully to a number of experimental results (Black et al., 1979; Sudbo et al., 1979). However, several experiments examining product branches ratios in relatively complex molecules, e.g., cyclopropane, have shown that the laser selectivity could not be explained by the statistical theory (Hall and Kaldor, 1979). The evidence suggests that the laser-induced reactions could result from a nonergodic or partially mode-selective excitation. Several research groups have also examined the bond localization character of large molecules, e.g., benzene, both experimentally and theoretically (Zewail, 1980; Bray and Berry, 1979; Heller and Mukamel, 1979; Thiele et al., 1980). This nonstatistical behavior is explained by a local mode model in which the total system is divided into two groups of vibrational modes. Within each group, the vibrational modes are strongly coupled and the photon energy quickly randomized. Between groups, however, the coupling is considerably weaker, and intergroup randomization rates are therefore appreciably slower by the concept of the energy-gap law (Nitzan et al., 1975). Another theory called the "restricted IVR model" has been proposed for the possibility of laser-selective photochemistry (Thiele et al., 1980).

ii. Multistep excitations. It has been experimentally shown that the dissociation yield of a molecule may be greatly enhanced via a two-step excitation (IR + UV) combining a single-step IR process with a single-step UV process. Strictly speaking, this is not a multiphoton process, but is rather a sum of two uncorrelated single-photon events, as mentioned in Section II,A. However, there is the possibility, in this situation, of multiphoton excitation, particularly from the IR laser. The UV laser could be replaced with several IR lasers, e.g., two, to achieve a multistep (three-frequency) excitation. Using a set of three phased pulses with frequencies resonant to the energy separations of three anharmonic level pairs and with pulse durations equal to different integral multiples (n) of π, efficient photon energy absorption can be achieved. In this process, the successive lasers interact with the molecule when the population is completely inverted by the first, i.e., at the moment of time $t = \tau$, such that $\Omega\tau = (2n + 1)\pi$, where Ω is the Rabi frequency (Oraevski et al., 1976).

A classical treatment of the dissociation of HF by two different IR lasers has shown (not surprisingly) that lower laser power densities of each laser are required for molecular dissociation than for a corresponding single-laser process (Stine and Noid, 1979). A quantum model of two-laser excitation of SF_6 has also suggested the advantage of multistep excitation (Narducci and Yuan, 1980).

III. Molecular Interactions and Reaction Dynamics

A. GAS-PHASE PROCESSES

The main feature of laser-induced dynamical processes is that laser photons do not have to be in resonance with the asymptotic separations in energy levels of the collision species (DeVries *et al.*, 1980). This means that light absorption and collisional energy transfer do not take place as independent events. There are two viewpoints of the situation: atomic and molecular. For simplicity, consider atom–atom collisions A + B. The atomic description makes use of asymptotic atomic energy levels of the A + B system. For example, reference can be made to the Na(3p) + Na(5s) energy level of the Na–Na system in the separated-atom limit. A photon slightly detuned from resonance (with respect to the atomic energy levels) will still be absorbed, the difference in energy being made up for by the collision. In the molecular picture, however, we can still speak of a resonance absorption process; however, in this case, the electronic energy levels are changing as a function of internuclear distance (as represented by interatomic potential energy curves), and the absorption takes place at some finite internuclear separation where the resonance condition is fulfilled. The two points of view are illustrated in Fig. 6. The second point of view elucidates the inseparability of radiative interaction and collision dynamics much more succinctly.

Theoretically, this inseparability may be exploited to introduce a mode of description of laser-influenced collision processes known as the electronic-field representation (George *et al.*, 1977). This representation is based on the idea of "dressing" molecular quantum states with photon quantum states. The resulting "dressed" states allow description of colli-

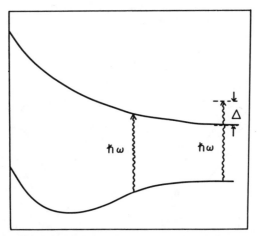

FIG. 6. Two viewpoints of looking at the collision-induced absorption of a photon. Δ is the detuning. In the separated-atom-limit picture (represented by the wavy line on the right), the photon is not in resonance with the energy levels; collisional effects make up for the detuning energy. In the molecular picture (represented by the wavy line on the left), the resonance requirement is fulfilled at a finite internuclear separation. Note that the wavy lines are of the same length—representing a photon of the same frequency.

sional and radiative interactions on the same footing and clearly reveal the altered dynamics of a collision system due to the presence of the radiation. The simplest way to introduce the electronic-field representation is as follows. We begin with a set of potential curves corresponding to different electronic energy levels of the collision system. Suppose an intense coherent laser of frequency ω is applied. All of the curves are then shifted by an amount $\hbar\omega$. The set of shifted and unshifted curves may cross each other at several points. At every crossing between a shifted and unshifted curve, radiative interaction is effective and the curves "couple" with each other to generate an avoided crossing. These avoided crossings change the shapes of the curves and thus effectively change the dynamics of the system. The radiatively altered curves are called electronic-field curves, and the quantum mechanical representation that they generate is known as the electronic-field representation.

The use of these curves is illustrated in Figs. 7 and 8. Figures 7 and 8 represent a nonreactive and reactive situation, respectively. Figures 7a and 8a represent the field-free and field-shifted potential curves with real crossings, whereas Figs. 7b and 8b represent the electronic-field curves with avoided crossings. The relevant curves for the description of the dynamics in Fig. 7a are E_1 and E_2, which correspond asymptotically to $W_1 + \hbar\omega$ and W_2, respectively. In this example, the barrier in the field-

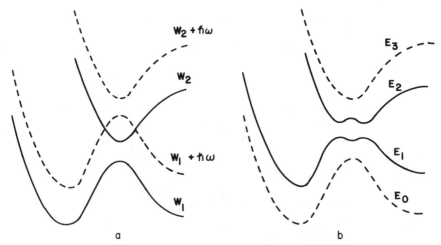

FIG. 7. A sketch of two model nonreactive electronic surfaces W_1 and W_2 plus their shifted images $W_1 + \hbar\omega$ and $W_2 + \hbar\omega$ (a), and the resulting electronic-field surfaces E_0, E_1, E_2, and E_3 obtained from them (b). This figure illustrates the utility of the electronic-field approach in clarifying the dynamical behavior of a more complicated scattering system undergoing irradiation.

free curve W_1 has been turned into a shallow valley (in E_1), and the valley in the field-free curve W_2 has been turned into a barrier (in E_2), thus clearly showing the effect of the radiation field in altering the dynamics of the system. In Fig. 8a, W_1 represents a reaction curve, exothermic with respect to reaction (reaction coordinate running from right to left), whereas W_2 and W_3 both exhibit steep reaction barriers. The presence of a strong field changes the situation altogether (Fig. 8b). A reaction starting on the surface of $W_1 + \hbar\omega$ could now be diminished, while one starting on surface W_2 would actually become exothermic. A reaction starting on W_3 could also be enhanced. This example again illustrates the significant effects on dynamics due to the presence of the field.

The following presents some actual systems illustrating collisional energy transfer in the presence of intense laser fields. The pioneering experiment of Falcone *et al.* (1977) provides

$$Sr^*(5p\ ^1P_0) + Ca(4s^2\ ^1S) + \hbar\omega(4976.8\ \text{Å}) \rightarrow Sr(5s^2\ ^1S) + Ca^*(4p^2\ ^1S)$$

This process can be described as an example of "cooperative collisional and optical pumping." In the atomic picture, the excitation of the $Sr^*(5p\ ^1P_0)$ is transferred to $Ca^*(4p\ ^1P_0)$ by collision and the photon then pumps the Ca to $Ca^*(4p^2\ ^1S)$. Pure optical pumping would not have been able to excite $Ca(4s^2\ ^1S)$ to $Ca^*(4p^2\ ^1S)$ because of unchanged parity between the

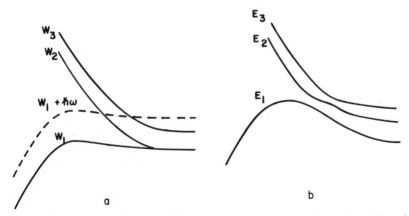

FIG. 8.   A sketch of three model electronic surfaces, one of which (W_1) is reactive, plus the shifted image $W_1 + \hbar\omega$ of the reactive surface (a). If the dynamical coupling of the corresponding electronic states is negligible, then the relevant electronic-field surfaces are those illustrated in b. Here, E_1 correlates to W_2 asymptotically far to the right, E_2 similarly correlates to W_3, and E_3 to $W_1 + \hbar\omega$. It is evident that the intense field can greatly inhibit reaction for this system when the initial electronic state corresponds to W_1, while assisting reaction from W_2 and (to a lesser extent) W_3.

initial and final states. The molecular picture would provide potential surfaces corresponding asymptotically to the energies $Sr^*(5p\ {}^1P_0)$ + $Ca(4s^2\ {}^1S)$ + $\hbar\omega$ and $Sr(5s^2\ {}^1S)$ + $Ca^*(4p^2\ {}^1S)$. Another example of laser-induced collisional energy transfer (Cahuzac and Toschek, 1978) is given by

$$Eu(^8P_J) + Sr(^1S_0) + \hbar\omega_2(\lambda_2 \simeq 658\ nm) \longrightarrow Eu(^8S_{7/2}) + Sr(^1D_2)$$

where $J = \frac{5}{2}, \frac{7}{2}$, or $\frac{9}{2}$. This process can be described analogously to the preceding example.

An example of laser-induced reaction (Hering et al., 1980) is given by

$$K(4^2S) + HgBr_2 + \hbar\omega\ (595\ nm) \longrightarrow KBr + HgBr^*(B^2\Sigma)$$
$$\longrightarrow KBr + HgBr(X^2\Sigma^+) + \hbar\omega_2(500\ nm)$$

No single-photon absorptions are known for $\lambda \sim 590$ nm for reagents or products of this system. Hence resonance for $\hbar\omega$ can only be fulfilled for some transition state (existing only for finite reaction coordinates) (Hering et al., 1980). The fact that $HgBr^*(B^2\Sigma)$ is formed is inferred from the fluorescence of 500 nm to the ground state $HgBr(X^2\Sigma^+)$. Another example (Wilcomb and Burnham, 1981) is the reaction

$$Xe + Cl_2 + \hbar\omega(193.3\ nm) \longrightarrow XeCl^*(B,C) + Cl$$

which does not proceed in the absence of the ArF laser.

Two dynamical processes resembling laser-induced collisional energy transfer (insofar as altered dynamics resulting from radiative interaction is concerned) are laser-induced Penning and associative ionization. Again, the laser does not have to come in resonance with asymptotic energy levels: it interacts with the collision system mainly at finite internuclear separations. The ionization usually takes place via cooperative optical and collisional excitation to some excited electronic state whose potential curve is either embedded in the ionization continuum (generated by the continuous range of free electronic energies) (Bellum et al., 1978; Bellum and George, 1979) or crosses the ionization threshold curve at some finite nuclear separation. Both kinds of ionization (Penning and associative) have been reported (Polak-Dingels et al., 1980) in the study of the Na–Na system

$$Na + Na \xrightarrow{\hbar\omega} Na_2^+ + e^- \quad \text{(associative)}$$
$$Na + Na \xrightarrow{\hbar\omega} Na^+ + Na + e^- \quad \text{(Penning)}$$

B. Condensed Phases

1. *Liquids*

Applications of lasers in the chemistry of pure liquids or solutions have exploited two characteristics of laser radiation: monochromaticity and ultrashort high-intensity pulses; the former involves excitation of specific vibrational modes whereas the latter is used in ultrafast, relatively non-specific heating of the sample in the so-called laser temperature-jump relaxation spectroscopy. The use of picosecond pulses of laser radiation allows a study of ultrafast (picosecond time scale) processes in liquids such as orientational or rotational relaxation, electron transfer, vibrational dephasing or energy transfer, electronic energy transfer, formation of charge-transfer complexes, photodissociation, recombination of ions to form molecules in a "cage" of solvent molecules, and a variety of other ultrashort transient phenomena.

Laser light-scattering experiments can be designed to measure either the polarized or the depolarized component of the output signal. Because the latter is often many orders of magnitude smaller than the former, intense laser sources are necessary for meaningful detection of the depolarized light scattering. A large amount of valuable information, especially about dynamical many-body effects in liquids, has been obtained from this new spectroscopy.

Liquid-phase laser chemistry can be divided into three categories, albeit somewhat arbitrarily because there are substantial overlaps between them; the categories are (1) light-scattering studies (polarized or depolarized), (2) laser-induced chemical reactions, and (3) time-resolved studies using ultrashort laser pulses.

a. Laser Light Scattering. Thermal fluctuations of the dielectric constant of a medium cause an incident polarized beam of laser light to scatter. The total scattering intensity can be thought of as a superposition of scattering from individual molecules, and the static and dynamic correlations between these are reflected in the output signal. In addition to the large component with a polarization similar to that of the incident light, a depolarized component due to intrinsic (or collision-induced) anisotropies in the polarizabilities of the molecules is present in the scattered light. The two provide information about different types of relaxation processes in the system (Bauer *et al.*, 1976; Berne and Pecora, 1976).

Experimental techniques in this field differentiate between processes faster than 10^{-6} sec and those slower than 10^{-6} sec (Berne and Pecora, 1974; Fleury and Boon, 1973). Filter methods (Benedek, 1968) involving frequency scanning by a grating monochromator (10^{10}–10^{14} Hz) or a Fabrey–Perot interferometer (10^6–10^{11} Hz) are used for the former case and provide the autocorrelation function $\langle E_s^*(0)E_s(t)\rangle$, where E_s is the amplitude of the electric field of the scattered electromagnetic radiation. In contrast, processes slower than 10^{-6} sec are investigated by techniques involving optical beats (Cummins, 1971) ("optical mixing"): the heterodyne method gives $\mathrm{Re}\langle E_s^*(0)E_s(t)\rangle$ whereas the homodyne method gives $\langle |E_s(0)|^2 \cdot |E_s(t)|^2\rangle$. The following discussion presents a few applications.

   i. Diffusion of small macromolecules (Ford, 1972). Time scales for these processes are 10^{-3}–10^{-1} sec and thus optical mixing methods are applicable. Polarized-scattering studies provide information on the dimensions and molecular weights of the macromolecules, the concentration dependence of diffusion coefficients, and aggregation effects. A striking example of the latter is a study of the onset of aggregation of hemoglobin S molecules (Wilson et al., 1971). The high molecular weight of an aggregate results in a larger contribution to the time correlation function (because of its smaller diffusion coefficient relative to that of a monomer). This results in a high sensitivity of light scattering to small amounts of aggregation. Exploitation of this sensitivity has allowed the detection of aggregation of hemoglobin S molecules.

   ii. Orientational relaxation of small molecules (Fujime, 1972). Two distinct situations exist in solutions of small molecules. The case where correlations between different molecules are negligible obtains single-molecule relaxation times. For strongly correlated molecules, however, collective orientational relaxation becomes important, and it is necessary to separate the degrees of freedom into slowly and quickly relaxing modes. A statistical mechanical formalism due to Mori (1965a,b), based on a calculation of the temporal behavior of the slowly relaxing modes, then provides an interpretation of both the static and the dynamic correlations in the fluid. These are contained in the linewidths and intensities of the Lorentzian curves, one for each different type of anisotropic molecule present.

   CW He–Ne, Ar^+, and Kr^+ lasers are used, and the filter method is necessary because of time scales of $\sim 10^8$–10^{11} Hz. Some general qualitative features have emerged from studies involving depolarized light scattering (Bauer et al., 1976). Thus, for pure, aromatic molecules (high concentration limit), orientational relaxation behavior is in qualitative agreement with the Stokes–Einstein theory, viz., $\tau \propto \eta/T$, η being the viscosity and T the temperature. For dilute solutions, however, $\tau = C\eta + \tau_0$, with substantial quantitative deviations of C from the Stokes–Einstein theory.

   Results of similar studies have provided information on the influence of solvent hydrogen bonding on reorientation (Alms *et al.*, 1973), as well as tests of various assumptions (Einstein, 1956; Adler *et al.*, 1970; Hu and Zwanzig, 1974; Youngren and Acrivos, 1975), e.g., "stick" or "slip" boundary conditions, made in connection with the viscosity dependence of reorientation times in hydrodynamical theories of fluid motion.

Particularly clear and detailed interpretations of the rotational dynamics are possible for systems where the collisional part of the correlational function is known to be small (Bruining and Clarke, 1975; Schoen *et al.*, 1975), for example, liquid CO, O_2, and N_2. The spectra of these molecules were found to be roughly Gaussian. Furthermore, the results for N_2 and CO are consistent with an average 50° rotation between collisions.

iii. Electrophoretic light scattering (Ware, 1974). For solutions of macroions, electrophoresis and the heterodyne method are combined to provide a powerful technique for measuring mobilities, diffusion coefficients, and relative concentrations in ionic mixtures. The very high resolution and speed (relative to conventional electrophoresis) as well as the ability to use very low concentrations make electrophoretic light scattering a very attractive technique for chemical kinetics. Systems to which this technique has been applied include (Bauer *et al.*, 1976; Ware, 1974; Bennett and Uzguris, 1973; Uzguris and Kaplan, 1974; Uzguris, 1974) polystyrene latex spheres, the bacterium *Staphyloccus epidermis*, human erythrocytes, and bovine serum albumin monomers and dimers.

iv. Intensity fluctuation spectroscopy (Schaefer and Berne, 1972; Schaefer, 1974). This technique applies particularly to very dilute solutions of large (micrometer-size) molecules which move independently. Two widely separated time scales are involved in the fluctuations of the system: τ_V, related to the volume defined by the intersection of the incident and detected beams, and τ_q, related to the scattering length q. These are referred to as the number and interference fluctuation times, respectively. Variations of τ_V provide an elegant method for studying a variety of processes which involve a net flux of particles.

b. Laser-Induced Chemical Reactions. A prominent technique for stimulating chemical reactions with lasers utilizes ultrashort pulses (~1 psec) of intense radiation in temperature-jump spectroscopy (Flynn and Sutin, 1974). This method belongs to the class of line-broadening or relaxation methods, the latter involving a very fast perturbation of the system away from equilibrium. The temperature-jump method is very popular because it requires only that the reaction being studied have a nonzero enthalpy change $\Delta H°$, as opposed to, for example, ultrasound absorption, which requires a finite volume change, or to dielectric relaxation, which requires a change in the effective electric moment. With a laser providing the temperature jump, very fast heating times (~ 1 psec) can be realized and very small samples can be used because of the focusing capabilities of laser beams.

Nd: glass and ruby lasers have sufficient power density ($\sim 10^8$ W/cm^2) for effective temperature jumps in many solvents other than water. Dyes are often used in aqueous solutions to overcome this problem. For systems of biological interest, however, techniques are available which exploit the stimulated Raman effect to red-shift the output of a Nd: glass or ruby laser to provide wavelengths absorbed by water.

The technique is most effective for studying reactions which are first order in the forward and/or backward direction. Typical applications are intersystem crossings in octahedral transition-metal complexes (Beattie *et al.*, 1973) and the triiodide ionic equilibrium (Turner *et al.*, 1972)

$$I_2 + I^- \rightleftharpoons I_3^-$$

The Raman laser temperature-jump study of this reaction (in aqueous solution) provides an understanding of the intermediate steps, including a water-exchange rate and numerical values of the rate constants.

Instead of inducing reactions via a temperature jump in the solvent, direct absorption of the laser energy by the solute molecules can be utilized to study mechanisms and reaction kinetics. A simple but important application of the method is the photoionization of water (Goodall and Greenhow, 1971). In a study of the wavelength and temperature dependence of the quantum yield for the laser-induced ionization, it was shown that a single photon is sufficient, and that the quantum yield increases dramatically with photon energy (Goodall *et al.*, 1979). The study also provided an understanding of the relative importance of decomposition and relaxation channels in the kinetic process.

In addition to pulsed lasers, CW lasers have also been used in the study of certain electronic excitations in solution. Examples include production of O_2 in the singlet state (Matheson and Lee, 1970), trans–cis isomerization of indigo dyes and subsequent cis–trans conversions (Giuliano *et al.*, 1968), and the study of photosynthesis (Weiss and Sauer, 1970; Witt *et al.*, 1961; Parsons, 1968).

c. Time-Resolved Studies Using Ultrashort Laser Pulses. Time-resolved spectroscopy with ultrashort laser pulses (Laubereau and Kaiser, 1978) involves the initial excitation followed by a series of weaker, appropriately time-delayed probe pulses which do not overlap the excitation pulse. The excitation is either via direct resonant absorption or a laser beam shifted into the anti-Stokes region by a stimulated Raman process (Flynn and Sutin, 1974; Eckhardt et al., 1962). The fluorescence or Raman scattering from the sample is then probed to derive information regarding vibrational dephasing, orientational relaxation, energy transfer or relaxation (population changes), collective beating due to isotopic interference, and a number of related dynamical processes in liquids.

The effects of hydrogen bonding of rhodamine 6G were studied in various liquids using a mode-locked Nd : glass laser (Chuang and Eisenthal, 1974). Orientational relaxation times were found to vary linearly with the viscosity; furthermore, the different hydrogen bonding in methanol and chloroform, which have the same viscosity, did not give rise to different relaxation rates, implying that, in the absence of aggregation effects, rotational motion of rhodamine 6G is independent of hydrogen bonding with the solvent.

Charge transfer is an important primary process in many reactions of biological interest. In a study of one such reaction

$$^1A + {}^1D \xrightarrow{h\nu} {}^1A^* + D \longrightarrow {}^1(A^-D^+)$$

(A, anthracene; D, N,N'-diethylaniline) measurement of the differently polarized outputs resulting from a ruby laser excitation allowed a separation of the effects due to orientational relaxation and complex formation (Chuang and Eisenthal, 1974, 1975). The results showed the transient portion of the electron transfer process to be in accord with the diffusion model of Smoluchowski or the Noyes molecular-pair model (Noyes, 1961) at the earliest and intermediate times, whereas the long-term behavior corresponded to a steady-state diffusion.

Evidence of the Franck–Rabinowitch cage effect in liquids was obtained in another application of time-resolved laser spectroscopy (Chuang et al., 1974). For I_2 molecules excited by a Nd : glass laser, measurements of the intensity of the parallel and perpendicular components of the transmitted light as a function of time yielded geminate (i.e., involving the original atomic partners of an I_2 molecule in the *same* solvent cage) recombination times of 70 and 140 psec in hexadecane and CCl_4, respectively.

The study of vibrational relaxation rates of excited states in liquids is generally quite difficult. The use of an incoherent-scattering technique makes such determinations possible, in some cases for the first time. Examples are the determination of the population lifetime T_1 of the 2939-cm^{-1} mode of 1,1,1-trichloroethane ($T_1 = 5.2$ psec) and a time-resolved study of intermolecular energy transfer from the above mode to the 2227-cm^{-1} mode of CD_3OD in a mixture of the two liquids (Laubereau, 1980).

2. Solids

Although the effects of laser radiation on homogeneous systems in the gas phase, liquid phase, or solid phase (Ready, 1971; White and Peercy, 1980) have been intensely studied, much less has been done on heterogeneous systems, e.g., gas/solid, gas/liquid, and liquid/solid (Djidjoev *et al.*, 1976; Karlov and Prokhorov, 1977; Goldanski *et al.*, 1976; Lin, 1980; George *et al.*, 1980; Lin and George, 1979a,b, 1980a–c, 1981a–c; Slutsky and George, 1978, 1979; Lin *et al.*, 1980).

The current understanding of the nature of heterogeneous catalysis involves one or more of the following processes (Thomas and Thomas, 1967): (1) adsorption (physical or chemical) and desorption of the species on the catalytic surface; (2) migration of adsorbed species and subsequent collisions; (3) interactions (via dipole–dipole, electron transfer, etc.) between the adspecies, either directly or surface mediated; and (4) scattering (reactive or nonreactive) of gas-phase species by the clean surface or adsorbed species.

The effect of laser radiation on these heterogeneous processes depends upon the nature of the surface (metal, insulator or semiconductor, smooth or rough, etc.), the electronic and vibrational structure of the adspecies/surface system, and, of course, the frequency, intensity, fluence, and polarization of the laser beam. Depending upon the physical and chemical state of the excited species, there are several possible ways in which laser radiation might influence heterogeneous processes. Some important types of laser-stimulated surface processes (LSSP) and related experimental and theoretical studies are discussed in the following.

1. Laser excitation of reactants in the gas phase. The vibrational excitation of a molecule in the gas phase before striking a catalytic surface could lead to an enhanced rate of reaction if the excitation energy can be used to overcome the reaction barrier, or to a reduced rate of reaction due to decreased adsorption of the species on the catalytic surface. For example, in the catalytic decomposition of formic acid over platinum (Umstead and Lin, 1978), the preexcitation of the gaseous formic acid molecules (by a 10-W/cm^2 CW CO$_2$ laser) resulted in a 50% increase in the ratio of the products CO$_2$/CO compared to that without the laser. Another study of LSSP of the first type (Tu *et al.*, 1981; Chuang, 1980, 1981) showed that SF$_6$, vibrationally excited by CO$_2$ laser radiation, was very reactive to silicon, and that the excitation of the silicon substrate alone could not cause the heterogeneous reaction to occur.

2. Laser excitation of reactants adsorbed on a solid surface. A different kind of LSSP (Djidjoev *et al.*, 1976) involves laser radiation directly incident on the surface of aerosil (SiO$_2$) in an ammonia atmosphere. Here the

rate of decomposition of chemisorbed NH_2 groups in the radiation field of a low-power CO_2 laser (30 W/cm^2) was found to be three orders of magnitude larger than that in the corresponding thermal reaction.

3. Laser heating of substrate. In addition to the selective-type excitations (types 1 and 2), where the laser photon energy is deposited in a specific mode (or functional group) of the reactant, laser radiation may also be used for local heating of the substrate, and may, in turn, influence the surface rate processes. A combination of the pyrolytic (substrate heating) and photolytic (dissociation of the reactant) reactions has been studied in laser-induced chemical vapor deposition (LCVD) with applications to microelectronics (Ehrlich *et al.*, 1979, 1980a,b,c, 1981).

The laser–matter interaction of heterogeneous systems involves addressing the questions on the following fundamental mechanisms:

How is the laser photon energy selectively transferred to the active mode of the species with subsequent relaxation to the bath modes (the inactive modes of the species plus the surface phonon modes)?

What are the energy relaxation dynamics and the nature (selective and nonselective) of the energy deposition in the adspecies/surface system?

How does the laser radiation affect the rate processes via field-induced adsorption, migration, diffusion, dissociation, and desorption?

How does the laser photon energy enhance the gas–solid interaction and change the thermal properties of the heated solid?

To describe theoretical techniques for addressing these questions, the following will focus first on LSSP selective types 1 and 2 just given, with a subsequent discussion on nonselective LSSP involving LCVD and laser annealing (Section III,B,2,b).

a. Selective Excitation and Heterogeneous Catalysis. Herein are discussed theoretical methods for describing energy transfer and bond breaking in terms of physical parameters, including desorption probabilities, pumping rates, damping factors, etc., in connection with possible mechanisms for selective processes.

Consider a model system consisting of a group of adspecies driven by IR radiation. The microscopic Hamiltonian for the entire adspecies/surface system (Lin and George, 1979a; Lin *et al.*, 1980) may be given as

$$H = H_A + H_B + H_{AB} + H_{AA} + H_{ABA} + H_{AF} \qquad (18)$$

where H_A and H_B are the unperturbed Hamiltonians of the active (A) and bath (B) modes, respectively, and H_{AB} is the Hamiltonian describing the multiphonon coupling between the active and bath modes. The terms H_{AA} and H_{ABA} represent direct and indirect interactions between the active modes, respectively. The last term, H_{AF}, represents the interaction between the active mode and the laser field. The direct-coupling term H_{AA} is related to the derivatives of the pairwise interaction potential between the adspecies with respect to the normal coordinates of the active modes (X_i). For example, in the case of a dipole–dipole interaction, the coupling constant (Lin et al., 1980) is given by

$$D_{ij} = (\alpha_i \alpha_j / 2\bar{R}^3)(\partial \mu_i / \partial X_i)_0 (\partial \mu_j / \partial X_j)_0 \tag{19}$$

where μ_i is the dipole moment of the ith active mode, $\bar{R}$ is the average distance between the dipoles (related to the adspecies coverage θ by $\bar{R} = 0.5\,\theta^{-2}$), and α_i is a quantization constant. The adspecies–field interaction Hamiltonian H_{AF} may be expressed in second-quantized notation (Lin and George, 1979a) as

$$H_{AF}(t) = [v_0(t) + v_1(t)(a\dagger + a) + v_2(t)(a\dagger + a)^2 + \cdots]\cos(\omega t) \tag{20}$$

where the first term v_0 is related to pure rotational transitions, while the remaining terms are related to rotation–vibration transitions for single-quantum, two-quanta, etc. processes, and are characterized by the derivatives of the dipole moment.

The Heisenberg–Markovian picture (Louisell, 1973) allows description of the quantum equations of motion for the average excitation of the active mode in which the many-body effects of the phonon bath modes (H_{AB} and H_{ABA}) and the active-mode interaction (H_{AA}) may be replaced by single-body parameters, namely, damping and dephasing factors and frequency shifts of the active mode. For a Markovian and adiabatic process, the coupled equations of motion governed by the total Hamiltonian in Eq. (18) reduce to a nonlinear differential equation (Lin et al., 1980) for the average excitation $\langle n \rangle$

$$\frac{d\langle n \rangle}{dt} = \frac{AI(\gamma_1 + \gamma_2)}{(\Delta - 2\varepsilon^*\langle n \rangle)^2 + (\gamma_1 + \gamma_2)^2/4} - \gamma_1(\langle n \rangle - \bar{n}) \tag{21}$$

where $\Delta = \omega_A - \omega$ is the detuning, ω_A and ω being the frequencies of the active mode and the laser field (with intensity I), respectively; A is a constant proportional to the square of the derivative of the active-mode dipole moment evaluated at the equilibrium position; γ_1 and γ_2 are T_1(energy) and T_2(phase) relaxation rates induced by the many-body effects of the Hamiltonians H_{AA}, H_{AB}, and H_{ABA}; $\bar{n}$ is a Bose–Einstein function for the bath modes; and $2\varepsilon^*\langle n \rangle$ is the anharmonic correction.

Equation (21) describes the conservation of energy of the total system (active and bath modes and laser field) with the following important features: (1) the pumping rate is linearly proportional to the laser intensity for low excitations, although for high excitations, the anharmonic term leads to nonlinear behavior; (2) the energy relaxation rate for the active mode is governed only by γ_1 and not by the dephasing factor γ_2, whereas the absorption cross section is governed by the total broadening ($\gamma_1 + \gamma_2$); (3) the steady-state average excitation $\langle n \rangle^{ss}$ is given by a cubic equation which yields the power-dependence law $\langle n \rangle^{ss} \propto I^\alpha$, with $\alpha = 1$ and <1 for low and high excitations, respectively. Figure 9 shows the average excitation as a function of laser intensity (Lin *et al.*, 1980).

Figure 10 shows the selective nature of LSSP via numerical solutions of the nonlinear Eq. (21). Another view of selective versus nonselective LSSP is presented in Fig. 11, where the results for a multilevel quantum system are plotted, obtained by solving for the energy populations in different modes (Lin and George, 1980b,c). In these two figures the competition between the selective and nonselective aspects of LSSP is portrayed for a range of values of parameters such as pumping rates, damping rates, and coupling constants.

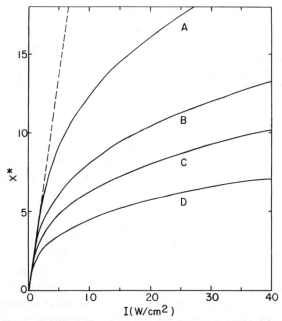

FIG. 9. Maximum excitation X^* vs laser intensity I under the optimal detuning condition for $\gamma_1 = 10^4 \ \text{sec}^{-1}$, $\gamma_2 = 10 \ \text{cm}^{-1}$, and (A) $\varepsilon^* = 1 \ \text{cm}^{-1}$, (B) $\varepsilon^* = 2 \ \text{cm}^{-1}$, (C) $\varepsilon^* = 3 \ \text{cm}^{-1}$, and (D) $\varepsilon^* = 5 \ \text{cm}^{-1}$. The resonant harmonic excitation ($\varepsilon^* = \Delta = 0$) is shown by the broken line. (From George *et al.*, 1980.)

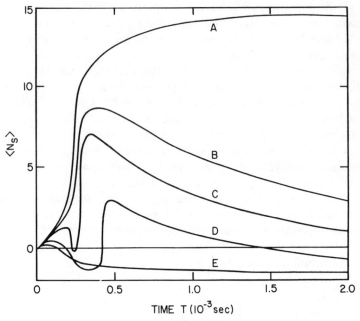

FIG. 10. Average selective excitation number $\langle N_s \rangle$ (the energy difference between the active mode and the bath modes) as a function of time t for various values of the energy relaxation rate γ_1: (A) 10^2 sec^{-1}, (B) 10^3 sec^{-1}, (C) 2×10^3 sec^{-1}, (D) 6×10^3 sec^{-1}, and (E) 10^4 sec^{-1}. Values of the other parameters: $\gamma_2 = 10$ cm^{-1}, $\Delta \equiv \omega_A - \omega = 24.9$ cm^{-1}, $\varepsilon^* = 2$ cm^{-1}, and the laser intensity $I = 10$ W/cm^2. Curves A and B represent the highly selective excitations and curve E represents the nonselective excitation. (From Lin et al., 1980.)

The characteristics of systems exhibiting selective effects and surface-enhanced bond breaking are now discussed. Consider a system whose vibrational degrees of freedom can be pictured as a set of distinct groups in the frequency domain. For simplicity, consider the case of only two groups, A and B, where group A is referred to as the excited group, consisting of the active mode plus other internal modes of the adspecies coupled strongly to the active mode, and group B consists of the remaining modes of the total system (adspecies plus solid). Depending on the relative magnitudes of the intragroup (R^{intra}) and the intergroup (R^{inter}) coupling rates, which characterize the energy randomization rate *within* and *between* the group(s), respectively, and the laser pumping rate (V), several types of laser excitations may be introduced: group selective, molecule selective, and purely thermal (nonselective). For LSSP to be characterized as group or molecule selective, the pumping rate must be greater than the energy relaxation rate(s), i.e., $V > R^{\text{intra}}$ and/or $V >$

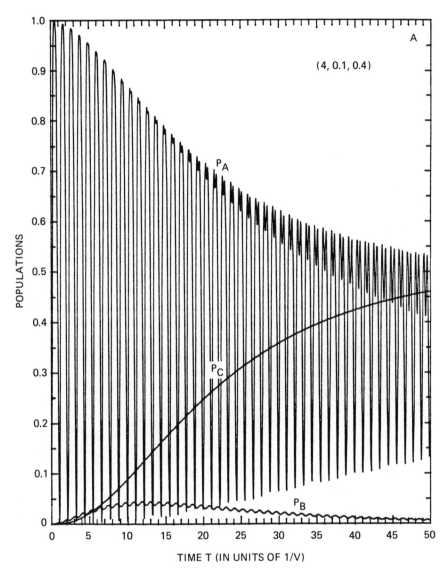

FIG. 11. The level populations of the active mode (P_A) and the bath modes (P_B) and (P_C) of two-photon multiphonon processes for the pumping rate (V), coupling factor (g), and damping rates (γ) given by (V, g, γ) = (A)(4, 0.1, 0.4) and (B) (4, 1, 1). Note that (A) shows the highly selective excitation of the active mode with high P_A and low P_C while (B) shows the nonselective thermal excitation of the C modes with high P_C and low P_A. (From Lin and George, 1980c.)

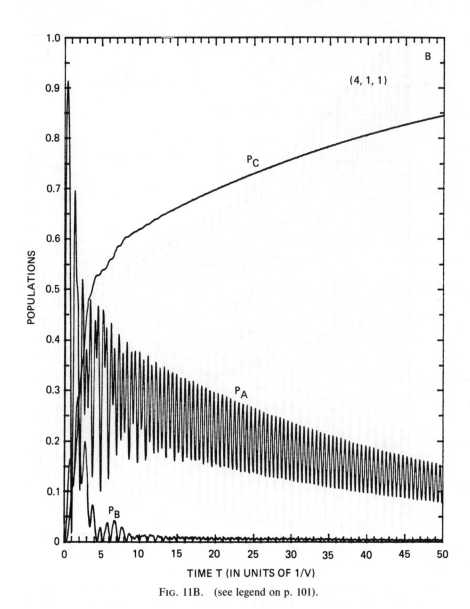

FIG. 11B.   (see legend on p. 101).

$R^{\text{inter}} \gg R^{\text{intra}}$. For the case of very high pumping or low relaxation rates such that $V \gg R^{\text{intra}}$ and R^{inter}, a group-selective excitation would be possible. However, for low-intensity excitations, group selectivity seems less likely than molecule selectivity because the only required condition is

V and $R^{\text{intra}} \gg R^{\text{inter}}$. According to the "energy-gap law", it is highly probable that a system will have very weak intergroup coupling but strong intragroup coupling, examples being large molecules adsorbed on solid surfaces (SF_6/metal) or long-chain adspecies (A–B–C–D/metal). In these systems the vibrational stretch of the excited species (which is not directly connected to the surface) may behave like a small subsystem, where the excited bond coordinate is composed of the normal coordinates of the related group. This selectively excited adspecies, although strongly coupled within the group, is weakly coupled to the remaining species (or functional group) of the system. Therefore, selective bond breaking of adspecies (or certain functional groups) will be possible (Lin and George, 1981b).

Another important feature of selective bond breaking in a heterogeneous system (and usually not present in a homogeneous gas-phase system) is the surface-enhanced local field acting on the adspecies. The pumping rate may be enhanced by a significant factor when the local field is increased due to surface effects such as roughness and electron transfer involving the substrate. Using a simple relation between the pumping rate and the local electric field, $I \propto (E_{\text{loc}})^2$, the magnitude of the laser intensity required for selective excitation may be reduced by a factor of 10^8 when the local electric field is enhanced by a factor of 10^4. The enhancement by the local surface field (which explains surface-enhanced Raman scattering) can play an essential role in LSSP, where considerably lower laser intensities ($10–10^3$ W/cm^2) are used compared to those for photodissociation of gaseous polyatomic molecules.

Given the absorption cross section [the first term of Eq. (21) divided by I], the master equation for the energy population can be solved to find physical quantities such as the laser-induced desorption probability, the desorption rate of the adspecies (or dissociation rate of the gaseous reactant), and the average excitations. Consider a heterogeneous rate process in which the overall reaction rate is limited by the removal of the product from the surface. The reaction rate may then be enhanced by laser-induced desorption of the product species. The desorption probability may be expressed in a simple Arrhenius form

$$P_D \propto \exp(-E_A/kT_{\text{eff}}) \qquad (22)$$

where E_A is the activation energy for bond breaking and T_{eff} is the vibrational temperature of the selectively excited adspecies (product) given by $kT_{\text{eff}} \approx \langle N \rangle / S$, $\langle N \rangle$ being the average number of photons absorbed by the adspecies with S vibrational modes.

Another example of LSSP is where the reactant is vibrationally excited before sticking to the substrate surface. Here the overall reaction probability may be expressed as

$$P = P_V P_S P_R \tag{23}$$

where P_V is the probability that the gaseous reactant is in the vibrationally excited state accessed by the laser radiation, P_S is the (laser-enhanced) sticking probability of the excited reactant, and P_R is the surface-catalyzed reaction probability. $P \approx P_S$ when the sticking of the excited reactant is the rate-limiting step, which may be greatly enhanced by the laser radiation. Furthermore, Eq. (23) can be used to describe a surface rate process where the dissociation of the gaseous reactant is the rate-limiting step. In this case the vibrational excitation probability is replaced by the dissociation probability given by Eq. (17).

There is another interesting example of laser/surface-catalyzed rate processes, namely, the recombination of atoms A and B on a catalytic surface (K):

$$
\text{A + B} \underset{}{\overset{k_1}{\rightleftharpoons}} \text{(AB)} \underset{}{\overset{k_2}{\rightleftharpoons}} \text{AB +}
$$
$$
\text{(K)} \qquad\qquad \text{(K)} \qquad\qquad \text{(K)}
$$

Laser radiation can influence this reaction (1) by increasing the mobility of the reactant atoms (A or B) through photon excitation of the A–K or B–K bond with subsequent enhancement of the reaction rate k_1; (2) by removal of the excess energy from the unstable complex (AB) on the substrate surface via laser-stimulated emission accompanied by surface-phonon-mediated relaxation, thereby increasing the reaction rate k_2; and (3) by breaking the AB–K bond either directly through laser excitation of the adspecies or indirectly through thermal desorption by laser heating of the surface. The direct desorption of chemisorbed species from a solid surface usually requires multiphoton absorption, necessitating the use of high-power radiation. However, much lower powers may be sufficient for the desorption of a diatomic molecule adsorbed on a solid surface (A–B–K) if the photon energy absorbed by the A–B molecule can be easily transferred to the surface to break the B–K bond via anharmonic coupling with the A–B–K system.

As a final example of LSSP, consider laser/surface-enhanced predissociation of a diatomic molecule adsorbed on a solid surface. It has been suggested that laser-induced predissociation of gaseous species such as NO, CO, and H_2 can be greatly enhanced by the presence of a surface magnetic field (Bhattacharyya et al., 1980; Lin and George, 1981c). In these cases the magnetic field splits the electronic curves (other than singlet) of the "adsorbed" species into multiple branches, giving rise to different dissociation channels and thus altering the predissociation probability. One specific example with a ground singlet and an excited triplet is shown in Fig. 12. The dynamical probabilities of predissociation by a two-step excitation, namely, the absorption of one IR and one UV photon, are shown in Fig. 13 for the surface-phonon-coupled and surface-phonon-free cases.

b. Laser Applications to Microelectronics Fabrication. A rapidly developing field of laser applications is microelectronics fabrication at the large-scale integration (LSI) and very-large-scale integration (VLSI) levels, the latter being associated with the development of very-high-speed

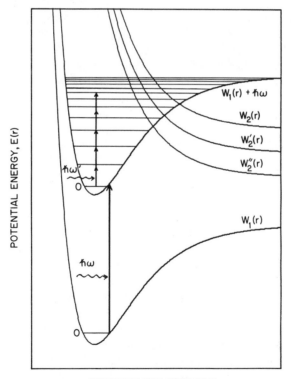

INTERNUCLEAR DISTANCE, r

FIG. 12.   Schematic energy diagrams of a multilevel–multistate system subject to two lasers with frequencies ω and ω', respectively. Note that the electronically excited triplet state is split by the surface magnetic field.

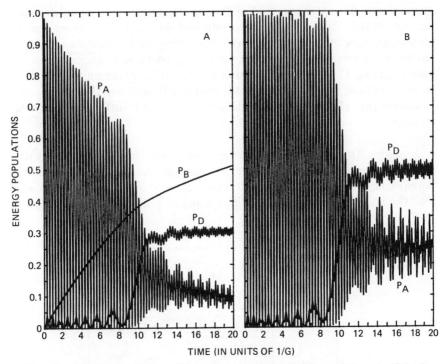

FIG. 13. The time evolution of the energy populations of the active mode (P_A), the phonon (bath) modes (P_B), and the predissociation probability (P_D) for ($V, G, \gamma, \Delta, \alpha$) = (10, 1, 0.1, 0, 1) for a phonon-coupled (A) and (10, 1, 0, 0, 1) a phonon-free (B) system. V, G, γ, Δ, and α are the pumping rate, Landau–Zener factor, phonon-induced damping, field detuning, and phase detuning, respectively. (From Lin and George, 1981c).

integrated circuits (VHSIC) (Capece, 1978, 1979; Avery, 1980; Murphy, 1979). VLSI technology requires fabrication of very-large-scale integrated circuits, where linewidths are reduced from the current 3 μm size of LSI to the submicron range and where it is possible to have over a million devices on one chip (Smith *et al.*, 1974; Lyman, 1980; Ratnakumar *et al.*, 1980).

Because this exciting field is in its infancy, much of the following discussion should be viewed as suggesting the "potential" of laser photochemistry in microelectronics, rather than reviewing well-established techniques. Fabrication methods for LSI/VLSI involve the following steps (Fogiel, 1972): (1) substrate preparation, (2) lithography, (3) oxidation/passivation, (4) diffusion/doping, (5) epitaxy, (6) chemical processing, and (7) interconnection, lead attachment, and packaging.

VLSI is still in the experimental phase, with a major portion of the effort aimed at solving the resolution problem at the lithography stage (Ratnakumar *et al.*, 1980; Smith *et al.*, 1974; Lyman, 1980; Decker and Ross, 1980). Some laboratories (Smith *et al.*, 1974; Lyman, 1980) believe that currently popular optical microlithography methods, which use masks in a contact or projection mode for LSI, can be improved for the submicron work of VLSI. Maskless techniques involving electron beams (and, less frequently, ion beams or X rays), however, seem to be more promising alternatives for VLSI (Brewer, 1971; Smith *et al.*, 1973; Spears and Smith, 1972; Broers and Hatzakis, 1972). No major changes in the technology for epitaxy are being proposed for VLSI, with chemical vapor deposition (CVD) being by far the most popular method for depositing thin films of a large variety of materials on substrates, most often semi-conductors or insulators (Feist *et al.*, 1969; Shaw, 1975; Grünbaum, 1975; Kern and Ban, 1978). Experiments involving CVD have utilized laser radiation to control the reaction at various stages (Deutsch *et al.*, 1979; Ehrlich *et al.*, 1980a,b; Baranauskas *et al.*, 1980; Allen and Bass, 1979; Christensen and Lakin, 1978; Steinfeld *et al.*, 1980; Fan *et al.*, 1979; Gat *et al.*, 1979; Williams *et al.*, 1978; Sandow, 1980; Tamaru *et al.*, 1980; Gibbons *et al.*, 1979; Bean *et al.*, 1978). The following will concentrate on the CVD aspect of the fabrication process. However, other aspects such as annealing will also be briefly discussed (Section III,B,2,b,ii).

i. Chemical vapor deposition (CVD). The generic processes occurring in CVD involve fundamental steps of gas–surface interactions (Kern and Ban, 1978) with the following typical sequence of events: (1) diffusional transfer of gas particles (reactants) to the surface; (2) adsorption of reactants on the surface; (3) events *on* the surface, e.g., reactions, migration, and lattice incorporation; (4) desorption of products from the surface; and (5) diffusion of desorbed species away from the surface.

Diffusion steps 1 and 5 are fairly well understood within frameworks such as the kinetic theory of gases (Geankopolis, 1972; Bird *et al.*, 1960; Vanderputte *et al.*, 1975) and generally do not represent critical points in the chemical process. Steps 2, 3, and 4 involving the actual gas–surface interaction are usually more important for the overall reaction, often include a rate-determining step, and form an area of active research (Kern and Ban, 1978; Bryant, 1977). In this context, results of model calculations of desorption dynamics (Doyen, 1980; De and Landman, 1980; Gortel *et al.*, 1980a; Bendow and Ying, 1973; Efrima *et al.*, 1980; Lin and Wolken, 1976a; Antoniewicz, 1980), phonon-induced surface migration (Efrima and Metiu, 1978; Prager and Frisch, 1980), elastic scattering (Lee and George, 1979a; Doll, 1974; Garcia *et al.*, 1979; Goodman and Tan, 1973), and inelastic scattering (Shugard *et al.*, 1977; Diebold *et al.*, 1979;

Tully, 1980; McCreery and Wolken, 1976; Lin and Wolken, 1976b; Diebold and Wolken, 1979; Milford and Novaco, 1971; Cole *et al.*, 1979; Chow and Thompson, 1976) have appeared in the literature. Experimental work in this area is dominated by surface spectroscopies such as IR (Eischens, 1972; Yates *et al.*, 1979; Pirug *et al.*, 1979; Evans and Weinberg, 1979), low-energy electron diffusion (LEED) (Ertl and Schillinger, 1977; Ibbotson *et al.*, 1980), photoelectron scattering (PES) (Eberhardt and Himpsel, 1979; Oshima *et al.*, 1979), Auger electron spectroscopy (AES) (Taylor *et al.*, 1978), and electron energy-loss spectroscopy (EELS) (Thiel *et al.*, 1979; Baro and Ibach, 1979), all of which provide information about the geometry of the adsorption sites and changes in core electronic structure due to adsorption. Information on dynamics (adsorption/desorption, migration) of adsorbed species is obtained from temperature-programmed desorption (Tamm and Schmidt, 1969; Barford and Rye, 1974; Fair and Madix, 1980; Cohen and King, 1973), inelastic electron-tunneling spectroscopy (IETS) (Eib *et al.*, 1979), field ionization microscopy (FIM) (Ertl, 1979), field emission microscopy (FEM) (Stewart and Ehrlich, 1975), and mass spectroscopy (Brumbach and Somorjai, 1974; Bernasek and Somorjai, 1975; Palmer and Smith, 1974).

CVD processes are of almost unlimited range and involve deposition of vitreous or crystalline forms of metallic, semiconducting, or insulating materials on solids which span a similar range. Consequently, the variety of physical processes encompassed is also very extensive. Typical examples of CVD reactions include the deposition of Si on Si by silane pyrolysis (Richman *et al.*, 1970); GaAs on spinel (MgOnAl$_2$O$_3$; $1.5 < n < 2.5$) by vapor-phase organometallic decomposition using trimethylgallium, (CH$_3$)$_3$Ga, and arsine (AsH$_3$) (Wang *et al.*, 1974); SiO$_2$ on Si by oxidation of tripropylsilane [(C$_3$H$_7$)$_3$SiH] (Avigal *et al.*, 1974); and W on Si by hydrogen reduction of WF$_6$ or WCl$_6$ (Melliar-Smith *et al.*, 1974). Other CVD reactions can be found in a review by Kern and Ban (1978). Although alternative methods of depositing thin films of different materials on solid substrates are available (Vossen, 1978), for example, molecular-beam epitaxy, electrodeposition, sputtering, direct transport of vaporized materials, plasma anodization, ion implantation, and liquid-phase epitaxy, CVD is by far the most important method for epitaxial growth from a commercial point of view because of atmospheric pressure operation, ease of doping, the almost unlimited number of reactions and starting materials available, and many other advantages (Feist *et al.*, 1969; Shaw, 1975). However, the high temperature often required for the reactions and the toxic, corrosive, or explosive nature of the reactants lead to problems such as undesirable diffusion, alloying or chemical reactions, and contamination or corrosion of substrate and deposit films. By-products of the reaction (such as polymers) may also attack or settle in the apparatus. In many cases there are also difficulties associated with the control of the growth rate and masking processes. Finally, the stresses imposed on the substrate by strong temperature gradients lead to softening, warping, and other defects which are responsible for an unusually high rejection rate of the mass-produced circuits (Lyman, 1977; Kamins, 1974; Auston *et al.*, 1979).

The recently initiated use of lasers in microelectronics fabrication presents some very exciting possibilities for overcoming some of the problems in conventional CVD. Thus, for example, the application of lasers in the annealing of amorphous films in some CVD reactions (Tamaru *et al.*, 1980; Gibbons *et al.*, 1979; Bean *et al.*, 1978) or of damaged surface layers resulting from ion implantation (Fan *et al.*, 1979; Gat *et al.*, 1979; Williams *et al.*, 1978; Auston *et al.*, 1979) via solid-phase or liquid-phase epitaxy seems to be fairly well established as an industrial process. A somewhat less established application involves the initial deposition of chemically produced elements or compounds in a laser-stimulated gas-phase reaction. In experiments by Deutsch *et al.* (1979) and Ehrlich *et al.* (1980a), a UV laser beam of wavelength 257.2 nm was used to break specific molecular bonds in gaseous trimethylaluminum [$(CH_3)_3Al$] and dimethylcadmium [$(CH_3)_2Cd$], from which Al or Cd, respectively, were deposited on a quartz substrate. These experiments represent a radical departure from the conventional laser processing techniques involving only thermal effects such as annealing or laser heating of the substrate (Baranauskas *et al.*, 1980; Allen and Bass, 1979; Christensen and Lakin, 1978) in CVD. The feasibility of combining both the thermal and nonthermal laser processing techniques in CVD has been clearly demonstrated (Ehrlich *et al.*, 1980b). The use of a focused laser beam (2–3 μm) in these experiments resulted in a strong localization of both the deposition process and the surface heating, thereby obviating the need for high temperatures throughout the reaction chamber. Some of the problems associated with the high temperature could thus be minimized by the use of lasers.

The revolutionary potential of the combined thermal plus nonthermal laser processing technique becomes clear when it is realized that the lithography and deposition steps, which have traditionally been independent and sequential (Fogiel, 1972), can in effect thereby be combined into one step. Serious development of this technique holds forth the promise of eliminating whole steps from the fabrication line, leading to substantial economies and simplifications in microelectronics technology.

Visible or UV (V/UV) lasers may possibly be used as lithographic tools (scribes). [Assuming the extent of the limiting focal region to be about the same as the wavelength of the light, infrared lasers would be quite ineffective for VLSI (submicron) lithography.] Finely focused V/UV laser beams of this nature will function mainly as sources of very high local temperatures, and lithography would primarily involve vaporization (or melting) on the solid surface.

An analysis of the common CVD reactions (Kern and Ban, 1978) provides meager but interesting insights into the nature of some important mechanisms involved in deposition. Growth rates are seen to be sensitive to temperature, orientation, the form of carrier gas, and of course the nature of the substrate; the sensitivity furthermore varies over a large range depending on other conditions. Thus, for example, the temperature dependence of the deposition rate is weak if the rate-determining step is the diffusional transport of reactants in the gas phase to the surface because the diffusion constants are almost independent of temperature (Geankopolis, 1972; Bird et al., 1960; Vanderputte et al., 1975). However, the temperature dependence is strong if the rate is determined by a step in a surface reaction (Joyce and Bradley, 1963; Farrow, 1974) such as, for example, the rate of desorption of hydrogen produced by the decomposition of silane on a silicon substrate. This is a particularly clear example of a situation where the selective laser stimulation of a single phase of a reaction could increase the overall rate. Thus, by pumping the hydrogen-to-surface bond with low-power infrared laser radiation of the "right" frequency, the rate of hydrogen desorption could be enhanced and higher deposition rates could be achieved at much lower temperatures. Desorption could conceivably also be enhanced by exciting the system electronically using higher frequency laser radiation. However, the mechanism in this case is not quite as clear, and the selectivity might not be as sharp.

The diverse dependencies of deposition rates on the nature of the substrate and the orientation of the crystal face represent a more difficult situation (Kern and Ban, 1978). A variety of mechanisms can be invoked to explain the variations in the deposition rate, some of them interrelated.

1. Lattice mismatch between the film and the substrate can lead to dislocations and poor growth rates.

2. The charge-density variation on the surface changes dramatically when different lattice planes are exposed, as does the concentration of dangling bonds (i.e., the degree of unsaturation on the surface), such that the layer thickness at which the influence of the substrate surface disappears and steady-state epitaxy sets in is also dictated by the exposed lattice plane.

3. Some crystal planes permit a larger number of intrinsic defects which can enhance deposition if they act as active sites for catalytic initiation of a reaction step, or de-enhance deposition if a smooth surface

is required for growing single-crystal films of products condensing from a gas-phase reaction.

4. In the final analysis, variations in the nature of the surface potential (both normal and parallel to the surface) at different crystal faces lead to differences in the dynamics of adsorption, desorption, migration, or even lattice incorporation or penetration. It is likely that one or more of these processes represent a bottleneck for the rate of deposition, and a clear understanding of the relationship could lead to improved efficiencies for the CVD reactions.

An important class of CVD reactions involves "selective" deposition (Berkenblit and Reisman, 1971; Engeler *et al.*, 1970), where the film growth takes place only on one part of the substrate due to different sticking coefficients on the different materials constituting the substrate. The depositions of silver and tungsten described earlier (Melliar-Smith *et al.*, 1974; Shaw and Amick, 1970; Voorhoeve and Merewether, 1972) are two examples of such a process; the metals deposit only on the *silicon* exposed through a layer of deposited silicon oxide, while the oxide is only mildly etched. Selective deposition of semiconductors has also been exploited for growing epitaxial germanium on small areas of Ge or GaAs substrates (Berkenblit and Reisman, 1971), and (in the "epicon" TV camera tube structure) for growing pyramidal-shaped structures of p-type silicon on top of a patterned oxide-covered wafer of n-type silicon (Engeler *et al.*, 1970). The latter structures are nucleated only on silicon and grow up through apertures in the oxide layer. Each structure forms a capacitor, i.e., an oxide layer sandwiched between layers of n- and p-type silicon.

Selective deposition is taken one step further in the so-called "electroless plating" by selective deactivation (Feldstein and Lancsec, 1970; Paunovic, 1980). Here the surface is chemically prepared using a two-step adsorption/chemical replacement technique which results in a layer of catalytic palladium on a photoresistively patterned surface. The activated substrate is then etched, causing selective deactivation, and finally plated along the active regions.

It is possible to conceive of a laser playing the role of selective activator or deactivator. By causing disturbances in specific vibrational or electronic energy transfer processes involved in adsorption, desorption, or migration of a gas atom or molecule on a surface, sticking probabilities can be altered. Use of high-power lasers as sources of local heating is not expected to be effective in this mode because all degrees of freedom would be equally favored. However, relatively low-power lasers with frequencies and linewidths chosen properly can be expected to interact strongly with a single vibrational or electronic degree of freedom (or, alternatively, with a very narrow band of such states). Whether this guarantees selective surface activation or deactivation depends crucially on an understanding of the nature of the time evolution of the gas–surface system in regard to energy transfer between the adspecies and the solid and among the adspecies. Other important elementary processes include migration, desorption, transitions of the adspecies between some precursor (perhaps physisorbed) state and a chemisorbed state, changes of adsorption site (for example, to one where the adspecies is not laser active), surface collisions and reactions, charge-transfer processes, and the arrival of more gas-phase species at the surface (leading eventually to a complete monolayer and, subsequently, multilayer formation).

The kinetics of processes taking place *on* the surface can be separated into two components: (1) the kinetics of adsorption of the first few atoms or molecules to reach the surface, and (2) the kinetics of subsequent multilayer formation. The theoretical problem associated with multilayer formation is not well defined at this stage. It is, of course, the next logical step after initial deposition begins, and as such is an important component of the overall CVD reaction. However, it represents a set of processes fundamentally different from those involved in the initial adsorption and is more closely related to homogeneous crystal growth (or agglomeration of microcrystallites, if actual epitaxy is not being considered).

In many cases of CVD, the fundamental reaction (for example, the decomposition of metal alkyls) actually takes place in the gas phase, with the desired material then condensing on the substrate. The stimulation of such gas-phase reactions by laser radiation has been and continues to be actively studied (George *et al.*, 1979; George, 1979; Zewail, 1980; Letokhov, 1980; Zare and Bernstein, 1980; Lee and Shen, 1980). Outstanding problems are associated with intermolecular energy-transfer processes which have a direct bearing on the ultimate destination of the laser energy and hence the nature of the final states of the gas molecules (dissociation, ionization, formation of excited intermediates, thermal randomization, etc.). A study of these gas-phase processes is important for a variety of reasons. The formalisms used for the study of gas–surface processes are often closely related to techniques used in gas-phase studies, often being extensions or modifications of the latter; also, the gas-phase reactions form an important part of the overall CVD process, which can easily be treated with present-day techniques.

The laser can interact with the gas–surface system in three ways: direct pumping of the gas-phase species alone, pumping of the adspecies–substrate system, and interaction where collisional dynamics lead to the actual adsorption. The last way is particularly intriguing from a theoretical point of view, but the laser intensities required are too high from a practical standpoint (Lee and George, 1979b,c; George *et al.*, 1980). The other two types of interaction are intimately involved in LCVD for reasonable laser intensities.

Laser interaction with the gas phase can involve the following processes: (1) excitation (rotational, vibrational, or electronic) and/or ionization, (2) dissociation into ionic or atomic fragments, and (3) thermal effects (e.g., increasing average kinetic energy). This aspect (gas–laser interaction) has received considerable theoretical attention in the past and continues to be an active field of endeavor, but with many outstanding problems.

Laser stimulation of adspecies–surface interactions includes analogous processes: (1) preparing the adspecies or the surface in a particular state, (2) adsorption/desorption of neutral or ionized fragments (dissociation of the adspecies), and (3) thermal effects (e.g., increased mobility on the surface).

Phonon band structure and infrared LCVD. When a molecule is adsorbed on a surface, it can exchange vibrational energy with the solid via the bond with the nearest atoms of the solid. The nature of this interaction will have a direct bearing on whether the molecule stays on the surface at a given site, moves around on the surface, or is desorbed. In conjunction with electronic mechanisms to be described later, this multiphonon vibrational relaxation dictates the nature of the kinetics of deposition of specific adspecies.

In general, the response of the system to infrared radiation is quite complex, involving a large number of degrees of freedom which are often quite different from those of the isolated adspecies or solid. The monochromatic nature of laser radiation holds the promise of simplifying the situation by separating out a few (often just one) degrees of freedom which are influenced more by the field than the others because of a resonance or near-resonance condition. An example is the case of an adsorbed molecule with one of its vibrational modes (A) having a fundamental frequency very close to that of the laser. If none of the other vibrational modes of the molecule (B) or phonon modes of the solid (C) is close to the laser frequency, the electromagnetic energy is primarily adsorbed by the A mode (the active mode), and the B and C modes act as "bath" modes from a thermodynamic point of view.

Theoretical treatments of desorption. A variety of methods have been applied to the problem of desorption, in particular phonon-stimulated desorption. Inclusion of the effects of laser radiation has been attempted (Lee and George, 1979b,c; George *et al.*, 1980; Lin and George, 1979a,b, 1980a–c; Lin *et al.*, 1980; Slutsky and George, 1978, 1979; Bhattacharyya *et al.*, 1980; Murphy and George, 1981). The theoretical treatment of relaxation phenomena (Zwanzig, 1961) is based on either (1) a classical Liouville equation, namely

$$i[\partial f(\mathbf{r},\mathbf{p},t)/dt] = \mathcal{L}f(\mathbf{r},\mathbf{p},t) \tag{24}$$

where *f* is the phase–space ensemble density; $\mathbf{r}$, $\mathbf{p}$, and *t* represent the totality of coordinates, momenta and time, respectively; and $\mathcal{L}$ is the classical Liouville operator

$$\mathcal{L} = i[(\partial \mathcal{H}/\partial \mathbf{r})(\partial/\partial \mathbf{p})] - i[(\partial \mathcal{H}/\partial \mathbf{p})(\partial/\partial \mathbf{r})] \tag{25}$$

with $\mathcal{H}$ the classical Hamiltonian for the system; or (2) a quantum mechanical equation of motion such as the Heisenberg equation

$$i(\partial \rho/\partial t) = \hbar^{-1}(\mathcal{H}\rho - \rho\mathcal{H}) \tag{26}$$

where ρ and $\mathcal{H}$ are the quantum mechanical density matrix and Hamiltonian operator, respectively. Within these basic frameworks, previous investigations are distinguished by the natures of the approximations employed in constructing $\mathcal{H}$ or $\mathcal{L}$ and in solving the equations of motion. The following are some of the most common techniques currently in use.

1. Direct solutions of Heisenberg's equation of motion (Doyen, 1980; De and Landman, 1980; Gortel *et al.*, 1980a,b; Lin and Wolken, 1976a; Antoniewicz, 1980; Shugard *et al.*, 1977; Diebold *et al.*, 1979; Tully, 1980; Beeby and Dobrzynski, 1971; Holloway and Jewsbury, 1976; Holloway *et al.*, 1977; Wolken and McCreery, 1978; Nitzan and Jortner, 1972; Nitzan *et al.*, 1974; Nitzan and Silbey, 1974), some of which assume random-phase or Wigner–Weisskopf approximations.

2. Solutions of Liouville's equation using projection operator techniques, developed by Mori, Kubo and Zwanzig, closely related to the generalized Langevin equation (Diestler and Wilson, 1975; Lin and Adelman, 1978; Shugard *et al.*, 1978; Nitzan *et al.*, 1978).

3. Solutions of a generalized master equation for the population of the various modes of the system subject to a stochastic force (Müller and Brenig, 1979). The assumption of Markovian statistics simplifies the solution, but non-Markovian problems have also been treated (Grigolini, 1979).

4. A formalism developed by van Hove and Glauber which leads to expressions involving "time- and position-dependent displacement–displacement correlation functions" for the phonons (van Hove, 1954; Glauber, 1955), but which is somewhat difficult for actual computational purposes (Bendow and Ying, 1973).

5. A Green's functions approach (Tapilin *et al.*, 1978; Allen, 1979).

6. The Cabrera–Celli–Goodman–Manson (CCGM) approach (Cabrera *et al.*, 1970; Garcia *et al.*, 1979; Goodman and Tan, 1973).

7. A surface band-structure approach (Milford and Novaco, 1971; Cole *et al.*, 1979; Chow and Thompson, 1976; Tsuchida, 1968; Wolken, 1974; Gerber *et al.*, 1978).

In almost all of these cases the Born approximation is invoked, transition rates are calculated using the Fermi Golden Rule expression, and the solid is treated within a Debye model. The aspects in which the above treatments differ include:

a. Type of surface potential used, e.g., truncated harmonic oscillator, square well, spherically symmetric Morse or Lennard–Jones, anisotropic (angle-dependent) Morse or Lennard–Jones, single-potential function for outermost surface atom and adsorbed atom, or a sum of pair potentials between the adatom and atoms of the solid.

b. One-dimensional (linear chain) or three-dimensional treatment.

c. Second-order perturbation theory or fourth-order perturbation theory.

d. No-phonon, one-phonon, two-phonon, or multiphonon processes included in gas–solid energy transfer.

e. Band-structure effects due to two-dimensional periodicity of the surface included or ignored.

f. Classical trajectory or classical normal-mode treatments versus quantum mechanical calculations of expectation values using suitable basis sets.

g. "Local oscillators" versus "delocalized oscillators" ("phonons") treatment of the solid.

h. Adsorption potential with one bound state or a large number of bound states.

Different combinations of these options are necessary or sufficient for different systems, and a broad consensus as to their relative merits is generally lacking. Thus, for example, a one-dimensional model seems to be sufficient for many problems involving physisorption (Beeby and Dobrzynski, 1971; Holloway and Jewsbury, 1976; Holloway *et al.*, 1977; Gortel *et al.*, 1980b; Goodman and Romero, 1978). However, due to the small number of investigations aimed at chemisorption, no such indication exists for the case of strong bonding. The situation regarding multiphonon effects is also fraught with controversy. In the past, the intuitively attractive idea that one-phonon processes suffice if $E_0 < \hbar\omega_D$, two-phonon processes if $\hbar\omega_D < E_0 < 2\hbar\omega_D$ (where ω_D is the Debye frequency and E_0 represents the position of the lowest bound state of the adspecies with respect to the adsorption continuum), and so on, was generally accepted. However, work by Jedrzejek et al. (1981) suggests that N-phonon processes ($N >> 1$) may always be important, pointing out a possible problem of convergence criteria and perhaps the necessity of abandoning perturbation theory approaches in favor of direct solutions of the problem. The latter is computationally very difficult, and it is necessary to adapt existing theories to each specific problem.

The choice between local oscillators versus lattice normal modes (phonons) represents a very fundamental problem (Lin and Wolken, 1976a; Antoniewicz, 1980). In the case of atom–surface scattering, a phonon treatment seems appropriate for collision times τ much larger than the characteristic time for energy dissipation in the solid, and a local-modes treatment for the opposite case. However, in the case of adsorbed atoms or molecules where a corresponding "residence time" is often unknown, such clear distinctions are difficult.

The foregoing represents a survey of theoretical techniques that are currently employed in treating the adspecies–surface interaction. Specific approaches in use in the authors' laboratory include methods 1, 2, and 3 mentioned earlier and discussed further in the following.

1. By employing the Wigner–Weisskopf (or random-phase) approximation (Louisell, 1973), the phonon modes can be decoupled from the active mode, and the many-body effects of the phonon couplings may be reduced to a Fermi-type damping factor which is characterized by the phonon density of states. The advantage of this technique is that the many-body quantum system is thereby reduced to a one-body problem where the average excitation of the active mode is calculated including nonlinear effects (Lin et al., 1980) and migration-induced broadening (Slutsky and George, 1978, 1979).

2. A classical approach may be used instead of the quantum approach in *1.* For example, the generalized Langevin equation (GLE) may be employed to describe the dynamical behavior of the excitation processes. Solution of the GLE provides a lineshape function (or an absorption cross section) in a more general form which involves T_1 (energy) and T_2 (phase) relaxations and memory effects (Lin and George, 1981c). This technique provides phenomenological results to which parameters such as the total width of the lineshape can be semiempirically fitted.

3. By knowing the transition rate or the absorption cross section, both of which are calculated by the techniques in *1* and *2,* it is possible to solve a generalized master equation (GME) in photon energy space (George et al., 1980). The energy populations (which are solutions of the GME)

provide important information such as the photodissociation probability
and the nature of the laser excitations—selective or nonselective (Lin and
George, 1980a–c; Lin et al., 1980).

Photon-stimulated charge transfer on surfaces. In semiconductors,
the electronic band structure can be very complicated. The effects of
exposure to laser radiation, however, can be readily evaluated using simplified wave functions and time-dependent perturbation theory (Pidgeon
et al., 1979; Lee and Fan, 1974). These techniques can also be applied to a
study of photon absorption and associated charge-transfer processes in
the surface region.

Surfaces introduce additional features (Lundquist, 1975) into the bulk
electronic band structure of solids. In addition to the bulk valence bands
and conduction bands, there can exist surface bands. Furthermore, defects and adsorbed species can introduce additional bands and localized
states. Figure 14 is a simple schematic of what this band structure might
look like.

There may actually be several bands introduced by the surface, and
these may fall in various places in the band structure. However, it can be

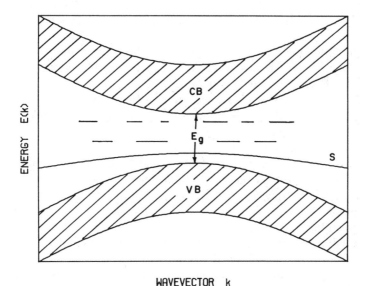

WAVEVECTOR k

FIG. 14.   Schematic representation of the electronic energy-level band structure of a
semiinfinite solid. Allowed energy levels are given as functions of wavevector **k** along a fixed
direction in the first Brillouin zone. CB, Conduction band; VB, valence band; S, surface
states; E_g, energy gap.

assumed in the solid under investigation that there is one surface band, S, and local states (represented in Fig. 14 by dashed lines) which fall in the bulk energy gap.

Electrons which have the energy associated with the surface band will be localized in the surface region. Similarly, electrons with energies corresponding to the local states will be localized in the region of the adsorbates or the defects which produced these states. On the other hand, the electrons located in the conduction and valence bands are essentially delocalized throughout the solid.

With a laser appropriately chosen (IR to visible) to excite electrons from a local adsorbate state to the conduction band or from the valence band to the local adsorbate state, the area around the adsorbed species can be made effectively more positive or negative. This charge difference can lead to a strong Coulomb interaction. By choosing the appropriate charge concentration, this process could be used to selectively clean the surface of contaminants. A laser can be used to excite the local states created by the contaminants without affecting the states associated with desired adsorbates. In fact, photon-stimulated desorption has been observed at the surface of a variety of metals (Woodruff *et al.*, 1980; Jaeger *et al.*, 1980). Although the desorption process is somewhat different from the one described here (Knotek and Feibelman, 1978), the main cause of the desorption is the Coulomb repulsion between the adsorbate and the surface.

In attempting to react different species on the surface, the solid can also be used as a charge bath. An electron can be excited into a local state induced by one species and an electron can be excited out of a local state induced by another species. Hence, even if the two species are far apart, using the solid bands as intermediaries can induce an effective charge transfer between species. The resultant adsorbates may then be in the ideal state for interaction with other species being deposited.

ii. Annealing. Annealing is a technologically important application of a nonselective thermal type of LSSP (see the beginning of Section III,B,2). The influence of high-power radiation on matter in the solid phase has been of interest for over a decade (Ready, 1971). However, laser annealing of semiconductors has only recently been explored in connection with microelectronics (White and Peercy, 1980). Laser annealing originally focused on the removal of defects introduced by ion implantation. Its scope has recently been broadened to include the transient heating of semiconductors, where lasers now provide a rich new range of opportunities for studying fast crystalline growth involving both solid- and liquid-phase epitaxy.

In an attempt to describe laser heating of solids, solutions are sought to the heat-conduction equation, which, for the case of one dimension (Carslaw and Jaeger, 1959), is usually given as

$$\partial T/\partial t = (\partial/\partial x)[D(T)(\partial T/\partial x)] + S/c\rho \qquad (27)$$

where $D(T)$ is the thermal diffusivity (in general, temperature dependent), and c and ρ are the specific heat and mass density, respectively. The local heating rate is governed by the source term S given by

$$S(x,t) = I(t)\,(1 - R)g(x) \qquad (28)$$

where $I(t)$ represents the laser intensity, and R and $g(x)$ are the reflectivity and diffusion factor, respectively. The diffusion factor accounts for the laser-generated charge carriers and reduces to the usual Beer's law form $g(x) = \exp(-\alpha x)$ for the case of long lifetimes of the carrier diffusion, where α is the optical absorption coefficient (Yoffa, 1980a,b).

This heat-transport equation cannot, in general, be solved analytically due to the temperature dependence of the thermal diffusivity $D(T)$. The thermal properties of the laser-heated solid are characterized not only by the solid but also by the types of laser radiation. For example, in pulsed laser annealing of semiconductors, a liquid layer is thought to form during the annealing process, and the redistribution of dopants is explained by using a model based on the diffusion of dopants in liquid silicon (Baeri *et al.*, 1978). On the other hand, a scanning CW laser can produce solid-phase recrystallization of the implanted layers, with no diffusion of implanted dopants during the annealing cycle (Liau *et al.*, 1979).

For the interaction of radiation with solids, there is controversy regarding the main physical mechanisms responsible for annealing, namely, thermal-melting models versus nonthermal models (Van Vechten *et al.*, 1979; Lo and Compaan, 1980; Kim *et al.*, 1981). The expression for the local lattice temperature of the solids consists of a laser heating term and a term corresponding to energy flux due to thermal diffusion. For materials with a large thermal conductivity which is weakly dependent on temperature, the effect of carrier diffusion on the rise of the lattice temperature may not be important. On the other hand, when the thermal conductivity decreases sharply with increasing temperature, the carrier diffusion plays an important role. In the case of silicon, the carrier diffusion does lower dramatically the heating efficiency near the surface. A more detailed understanding of transient laser-induced processes in semiconductors necessitates both a study of the relative time scales of processes such as electron–electron, electron–phonon, and electron–hole interactions and of the electronic structure of the material (Brown, 1980).

ACKNOWLEDGMENTS

This work was supported in part by the Air Force Office of Scientific Research (AFSC), United States Air Force, under Grant AFOSR-82-0046, the Office of Naval Research, and the National Science Foundation under Grant CHE-8022874. One of us (TFG) acknowledges the Camille and Henry Dreyfus Foundation for a Teacher–Scholar Award (1975–1984), and the John Simon Guggenheim Memorial Foundation for a Fellowship (1983–1984).

References

Adler, B. J., Gass, D. M., and Wainwright, T. E. (1970). *J. Chem. Phys.* **53**, 3813.
Allen, R. E. (1979). *Phys Rev. B* **19**, 917.
Allen, S. D., and Bass, M. (1979). *J. Vac. Sci. Technol.* **16**, 431.
Alms, G. R., Bauer, D. R., Brauman, J. I., and Pecora, R. (1973). *J. Chem. Phys.* **59**, 5310, 5321.
Ambartzumian, R. V., and Letokhov, V. S. (1972). *Appl. Opt.* **11**, 354.
Ambartzumian, R. V., Letokhov, V. S., Makorov, G. N., and Puretzky, A. A. (1978). *Opt. Commun.* **25**, 69.
Ambartzumian, R. V., Letokhov, V. S., Ryabov, E. A., and Chekalin, E. A. (1974). *JETP Lett. (Engl. Transl.)* **20**, 273.
Andreyev, S. V., Antonov, V. S., Knyazev, I. N., and Letokhov, V. S. (1977). *Chem. Phys. Lett.* **45**, 166.
Antoniewicz, P. R. (1980). *Phys. Rev. B* **21**, 3811.
Auston, D. H., Brown, W. L., and Celler, G. C. (1979). *Bell Lab. Rec.* July/August, 187.
Avery, G. E. (1980). *Mil. Electron./Countermeasures* **6**(7), 50.
Avigal, Y., Beinglass, I., and Schieber, M. (1974). *J. Electrochem. Soc.* **121**, 1103.
Baeri, P., Campisono, S. V., Foti, G., and Rimini, E. (1978). *Appl. Phys. Lett.* **33**, 137.
Baldwin, A. C., and Barker, J. R. (1981). *J. Chem. Phys.* **74**, 3813.
Baranauskas, V., Mammana, C. I. Z., Klinger, R. E., and Greene, J. E. (1980). *Appl. Phys. Lett.* **36**, 930.
Barford, B. D., and Rye, R. R. (1974). *J. Chem. Phys.* **60**, 1046.
Barker, J. R. (1980). *J. Chem. Phys.* **72**, 3686.
Baro, A. M., and Ibach, H. (1979). *J. Chem. Phys.* **71**, 4812.
Bassini, M., Biraben, F., Cagnac, B., and Grynberg, G. (1977). *Opt. Commun.* **21**, 263.
Bauer, D. R., Brauman, J. I., and Pecora, R. (1976). *Annu. Rev. Phys. Chem.* **27**, 443.
Bean, J. C., Leamy, H. J., Poate, J. M., Rozgonyi, G. A., Sheng, T. T., Williams, J. S., and Celler, G. K. (1978). *Appl. Phys. Lett.* **33**, 227.
Beattie, J. K., Sutin, M., Turner, D. H., and Flynn, G. W. (1973). *J. Am. Chem. Soc.* **95**, 2052.
Beeby, J. L., and Dobrzynski, L. (1971). *J. Phys. C* **4**, 1269.
Bellum, J. C., and George, T. F. (1979). *J. Chem. Phys.* **70**, 5059.
Bellum, J. C., Lam, K.-S., and George, T. F. (1978). *J. Chem. Phys.* **69**, 1781.
Bendow, B., and Ying, S.-C. (1973). *Phys. Rev. B* **7**, 622, 637.
Benedek, G. B. (1968). "Brandeis Lectures in Theoretical Physics" (M. Chrétien, E. P. Gross, and S. Deser, eds.). Gordon & Breach, New York.
Bennett, A. J., and Uzguris, E. E. (1973). *Phys. Rev. A* **8**, 2662.
Berkenblit, M., and Reisman, A. (1971). *Metall. Trans.* **2**, 803.
Bernasek, S. L., and Somorjai, G. A. (1975). *Prog. Surf. Sci.* **5**, 377.

Berne, B. J., and Pecora, R. (1974). *Annu. Rev. Phys. Chem.* **25**, 233.
Berne, B. J., and Pecora, R. (1976). "Dynamical Light Scattering with Applications to Chemistry, Biology and Physics." Wiley, New York.
Bhattacharyya, D. K., Lam, K.-S., and George, T. F. (1980). *J. Chem. Phys.* **73**, 1999.
Bird, R. B., Stewart, W. E., and Lightfoot, E. N. (1960). "Transport Phenomena." Wiley, New York.
Black, J. G., Kolodner, P., Shultz, M. J., Yablonovitch, E., and Bloembergen, N. (1979). *Phys. Rev. A* **19**, 704.
Bloembergen, N. (1975). *Opt. Commun.* **15**, 416.
Bray, R. G., and Berry, M. J. (1979). *J. Chem. Phys.* **71**, 4909.
Brewer, G. R. (1971). *IEEE Spectrum* **8**, 23.
Brinkmann, V., Hartig, W., Telle, H., and Walther, H. (1974). *Appl. Phys.* **5**, 109.
Broers, A. N., and Hatzakis, M. (1972). *Sci. Am.* **227**, 34.
Brown, W. L. (1980). *In* "Laser and Electron Beam Processing of Materials" (C. W. White and P. S. Peercy, eds.), p. 20. Academic Press, New York.
Bruining, J., and Clarke, J. H. R. (1975). *Chem. Phys. Lett.* **31**, 355.
Brumbach, S. G., and Somorjai, G. A. (1974). *CRC, Crit. Rev. Solid State Sci.* **4**, 429.
Bryant, W. A. (1977). *J. Mater. Sci.* **12**, 1285, and references therein.
Cabrera, N., Celli, V., Goodman, F. O., and Manson, R. (1970). *Surf. Sci.* **19**, 67.
Cahuzac, P., and Toschek, P. E. (1978). *Phys. Rev. Lett.* **40**, 1087.
Capece, R. P. (1978). *Electronics* **51**(24), 111.
Capece, R. P. (1979). *Electronics* **52**(19), 109.
Carmeli, B., and Jortner, J. (1980). *J. Chem. Phys.* **72**, 2054, 2070.
Carslaw, M. S., and Jaeger, J. C. (1959). "Conduction of Heat in Solids," 2nd ed. Oxford Univ. Press, London and New York.
Chow, H., and Thompson, E. D. (1976). *Surf. Sci.* **59**, 225.
Christensen, C. P., and Lakin, K. M. (1978). *Appl. Phys. Lett.* **32**, 254.
Chuang, T. J. (1980). *J. Appl. Phys.* **51**, 2614.
Chuang, T. J. (1981). *J. Chem. Phys.* **74**, 1453.
Chuang, T. J., and Eisenthal, K. B. (1974). "Laser Spectroscopy" (R. G. Brewer and A. Mooradian, eds.). Plenum, New York.
Chuang, T. J., and Eisenthal, K. B. (1975). *J. Chem. Phys.* **62**, 2213.
Chuang, T. J., Hoffman, G. W., and Eisenthal, K. B. (1974). *Chem. Phys. Lett.* **25**, 201.
Cohen, S. A., and King, J. G. (1973). *Phys. Rev. Lett.* **31**, 703.
Cole, W. E., Derry, G., and Frankl, D. R. (1979). *Phys. Rev. B* **19**, 3258.
Cosby, P. C., Bennett, R. A., Peterson, J. R., and Moseley, J. T. (1975). *J. Chem. Phys.* **63**, 1612.
Cotter, T. P., Fuss, W., Kompa, W., and Stafast, K. L. (1976). *Opt. Commun.* **18**, 220.
Cummins, H. Z. (1971). "Light Scattering in Solids" (M. Balkanski, ed.). Flammarion, Paris.
De, G. S., and Landman, U. (1980). *Phys. Rev. B* **21**, 3256.
Decker, C. A., and Ross, D. L. (1980). *J. Electrochem. Soc.* **127**, 45C.
Delone, B. (1975). *Sov. Phys.—Usp.* (*Engl. Transl.*) **18**, 169.
Deutsch, T. F., Ehrlich, D. J., and Osgood, R. M., Jr. (1979). *Appl. Phys. Lett.* **35**, 125.
DeVries, P. L., Lam, K.-S., and George, T. F. (1980). *In* "Electronic and Atomic Collisions" (N. Oda and K. Takayanagi, eds.), p. 683. North-Holland Publ., Amsterdam.
Diebold, A. C., and Wolken, G., Jr. (1979). *Surf. Sci.* **82**, 245.
Diebold, A. C., Adelman, S. A., and Mou, C. Y. (1979). *J. Chem. Phys.* **71**, 3236.
Diestler, D. J., and Wilson, R. S. (1975). *J. Chem. Phys.* **62**, 1572.
Djidjoev, M. S., Khokhlov, R. V., Kiselev, A. V., Lygin, V. I., Namiot, V. A., Osipov,

A. I., Panchenko, V. I., and Provotorov, B. I. (1976). *In* "Tunable Lasers and Applications" (A. Mooradian, T. Jaeger, and P. Stokseth, eds.), p. 100. Springer-Verlag, Berlin and New York.

Doll, J. D. (1974). *Chem. Phys.* **3**, 257.

Doyen, G. (1980). *Phys. Rev. B* **22**, 497.

Eberhardt, W., and Himpsel, F. J. (1979). *Phys. Rev. Lett.* **42**, 1375.

Eberly, J. H., and Lambropoulos, P., eds. (1978). *Multiphoton Processes, Proc. Int. Conf.*, 1977.

Eberly, J. H., Gallagher, J. W., and Beaty, E. C., eds. (1979). "Multiphoton Bibliography." Wiley, New York.

Eckhardt, G., Hellwarth, R. W., McClung, F. J., Schwartz, S. E., Weiner, D., and Woodbury, E. J. (1962). *Phys. Rev. Lett.* **9**, 455.

Efrima, S., and Metiu, H. (1978). *J. Chem. Phys.* **69**, 2286.

Efrima, S., Freed, K. F., Jedrzejek, C., and Metiu, H. (1980). *Phys. Lett.* **74**, 43.

Ehrlich, D. J., Osgood, R. M., Jr., and Deutsch, T. F. (1979). *Appl. Phys. Lett.* **35**, 175.

Ehrlich, D. J., Osgood, R. M., Jr., and Deutsch, T. F. (1980a). *Appl. Phys. Lett.* **36**, 698.

Ehrlich, D. J., Osgood, R. M., Jr., and Deutsch, T. F. (1980b). *Appl. Phys. Lett.* **36**, 916.

Ehrlich, D. J., Osgood, R. M., Jr., and Deutsch, T. F. (1980c). *IEEE J. Quantum Electron* **QE-16**, 1233.

Ehrlich, D. J., Osgood, R. M., Jr., and Deutsch, T. F. (1981). *Appl. Phys. Lett.* **38**, 399.

Eib, N. K., Gent, A. N., and Henriksen, P. N. (1979). *J. Chem. Phys.* **70**, 4288.

Einstein, A. (1956). "Investigations on the Theory of the Brownian Movement." Dover, New York.

Eischens, R. P. (1972). *Acc. Chem. Res.* **5**, 74.

Engeler, W. E., Blumenfeld, M., and Taft, E. A. (1970). *Appl. Phys. Lett.* **16**, 202.

Ertl, G. (1979). *Gazz. Chim. Ital.* **109**, 217.

Ertl, G., and Schillinger, D. (1977). *J. Chem. Phys.* **66**, 2569.

Evans, H. E., and Weinberg, W. H. (1979). *J. Chem. Phys.* **71**, 4789.

Fair, J., and Madix, R. J. (1980). *J. Chem. Phys.* **73**, 3480.

Falcone, R. W., Green, W. R., White, J. C., Young, J. E., and Harris, S. E. (1977). *Phys. Rev. A* **15**, 1333.

Fan, J. C. C., Chapman, R. L., Donnelly, J. P., Turner, G. W., and Bozler, C. O. (1979). *Appl. Phys. Lett.* **34**, 780.

Farrow, R. F. C. (1974). *J. Electrochem. Soc.* **121**, 899.

Feist, W. M., Steele, S. R., and Ready, D. W. (1969). *Phys. Thin Films* **5**, 237.

Feldstein, N., and Lancsec, T. S. (1970). *RCA Rev.* **31**, 439.

Fleury, P. A., and Boon, J. P. (1973). *Adv. Chem. Phys.* **24**, 1.

Flynn, G. W., and Sutin, N. (1974). *In* "Chemical and Biochemical Applications of Lasers" (C. B. Moore, ed.), Vol. 1, p. 309. Academic Press, New York.

Fogiel, M. (1972). "Modern Microelectronics." Research and Education Association, New York.

Ford, N. C. (1972). *Chem. Scr.* **2**, 193.

Forst, W. (1973). "Theory of Unimolecular Reactions." Academic Press, New York.

Fujime, S. (1972). *Adv. Biophys.* **3**, 1.

Fuss, W. (1979). *Chem. Phys.* **36**, 135.

Gan, Z., Yang, G., Feng, K., and Huang, X. (1978). *Acta Physiol. Sin.* **27**, 664.

Garcia, N., Goodman, F. O., Celli, V., and Hill, N. R. (1979). *Phys. Rev. B* **19**, 1808.

Gat, A., Lietoila, A., and Gibbons, J. F. (1979). *J. Appl. Phys.* **50**, 2926.

Geankopolis, C. J. (1972). "Mass Transport Phenomena." Holt, New York.

George, T. F. (1979). *Opt. Eng.* **18**, 167.

George, T. F., Zimmerman, I. H., Yuan, J. M., Laing, J. R., and DeVries, P. L. (1977). *Acc. Chem. Res.* **10**, 449.

George, T. F., Zimmerman, I. H., DeVries, P. L., Yuan, J. M., Lam, K.-S., Bellum, J. C., Lee, H. W., Slusky, M. S., and Lin, J. (1979). *In* "Chemical and Biochemical Applications of Lasers" (C. B. Moore, ed.), Vol. 4, p. 253. Academic Press, New York.

George, T. F., Lin, J., Lam, K.-S., and Chang, C. H. (1980). *Opt. Eng.* **19**, 100.

Georges, A. T., and Lambropoulos, P. (1977). *Phys. Rev. A* **15**, 727.

Gerber, R. B., Yinnon, A. T., and Murrell, J. N. (1978). *Chem. Phys.* **31**, 1.

Gibbons, J. F., Lee, K. F., Magee, T. J., Peng, J., and Ormond, R. (1979). *Appl. Phys. Lett.* **34**, 831.

Giuliano, C. R., Hess, L. D., and Margerum, J. D. (1968). *J. Am. Chem. Soc.* **90**, 587.

Glauber, R. J. (1955). *Phys. Rev.* **98**, 1692.

Goldanski, V. I., Namiot, V. A., and Khokhlov, R. V. (1976). *Sov. Phys.—JETP (Engl. Transl.)* **43**, 226.

Goodall, D. M., and Greenhow, R. C. (1971). *Chem. Phys. Lett.* **9**, 583.

Goodall, D. M., Greenhow, R. C., and Knight, B. (1979). "Laser-Induced Processes in Molecules" (K. L. Kampa and S. D. Smith, eds.). Springer-Verlag, Berlin and New York.

Goodman, F. O., and Romero, I. (1978). *J. Chem. Phys.* **69**, 1086.

Goodman, F. O., and Tan, W.-K. (1973). *J. Chem. Phys.* **59**, 1805.

Gortel, J. W., Kreuzer, H. J., and Teshima, R. (1980a). *Phys. Rev. B* **22**, 512.

Gortel, J. W., Kreuzer, H. J., and Spaner, D. (1980b). *J. Chem. Phys.* **72**, 234.

Grant, E. R., Shulz, P. A., Sudbo, Aa. S., Shen, Y. R., and Lee, Y. T. (1978). *Phys. Rev. Lett.* **40**, 115.

Grigolini, P. (1979). *Chem. Phys.* **38**, 389.

Grünbaum, E. (1975). *In* "Epitaxial Growth" (J. W. Matthews, ed.), Part B, p. 611. Academic Press, New York.

Hall, R. B., and Kaldor, A. (1979). *J. Chem. Phys.* **70**, 4027.

Heller, D., and Mukamel, S. (1979). *J. Chem. Phys.* **70**, 463.

Hering, P., Brooks, P. R., Curl, R. F., Jr., Judson, R. S., and Lowe, R. S. (1980). *Phys. Rev. Lett.* **44**, 687.

Holloway, S., and Jewsbury, P. (1976). *J. Phys. C* **9**, 1907.

Holloway, S., Jewsbury, P., and Beeby, J. L. (1977). *Surf. Sci.* **63**, 339.

Hu, C., and Zwanzig, R. (1974). *J. Chem. Phys.* **60**, 4354.

Ibbotson, D. E., Wittrig, T. S., and Weinberg, W. H. (1980). *J. Chem. Phys.* **72**, 4885.

Jaeger, R., Feldhaus, J., Haase, J., Stöhr, J., Hussain, Z., Menzel, D., and Norman, D. (1980). *Phys. Rev. Lett.* **45**, 1870.

Jedrzejek, C., Efrima, S., Metiu, H., and Freed, K. F. (1981). *Chem. Phys. Lett.* **79**, 227.

Joyce, B. A., and Bradley, R. R. (1963). *J. Electrochem. Soc.* **110**, 1235.

Kamins, T. I. (1974). *J. Electrochem. Soc.* **121**, 681.

Karlov, N. V., and Prokhorov, A. M. (1977). *Sov. Phys.—Usp. (Engl. Transl.)* **20**, 721.

Kern, W., and Ban, V. S. (1978). *In* "Thin Film Processes" (J. L. Vossen and W. Kern, eds.), p. 257. Academic Press, New York.

Kim, D. M., Kwong, D. L., Shah, R. R., and Crosthwait, D. L. (1981). *J. Appl. Phys.* **52**, 4995.

Knotek, M. L., and Feibelman, P. J. (1978). *Phys. Rev. Lett.* **40**, 964.

Laubereau, A. (1980). "Vibrational Spectroscopy of Molecular Liquids and Solids" (S. Bratos and R. M. Pick, eds.). Plenum, New York.

Laubereau, A., and Kaiser, W. (1978). *Rev. Mod. Phys.* **50**, 607.

Lee, C. C., and Fan, H. Y. (1974). *Phys. Rev. B* **9**, 3502.

Lee, H. W., and George, T. F. (1979a). *J. Chem. Phys.* **70**, 3685.
Lee, H. W., and George, T. F. (1979b). *Theor. Chim. Acta* **53**, 193.
Lee, H. W., and George, T. F. (1979c). *J. Chem. Phys.* **70**, 4220.
Lee, Y. T., and Shen, Y. R. (1980). *Phys. Today* **33**(11), 52.
Letokhov, V. S. (1973). *Science* **180**, 451.
Letokhov, V. S. (1980). *Phys. Today* **33**(11), 34.
Liau, Z. L., Tsaur, B. Y., and Mayer, J. W. (1979). *Appl. Phys. Lett.* **34**, 221.
Lin, J. (1979). *Phys. Lett.* **70A**, 195.
Lin, J. (1980). Ph.D. Thesis, University of Rochester, Rochester, New York.
Lin, J., and George, T. F. (1979a). *Chem. Phys. Lett.* **66**, 5.
Lin, J., and George, T. F. (1979b). *J. Chem. Phys.* **72**, 2554.
Lin, J., and George, T. F. (1980a). *Phys. Lett.* **80A**, 296.
Lin, J., and George, T. F. (1980b). *Surf. Sci.* **100**, 381.
Lin, J., and George, T. F. (1980c). *J. Phys. Chem.* **84**, 2957.
Lin, J., and George, T. F. (1981a). *Phys. Rev. B* **24**, 64.
Lin, J., and George, T. F. (1981b). *Surf. Sci.* **107**, 417.
Lin, J., and George, T. F. (1981c). *Surf. Sci.* **108**, 340.
Lin, J., Beri, A. C., Hutchinson, M., Murphy, W. C., and George, T. F. (1980). *Phys. Lett.* **79A**, 233.
Lin, Y.-W., and Adelman, S. A. (1978). *J. Chem. Phys.* **68**, 9.
Lin, Y.-W., and Wolken, G., Jr. (1976a). *J. Chem. Phys.* **65**, 2634.
Lin, Y.-W., and Wolken, G., Jr. (1976b). *J. Chem. Phys.* **65**, 3729.
Lo, H. W., and Compaan, A. (1980). *Phys. Rev. Lett.* **44**, 1604.
Louisell, W. H. (1973). "Quantum Statistical Properties of Radiation." Wiley, New York.
Lundquist, S. (1975). *In* "Surface Science," Vol. 1, p. 331. IAEA, Vienna.
Lyman, J. (1977). *Electronics* **50**(15), 81.
Lyman, J. (1980). *Electronics* **53**(14), 115.
Lyman, J., Rockwood, S. D., and Freund, S. M. (1977). *J. Chem. Phys.* **67**, 4545.
McCreery, J. H., and Wolken, G., Jr. (1976). *J. Chem. Phys.* **64**, 2845.
Matheson, I. B. C., and Lee, J. (1970). *Chem. Phys. Lett.* **7**, 475.
Melliar-Smith, C. M., Adams, A. C., Kaiser, R. H., and Kushner, R. A. (1974). *J. Electrochem. Soc.* **121**, 298.
Milford, F. J., and Novaco, A. D. (1971). *Phys. Rev. A* **4**, 1136.
Minorsky, N. (1962). "Nonlinear Oscillators." Van Nostrand-Reinhold, Princeton, New Jersey.
Mori, H. (1965a). *Prog. Theor. Phys.* **33**, 423.
Mori, H. (1965b). *Prog. Theor. Phys.* **34**, 399.
Müller, H., and Brenig, W. (1979). *Z. Phys. B* **34**, 165.
Murphy, B. T. (1979). *Eur. Solid State Circuits Conf., 5th, 1979*, p. 70.
Murphy, W. C., and George, T. F. (1981). *Surf. Sci.* **102**, L46.
Narducci, L. M., and Yuan, J. M. (1980). *Phys. Rev. A* **22**, 261.
Narducci, L. M., Mitra, S. S., Shatas, R. A., and Coulter, C. A. (1977). *Phys. Rev. A* **16**, 247.
Nitzan, A., and Jortner, J. (1972). *Mol. Phys.* **25**, 713.
Nitzan, A., and Silbey, R. J. (1974). *J. Chem. Phys.* **60**, 4070.
Nitzan, A., Mukamel, S., and Jortner, J. (1974). *J. Chem. Phys.* **60**, 3929.
Nitzan, A., Mukamel, S., and Jortner, J. (1975). *J. Chem. Phys.* **63**, 200.
Nitzan, A., Shugard, M., and Tully, J. C. (1978). *J. Chem. Phys.* **69**, 2525.
Noyes, R. M. (1961). *Prog. React. Kinet.* **1**, 129.

Oraevski, A. N., Stepanov, A. A., and Shcheglov, V. A. (1976). *Sov. Phys.—JETP (Engl. Transl.)* **42,** 1012.

Oshima, C., Tanaka, T., Aono, M., Nishitani, R., Kawai, S., and Yajima, F. (1979). *Appl. Phys. Lett.* **35,** 822.

Palmer, R. L., and Smith, J. N., Jr. (1974). *J. Chem. Phys.* **60,** 1453.

Parsons, W. W. (1968). *Biochim. Biophys. Acta* **153,** 248.

Paunovic, M. (1980). *J. Electrochem. Soc.* **127,** 441C.

Pidgeon, C. R., Wherrett, B. S., Johnston, A. M., Dempsey, J., and Miller, A. (1979). *Phys. Rev. Lett.* **42,** 1785.

Pirug, G., Bonzel, H. P., Hopster, H., and Ibach, H. (1979). *J. Chem. Phys.* **71,** 593.

Polak-Dingels, P., Delpeck, J.-F., and Weiner, J. (1980). *Phys. Rev. Lett.* **44,** 1663.

Poppe, D. (1980). *Chem. Phys.* **45,** 371.

Prager, S., and Frisch, H. L. (1980). *J. Chem. Phys.* **72,** 2941.

Quack, M. (1978). *J. Chem. Phys.* **69,** 1282.

Ramaswamy, R., Siders, P., and Marcus, R. A. (1980). *J. Chem. Phys.* **74,** 4418.

Ratnakumar, K. N., Bartelink, D. J., and Meindl, J. D. (1980). *Dig. Tech. Pap., IEEE Int. Solid-State Circuits Conf., 1980,* p. 72.

Ready, J. F. (1971). "Effects of High-Power Laser Radiation." Academic Press, New York.

Richman, D., Chiang, Y. S., and Robinson, P. H. (1970). *RCA Rev.* **31,** 613.

Sandow, P. M. (1980). *Solid State Technol.* **23**(7), 74.

Sargent, M., Scully, M. O., and Lamb, W. E. (1974). "Laser Physics." Addison-Wesley, Reading, Massachusetts.

Schaefer, D. W. (1974). "Laser Applications to Optics and Spectroscopy" (S. F. Jacobs, M. Sargent, III, J. F. Scott, and M. O. Scully, eds.). Addison-Wesley, Reading, Massachusetts.

Schaefer, D. W., and Berne, B. J. (1972). *Phys. Rev. Lett.* **28,** 475.

Schoen, P. E., Cheung, P. S. Y., Jackson, D. A., and Powles, J. G. (1975). *Mol. Phys.* **29,** 1197.

Schulz, P. A., Sudbo, Aa. S., Krajnovich, D. J., Kwok, H. S., Shen, Y. R., and Lee, Y. T. (1979). *Annu. Rev. Phys. Chem.* **30,** 379.

Shaw, D. W. (1975). *In* "Epitaxial Growth" (J. W. Matthews, ed.), Part A, p. 89. Academic Press, New York.

Shaw, J. M., and Amick, J. A. (1970). *RCA Rev.* **31,** 306.

Shugard, M., Tully, J. C., and Nitzan, A. (1977). *J. Chem. Phys.* **66,** 2534.

Shugard, M., Tully, J. C., and Nitzan, A. (1978). *J. Chem. Phys.* **69,** 336.

Shultz, M. J., and Yablonovitch, E. (1978). *J. Chem. Phys.* **68,** 3007.

Slutsky, M. S., and George, T. F. (1978). *Chem. Phys. Lett.* **57,** 473.

Slutsky, M. S., and George, T. F. (1979). *J. Chem. Phys.* **70,** 1231.

Smith, H. E., Spears, D. L., and Bernacki, S. E. (1973). *J. Vac. Sci. Technol.* **15,** 21.

Smith, H. I., Efremow, N., and Kelly, P. L. (1974). *J. Electrochem. Soc.* **121,** 1503.

Spears, D. L., and Smith, H. I. (1972). *Solid State Technol.* **15,** 21.

Steinfeld, J. I., Anderson, T. G., Reiser, C., Denison, D. R., Hartsough, L. D., and Hollahan, J. R. (1980). *J. Electrochem. Soc.* **127,** 514.

Stenholm, S. (1979). *Contemp. Phys.* **20,** 37.

Steverding, B., Dudel, P., and Gibson, F. P. (1977). *J. Appl. Phys.* **48,** 1195.

Steverding, B., Dudel, P., and Gibson, F. P. (1978). *J. Appl. Phys.* **49,** 1260.

Stewart, C. N., and Ehrlich, G. (1975). *J. Chem. Phys.* **62,** 4672.

Stine, J. R., and Noid, D. W. (1979). *Opt. Commun.* **31,** 161.

Stone, J., and Goodman, M. F. (1979). *J. Chem. Phys.* **71,** 408.

Stone, J., Thiele, E., and Goodman, M. F. (1980). *J. Chem. Phys.* **73**, 2259.
Sudbo, A. S., Schulz, P. A., Grant, E. R., Shen, Y. R., and Lee, Y. T. (1979). *J. Chem. Phys.* **70**, 912.
Tamaru, M., Tamaru, H., and Tokuyama, T. (1980). *Jpn. J. Appl. Phys.* **19**, L23.
Tamm, P. W., and Schmidt, L. D. (1969). *J. Chem. Phys.* **51**, 5352.
Tapilin, V. M., Cunningham, S. L., and Weinberg, W. H. (1978). *Phys. Rev. B* **18**, 2656.
Tardy, D. C., and Rabinovitch. B. S. (1977). *Chem. Rev.* **77**, 369.
Taylor, J. L., Ibbotson, D. E., and Weinberg, W. H. (1978). *J. Chem. Phys.* **69**, 4298.
Thiel, P. A., Weinberg, W. H., and Yates, J. T., Jr. (1979). *J. Chem. Phys.* **71**, 1643.
Thiele, E., Goodman, M. F., and Stone, J. (1980). *Opt. Eng.* **19**, 10.
Thomas, J. M., and Thomas, W. J. (1967). "Introduction to the Principles of Heterogeneous Catalysis." Academic Press, New York.
Troe, J. (1977). *J. Chem. Phys.* **66**, 4745.
Tsuchida, A. (1968). *Surf. Sci.* **14**, 375.
Tu, Y. Y., Chuang, T. J., and Winters, H. F. (1981). *Phys. Rev. B* **23**, 823.
Tuccio, S. A., Foley, R. J., Dubrin, J. W., and Krikorian, O. (1975). *IEEE J. Quantum Electron.* **QE-11**, 101D.
Tully, J. C. (1980). *J. Chem. Phys.* **73**, 1975.
Turner, D. H., Flynn, G. W., Sutin, N., and Beitz, J. V. (1972). *J. Am. Chem. Soc.* **94**, 1554.
Umstead, M. E., and Lin, M. C. (1978). *J. Phys. Chem.* **82**, 2047.
Uzguris, E. E. (1974). *Rev. Sci. Instrum.* **45**, 74.
Uzguris, E. E., and Kaplan, J. H. (1974). *Anal. Biochem.* **60**, 455.
Vanderputte, P., Giling, L. J., and Bloem, J. (1975). *J. Cryst. Growth* **31**, 299.
van Hove, L. (1954). *Phys. Rev.* **95**, 249, 1374.
Van Vechten, J. A., Tsu, R., and Saris, F. W. (1979). *Phys. Lett.* **74A**, 422.
Voorhoeve, R. J. H., and Merewether, J. W. (1972). *J. Electrochem. Soc.* **119**, 364.
Vossen, J. L. (1978). *In* "Thin Film Processes" (J. L. Vossen and W. Kern, eds.), p. 3. Academic Press, New York.
Walker, R. B., and Preston, R. K. (1977). *J. Chem. Phys.* **67**, 2017.
Wang, C. C., Doughtery, F. C., Zanzucchi, P. J., and McFarlane, S. H., III. (1974). *J. Electrochem. Soc.* **121**, 571.
Ware, B. R. (1974). *Adv. Colloid Interface Sci.* **4**, 1.
Weiss, C., Jr., and Sauer, K. (1970). *Photochem. Photobiol.* **11**, 495.
White, C. W., and Peercy, P. S., eds. (1980). "Laser and Electron Beam Processing of Materials." Academic Press, New York.
Wilcomb, B. E., and Burnham, R. (1981). *J. Chem. Phys.* **74**, 6784.
Williams, J. W., Brown, W. L., Leamy, H. J., Poate, J. M., Rodgers, J. W., Rousseau, D., Rozgonyi, G. A., Shelnutt, J. A., and Sheng, T. T. (1978). *Appl. Phys. Lett.* **33**, 542.
Wilson, W. W., Luzzana, M. R., Penniston, J. T., and Johnson, C. S. (1971). *Proc. Natl. Acad. Sci. U.S.A.* **71**, 1260.
Witt, H. T., Rumberg, B., Junge, W., Doering, G., Stiehl, H. H., Weikard, J., and Wolff, C. (1961). *Prog. Photosynth. Res.* **3**, 1361.
Wolken, G., Jr. (1974). *J. Chem. Phys.* **60**, 2210.
Wolken, G., Jr., and McCreery, J. H. (1978). *Chem. Phys. Lett.* **54**, 35.
Woodruff, D. P., Traum, M. M., Farrell, H. H., Smith, N. V., Johnson, P. D., King, D. A., Benbow, R. L., and Hurych, Z. (1980). *Phys. Rev. B* **21**, 5642.
Yates, J. T., Jr., Duncan, T. M., Worley, S. D., and Vaughan, R. W. (1979). *J. Chem. Phys.* **70**, 1219.
Yeung, E. S., and Moore, C. B. (1972). *Appl. Phys. Lett.* **21**, 109.
Yoffa, E. J. (1980a). *Appl. Phys. Lett.* **36**, 37.

Yoffa, E. J. (1980b). *Phys. Rev. B* **21,** 2415.
Youngren, G., and Acrivos, A. (1975). *J. Chem. Phys.* **63,** 3846.
Zare, R. N. (1977). *Sci. Am.* **236,** 86.
Zare, R. N., and Bernstein, R. B. (1980). *Phys. Today* **33**(11), 43.
Zewail, A. H. (1980). *Phys. Today* **33**(11), 27.
Zwanzig, R. (1961). *Lect. Theor. Phys.* **3,** 106.

COMBUSTION RESEARCH WITH LASERS

James H. Bechtel, Cameron J. Dasch, and Richard E. Teets

General Motors Research Laboratories
Warren, Michigan

I. Introduction

Combustion processes have always been important for the production of light, heat, and other derived forms of energy. The combustion of hydrocarbon fuels is one of the most extensive, controlled chemical processes on the earth today. Man's interest in flames and combustion has evolved over many millennia, and during that period many properties of combustion have been discovered and documented. In order to better understand the combustion process today, it is necessary to measure both the physical and chemical attributes of flame structure and propagation.

Flames involve a complex interplay of many different chemical reactions, heat transfer, fluid flow, and species diffusion (see Fig. 1). Only

LASER APPLICATIONS, VOLUME 5

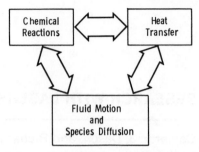

FIG. 1. The relation between heat transfer, mass transfer, and chemical reactions in combustion.

within the last decade or two have detailed theoretical calculations been attempted to describe the interaction of these chemical and physical processes. Advances in computer science now allow the numerical solution of systems of coupled differential equations that are mathematical statements of the laws of conservation of total mass, momentum, energy, and individual species. These equations are subject to a specific set of chemical reactions that are determined by the fuel, oxidizer, and diluent that comprise the flame reactants. The differential equations are also subject to the initial and boundary conditions of the combustion system that is being examined. The solutions to these equations provide theoretical predictions for temperature, pressure, composition, burning velocity, and fluid motion.

The science of optics has always played an important role in the measurement and understanding of combustion phenomena, including not only laboratory flames, but also practial combustion devices such as the internal combustion engine (see Fig. 2). For example, Nicholaus Otto, who might be called the father of the four-stroke spark-ignition engine, was convinced that his engine operated with an inhomogeneous, stratified fuel–air mixture (Bryant, 1967). This belief was later challenged by experiments that allowed an optical visualization of the fluid flow. The result was that Otto lost his patent protection in several countries (Amann, 1982).

Later optical experiments with internal combustion engines allowed researchers to infer new information about both the chemical processes and the fluid motion in an engine. Withrow and Rassweiler (1938) used high-speed cinematography to measure flame motion, sodium line reversal to measure temperature, and both emission and absorption spectroscopy to obtain qualitative measurements of flame composition.

The advent of the laser gave new momentum to optical experiments for combustion research. Laser experiments provide new ways of accurately

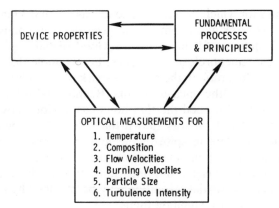

FIG. 2.   The importance of optical diagnostics for combustion research.

measuring temperature, species concentration, fluid flow velocities, and particle sizes in flames. The nonperturbing nature of laser experiments, as with all optical experiments, is a major advance over nonoptical experiments such as those that have thermocouples or gas sampling tubes. These probes may disturb the fluid flow, temperature, or chemistry, and may lead to uncertainties in the interpretation of the results.

Lasers are now used to probe many types of flames or devices and to determine many different combustion properties. Examples include laboratory flames that are tailored to provide new knowledge of specific aspects of flame chemistry, fluid flow, heat transfer, or molecular diffusion properties. These include flat flames, curved flames, premixed or diffusion flames, and laminar or turbulent flames. Examples of laser-probed devices include reciprocating internal combustion engines similar to those used to power automobiles, or gas turbine engines similar to those used to power jet aircraft.

Laser measurements can have very good temporal or spatial resolution compared to other techniques. Many lasers have pulse lengths in the 10–1000 nsec range, which is short compared to combustion time scales. This allows nearly instantaneous measurements that are useful for the understanding of turbulent combustion. Good spatial resolution is also possible because of the excellent focusing properties of laser beams. In a typical experiment, the measurement volume might be a near-cylindrical-shaped volume with a diameter of 50 μm and a length of 1 mm. This spatial resolution is very useful for probing the thin, primary reaction zone of flames. At atmospheric pressure this zone is typically a few tenths of a millimeter thick. For example, Bechtel (1980) and Bechtel et al. (1981) have measured temperature and species concentration profiles in laminar, premixed flames for which the reaction zone was about 0.3 mm thick. The resolution of the laser measurements allowed detailed comparison with theoretical calculations of flame structure. Good spatial resolution is also needed for measurements close to surfaces. For example, Johnston (1980) measured the air–fuel ratio in the gap of a spark plug as part of a study of misfire in an engine. Other measurements (Bechtel and Blint, 1980; Blint and Bechtel, 1982a,b) probed the thin boundary region between a flame and a cooled metal wall. It was shown that the fuel combustion in this region became complete in spite of the low temperature. All of these measurements are quasi-point measurements and are obtained by imaging the waist of a focused laser beam onto some type of detector. In some other spectroscopies, spatial resolution may be obtained by crossing two or more laser beams at a common focus.

There are many different types of laser measurements that may be used to probe flames or combustion devices. Examples of laser spectroscopies that are currently gaining much attention for combustion applications include spontaneous Raman spectroscopy, coherent anti-Stokes Raman spectroscopy, Rayleigh scattering, laser-induced fluorescence spectroscopy, stimulated Raman gain or loss spectroscopy, optogalvanic spectroscopy, optoacoustic (photoacoustic) spectroscopy, absorption spectroscopy, multiphoton absorption or fluorescence spectroscopy, laser-induced breakdown spectroscopy, and Mie scattering. The list is long and growing.

For these different spectroscopies, the scattered, absorbed, or emitted frequency is determined by both the incident laser frequency and the atomic and molecular species present. The output frequencies are thus a signature of what is present in the laser-probed region. The intensity of the output at a frequency or set of frequencies is related to the concentration of the specific atomic or molecular species within the probed volume, and the spectral distribution is related to the temperature.

Bechtel (1979) has exploited this to derive a temperature for hydroxyl from the populations of the individual quantum states. Numerous other researchers have also made similar types of determinations by either laser fluorescence, laser absorption, spontaneous Raman, or coherent anti-Stokes Raman spectroscopies. (The references to these additional research efforts will be given later in this article.) Consequently, laser spectroscopies now allow the determination of individual species temperatures if the population distributions exhibit Boltzmann distributions. For a specific molecular species there are vibrational, rotational, and translational temperatures that may be determined at different space–time points in a flame.

In addition to quasi-point measurements, there are other types of laser measurements, including multipoint or whole-field measurements. Although photography is one of the oldest applications of optics to combustion research, laser visualization experiments are of much current interest. Examples of these visualization experiments include holography, laser schlieren photography, time-resolved interferometry, laser tomography, Moiré deflection mapping as well as two-dimensional visualization of laser-induced fluorescence, and Rayleigh and Raman spectroscopies. For these types of measurements, as with many spectroscopic measurements, a laser may not be essential for the experiment, but a laser is often a much more convenient radiation source than a spectrally and spatially filtered incoherent source.

The applications of lasers to combustion research have grown remarkably during the last 10 years. Many of the spectroscopic experiments have become increasingly attractive because of developments in laser technology. Higher power lasers give improved detectivities for spectroscopic species concentration measurements. The improvements of tunable dye lasers, infrared-tunable semiconductor lasers, and other frequency-tunable sources have allowed researchers to devote more energy to understanding combustion and less to problems of laser engineering.

It is impossible to give here a comprehensive review of all aspects of how laser technology is impacting on combustion research. Rather, this article focuses on identifying the many applications of lasers to combustion research, with reference not only to the research done in the authors' laboratory, but also to other research efforts. There have been several reviews and conference proceedings, and the reader may find references to additional literature in these reports. Reviews of general laser applications in combustion research include those by Bechtel and Chraplyvy (1982), Bechtel (1981), Eckbreth *et al.* (1979), Penner and Jerskey (1973), Eckbreth (1981a,b), and McDonald (1980). Specific reviews on laser-induced fluorescence have been given by C. P. Wang (1976), Crosley (1980b, 1981, 1982), and Schofield and Steinberg (1981); reviews on spontaneous Raman spectroscopy have been reported by Lapp and Hartley (1976) and Lederman (1977). Reviews of coherent anti-Stokes Raman spectroscopy for combustion applications include those by Hall and Eckbreth (1981; see also Hall and Eckbreth, this volume), Attal *et al.* (1980), Tolles *et al.* (1977), Nibler and Knighten (1979), and Pealet *et al.* (1980). Review articles on laser Doppler velocimetry have been given by Self and Whitelaw (1976) and Stevenson (1982). Reviews on particle sizing have been presented by Stevenson (1978), Penner and Chang (1981), and Azzopardi (1979).

Several conference proceedings have also been published and provide numerous details of laser applications. These proceedings have been reviewed by Goulard (1976), Lapp and Penney (1974), Thompson and Stevenson (1979), Zinn (1977), and Crosley (1980a).

Other references to combustion, although not necessarily to optical and laser diagnostics, include the journals *Combustion and Flame, Combustion Science and Technology,* and *Progress in Energy and Combustion Science.* The Combustion Institute, which sponsors the journal *Combustion and Flame,* also sponsors biennial international symposia whose proceedings are published.

Section II discusses background information on the fundamentals of combustion. In Section III, molecular energy levels and spectra are briefly reviewed, and Section IV discusses laser absorption spectroscopy. Section V gives applications of laser-induced fluorescence spectroscopy to combustion research, and Section VI reviews spontaneous Raman spectroscopy. Coherent Raman spectroscopies are discussed in Section VII, and Section VIII reviews other spectroscopies such as Rayleigh, optogalvanic, optoacoustic, and laser-induced breakdown spectroscopy. Section IX will give applications of optical refraction, holography, and tomography to flame visualization and to the determination of temperature, species concentrations in flames, and vapor concentrations in jets. Particle sizing methods are discussed in Section X, and laser velocimetry measurements are given in Section XI.

II. Combustion Research: General Considerations

This section discusses the practical goals of combustion research and provides an introduction to the fundamentals of combustion. Several introductory books on combustion are available including Gaydon and Wolfhard (1979), Glassman (1977), Williams (1965), and Fristrom and Westenberg (1965). The latter includes useful discussions of common flame measurement techniques.

Combustion involves an interplay between chemistry and transport processes. The chemistry includes kinetics (rates of reactions) and thermodynamics (heat release and equilibrium concentrations). The transport processes include bulk flow (convection), turbulence, heat conduction, and species diffusion. For solid and liquid fuels, melting and vaporization are also important. Combustion is often categorized by whether the fuel and oxidizer are premixed or must diffuse together and by whether the flows are laminar or turbulent. Spark-ignition engines use turbulent, premixed combustion, whereas diesels use turbulent diffusion flames. A candle burns with a laminar diffusion flame.

A laminar premixed flame can be divided into three zones: (1) the cool preflame zone containing the fuel and oxidizer, (2) the thin flame front where fast reactions consume the fuel and increase the temperature, and (3) the burned-gas region where slower reactions change the species concentrations toward their equilibrium values. (These are also called the preheat, reaction, and recombination zones, respectively.) Reactive species (radicals) produced in the flame front and burned-gas region diffuse into the preflame zone and initiate the breakdown of the fuel. Conduction of heat from the burned-gas region to the preflame region is important because it greatly accelerates the reactions. Thus the laminar burning rate or flame speed depends on species diffusion and heat conduction as well as the chemical reaction rates. The same processes occur in laminar diffusion flames, but with the important difference that the fuel and oxidizer are on opposite sides of the flame front and must diffuse together to react. The mixture is thus fuel rich on the fuel side, near stoichiometric in the flame front, and fuel lean on the oxidizer side. Heat and radicals transported from the flame front into the fuel region can lead to thermal pyrolysis. (In premixed flames, oxidation rates are typically faster than pyrolysis rates, thus the latter are less important.)

In laminar flames, diffusion and conduction control the transport of reactive species and heat. In most practical combustion systems, this transport is augmented by turbulence. Turbulent structures such as vortices and eddies can rapidly mix fuel, oxidizer, and hot gases, leading to much higher burning rates than occur in laminar systems. A recent book by Libby and Williams (1980) provides a good introduction to turbulence in combustion.

To understand combustion chemistry, it must be determined which species and reactions are important, and how fast the reactions occur. Thermodynamics can be used to calculate the equilibrium concentrations and temperature, but some of the reactions may be too slow to reach equilibrium on the time scales of interest. Reaction rates are usually measured in systems such as shock tubes, fast-flow reactors, and well-stirred reactors where transport is not important and there is some freedom to vary the concentrations of various reactants. Measurements of temperature and species concentrations are needed to determine reaction rates and thermochemistry. A comparison of measured and calculated concentration profiles in premixed flames of various stoichiometries can be used to test the validity of a reaction mechanism, provided that transport is properly treated in the calculation. It should be emphasized that an incomplete mechanism may work well at one pressure and stoichiometry, and yet be inaccurate under other conditions. The chemistry of air flames with fuels such as hydrogen (Dixon-Lewis, 1979) and C_1 and C_2 hydrocarbons (Warnatz, 1979, 1981; Westbrook and Dryer, 1979; Westbrook,

1982) is reasonably well understood, although some questions remain. Reaction mechanisms for propane are currently under development. There has also been much research on the chemistry of pollutant formation. The chemistry of CO and of NO (formed from N_2) is reasonably well understood. Recent work has partially elucidated the chemistry of sulfur and of fuel-bound nitrogen. Much has also been learned about soot formation, but the fundamental chemistry is still unclear. Other examples where chemistry is very important for combustion include flame inhibition (for fire retardants) and autoignition in engines.

The theory of laminar transport (species and heat diffusion) is well developed (Hirschfelder *et al.*, 1964). Generally, only Fickian diffusion and heat conduction are considered; the Dufour and Soret effects are negligible except in some hydrogen flames. For stable species, the transport coefficients can be determined experimentally (e.g., from viscosity measurements), whereas for radicals these coefficients are often estimated from approximate Lennard–Jones parameters. The transport coefficients are not known with high accuracy, but this is usually not a serious limitation. This is partially because the diffusion term in the species conservation equation is often smaller than many of the reaction terms. Also, for most combustion models, uncertainties in the reaction rates (which can vary over a large range) are much more of a problem than uncertainties in the transport coefficients. Transport affects burning rates and flame widths, and is most important near boundaries, flame fronts, or where there are large gradients.

In turbulent flames, an understanding of the turbulent mixing is often more important than a detailed knowledge of the chemistry or diffusion. For example, in a highly turbulent hydrocarbon–air flame, the fuel oxidation reactions are fast compared to mixing times. Mixing due to laminar diffusion is negligible compared to turbulent mixing. Thus, a good description of the turbulence combined with a simple approximation for the chemistry (e.g., a one-step global reaction) will provide a good description of the combustion. Most turbulence models attempt to describe the flow, mean temperature, pressure, and concentrations and fluctuations of these quantities around the mean. Turbulence theory is based on the (nonlinear) Navier–Stokes equations combined with additional (closure) equations which describe the correlation of the fluctuations of various quantities. Because the correct formulation of these closure relations is still unclear, simultaneous time-resolved measurements of two quantities (such as temperature and velocity) are particularly useful. Turbulent velocity measurements, although nontrivial, are becoming common (see Section XI for more details). Techniques for measuring temperature and concentration fluctuations are being actively pursued, and several examples will be discussed in the sections on fluorescence, Raman, and Rayleigh scattering. The excellent time resolution and nonintrusive nature of lasers make them invaluable for turbulence measurements.

The preceding has dealt primarily with combustion fundamentals, but empirical measurements in practical systems are also important. For example, measuring the temperature near a turbine blade can lead to design changes which extend the turbine life. Measuring turbulence levels in internal combustion engines can lead to designs with faster burning rates. Other parameters of practical importance include emission levels, fuel distribution in the combustor, and droplet sizes in sprays. Specific applications will be discussed in the following sections.

III. Molecular Energy Levels and Spectra

A brief discussion of molecular spectroscopy is given here to provide an introduction to laser spectroscopy for combustion measurements. More detail is provided by Herzberg (1945, 1950, 1966), Huber and Herzberg (1979), and Wilson et al. (1955). Gaydon (1974) discusses the spectra of species commonly found in flames.

Each molecular level has a unique set of quantum numbers which describe its electronic state, vibration, rotation, electronic spin, and nuclear spin. Typically, the electronic states are widely separated in energy $(10^4-10^5$ cm$^{-1})$. [Energies and frequencies are conveniently expressed in wavenumbers (cm^{-1}) using E/hc or frequency/c, where h is Planck's constant and c is the speed of light.] Each electronic state has vibrational sublevels separated by 500–4000 cm^{-1} which in turn have rotational levels separated by 0.1–1000 cm^{-1}. For many molecules the electronic spin is zero, but if it is not, interactions between the spin and electronic and rotational motion will produce additional small energy splittings. The energy associated with nuclear spin interactions is usually negligibly small, but the nuclear spin affects the degeneracy g_i of a state and therefore affects the population n_i of the state

$$n_i = ng_i \exp(-E_i/kT) \Big/ \Big[\sum g_j \exp(-E_j/kT)\Big] \qquad (1)$$

Here k is Boltzmann's constant and n is the total number density for the molecule. Note that 1 K corresponds to 0.695 cm^{-1}, thus at room temperature only the ground electronic and vibrational states will be thermally occupied, whereas many rotational states will be occupied. At flame temperatures, several vibrational states will be occupied.

To a good approximation, the rotations and vibrations can be treated separately. The quantum mechanical energy levels for a rotating symmetric top are given by

$$E = hc[BJ(J + 1) + (A - C)K^2]$$ (2)

where A, B, and C are constants which depend on the moments of inertia, and J and K are rotational quantum numbers. (For a diatomic, A and C are zero.) The vibrations are approximately described as simple harmonic oscillations with energies given by

$$E = hc\omega(v + \tfrac{1}{2})$$ (3)

where v is the vibrational quantum number and ω is the vibrational frequency. This simple model is qualitatively correct, but it is usually necessary to know the energy levels more accurately. Correction terms due to anharmonicity of the vibrations and centrifugal distortion of the rotating molecule are usually included in the expression for the energy. For many molecules (especially diatomics), these levels have been measured accurately and formulas have been developed to calculate E as a polynomial function of v and J (Huber and Herzberg, 1979). However, for complicated molecules like octane, the spectroscopic data are incomplete.

IV. Laser Absorption Spectroscopy

Absorption spectroscopy, both experimentally and theoretically, is a relatively simple technique for making combustion measurements. The light transmitted through a sample of length s is given by

$$I(s) = I(0) \int dv\, G(v - v_L)exp\left[- \int dx\, g(v - v_i)\alpha \right]$$ (4)

where $I(0)$ is the input laser intensity, α is the absorption coefficient of the line, and G and g are the lineshapes of the laser and the absorption line centered at frequencies v_L and v_i, respectively. The integral of the lineshape over frequency is unity. The observed absorption is integrated over the line of sight, thus there is no spatial resolution in this direction. (Both g and α are functions of x.) This can be a serious limitation if the combustion medium is not homogeneous (Grabner and Hastie, 1982). In some cases, fluorescence can be used to determine the distribution of absorbers along the line of sight. Tomography and saturated absorption are two other techniques for improving the spatial resolution and will be discussed later.

The absorption coefficient α depends on the populations of the lower and upper levels of the transition and on the quantum numbers of these levels. For visible and ultraviolet absorptions, α is often expressed as

$$\alpha = n_i[1 - \exp(-hc\nu_i/kT)](e^2/4\varepsilon_0 mc^2)f \tag{5}$$

Here n_i is the number density of the lower state, the exponential gives the contribution from the upper state, e and m are the electron charge and mass, ε_0 is the free-space permittivity, and f is the oscillator strength, which is a dimensionless measure of the strength of the absorption. [The units in Eq. (5) are correct for frequencies in cm^{-1} such that g has units of centimeters.] For strong atomic transitions f is near 1, whereas for strong molecular transitions f is near 10^{-3} (C. P. Wang, 1976). For simple molecules the dependence of f on the rotational quantum number can be calculated using quantum mechanics (see the Hönl–London formulas given by Herzberg, 1950). However, when high accuracy is needed (especially for temperature measurements from relative rotational populations), simple quantum calculations may not be sufficient (Wang *et al.*, 1981; Learner, 1962). For infrared (vibrational) transitions, a slightly different notation is often used (see Hanson *et al.*, 1977, 1980).

Accurate absorption measurements require consideration of the lineshape functions in Eq. (5). The molecular lineshape will be broadened by collisions which lead to a Lorentzian lineshape and by the Doppler effect which has a Gaussian lineshape. The convolution of the two is a Voigt profile which depends on the parameter $a = (\Delta\nu_L/\Delta\nu_D)\sqrt{\ln(2)}$, where $\Delta\nu_D$ is the linewidth due to Doppler broadening and $\Delta\nu_L$ is the linewidth due to collision and natural lifetime broadening. The Voigt profile has been described by many workers (Penner, 1959; Lück and Muller, 1977), and simple, accurate algorithms exist for calculating the profile (Hui *et al.*, 1978; Klim, 1981; Humlicek, 1982). The linewidths will depend on temperature, pressure, and composition. Typical values are near $0.1 \ cm^{-1}$ at atmospheric pressure. If the laser linewidth is much less than the molecular linewidth, then G can be replaced by a delta function in Eq. (4). The lineshape g can then be determined by scanning the laser across the line. Figure 3 shows how well the Voigt profile fits the measured lineshape for CO in an atmospheric pressure flame (Varghese and Hanson, 1981; see also Hanson *et al.*, 1977, 1980; Lück and Thielen, 1978). If the oscillator strength is known, the number density n_i can then be accurately determined. Also, measuring n_i for different rotational or vibrational levels can be used to determine the temperature. An example from Hanson and Falcone (1978) is shown in Fig. 4.

If the laser linewidth is comparable or larger than the molecular linewidth, then g cannot be determined directly. The curve-of-growth technique is often used in this case (Bechtel and Teets, 1979; Cattolica, 1982).

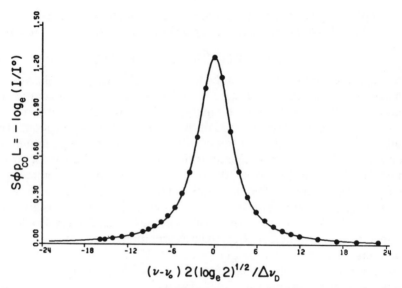

FIG. 3.   The absorption profile of a CO line in a methane–air flame. The points were
measured using a diode laser and the solid line is a Voigt profile with $a = 2.55$. [From
Varghese and Hanson (1981) with permission.]

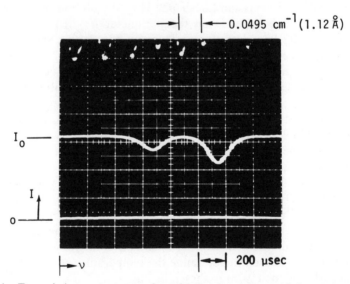

FIG. 4.   Transmission measurement of two CO lines in a propane–air flame using a single
sweep of a diode laser. The temperature determined from the ratio of the two peaks was 2100
K; the thermocouple temperature was 2070 K. Similar results were obtained with faster
scans which covered the two lines in 20 μsec. [From Hanson and Falcone (1978) with
permission.]

The integrated absorption [I_0 minus the integral of Eq. (4) over frequency ν_L] is a function called the curve-of-growth which is independent of laser lineshape but depends on the Voigt parameter "*a*." This function is evaluated for various "*a*" values by Penner (1959), Penner and Kavanagh (1953), Carrington (1959), and Hill (1979). For small absorption this function is independent of the molecular lineshape, as can easily be seen by expanding the exponential in Eq. (4). For large absorption, the integrated transmission depends strongly on the Voigt parameter. This is because the absorption by the wings of the line is important, and the wings of a Lorentzian and a Gaussian are very different. The curve-of-growth can be used to evaluate "*a*" by determining how the integrated transmission changes with optical depth. That is, a comparison of the signals for strong and weak lines or for multiple passes through the flame can be used to determine "*a*."

The curve-of-growth is applicable only to resolved lines. Also, because "*a*" depends on the temperature, pressure, and composition, it will vary with position in the flame. Finally, although the Voigt profile usually provides a very accurate description of the lineshape, some lines exhibit collisional narrowing (Dicke, 1953; Galatry, 1961). For these cases the usual curve-of-growth analysis is not correct. For all of these reasons, narrowband measurements which give the lineshape directly are preferable.

Both lasers and conventional light sources are useful for combustion measurements, although conventional sources are often cheaper or cover a wider spectral range. For example, Ottesen and Stephenson (1982) used Fourier transform infrared absorption to examine many species in sooting flames. Lasers typically provide better spectral resolution, spatial resolution, and higher data rates. Hanson (1977) and Hanson et al. (1977) have used semiconductor diode lasers to record absorption profiles on submillisecond time scales. The sensitivity of absorption measurements is usually limited by the stability of the light source, and absorptions below the 0.1–1% range are hard to measure. A wide variety of lasers have been used for absorption measurements, but the most generally useful are the tunable lasers, especially infrared diode lasers and the visible dye lasers. The latter can also produce infrared or ultraviolet light by such processes as second-harmonic generation in crystals or stimulated Raman shifting in hydrogen gas.

Several laser absorption measurements of OH in flames have been carried out. The near-UV spectrum of OH has been thoroughly studied (Dieke and Crosswhite, 1961; Goldman and Gillis, 1981; Wang et al., 1981; Smith and Crosley, 1981; Chidsey and Crosley, 1980). Teets and Bechtel (1981a) used OH flame measurements to evaluate the rate constant for the reaction $H + O_2 + H_2O \rightarrow HO_2 + H_2O$, which was shown to control the equilibration of the radicals above the flame. Lück and Tsatsaronis (1979) and Cattolica (1982) measured OH in investigations of methane/air flames. Dean *et al.* (1982) used OH absorption as part of a study of NH_3/O_2 flame chemistry. This chemistry is relevant to the combustion of fuels derived from coal or oil shale containing nitrogen impurities. It is also related to the "thermal denox" process (Lyon and Benn, 1979) in which ammonia added to the burned gas of an industrial burner reacts with NO to form N_2. Roose *et al.* (1981) have used diode laser measurements of NO in a shock tube to study this thermal denox process. Other studies of nitrogen chemistry in flames have involved absorption measurements of NH_2 (Green and Miller, 1981) and NH (Dean *et al.*, 1982; Anderson *et al.*, 1982b). Kychakoff and Hanson (1981) have used optical fibers to carry the laser beam into and out of the flame. This can provide improved spatial resolution and also better optical access. As a final example, Cattolica *et al.* (1982b) and Revet *et al.* (1978) used OH absorption measurements to study the perturbations of a sampling probe in a low-pressure flame (see also Stepowski and Cottereau, 1981b).

Other molecules detected using laser absorption include NO (Hanson *et al.*, 1980; Sell, 1981) and CO (Hanson, 1977, 1980; Hanson *et al.*, 1977, 1980; Schoenung and Hanson, 1981a; Sell *et al.*, 1980, 1981). Schoenung and Hanson (1981b) used a diode laser to detect CO in gas extracted from a turbulent flame with a sampling probe. The fast system time response allowed measurement of turbulence frequencies up to 1 kHz. Cuzillo *et al.* (1982), Olsen *et al.* (1978), Chraplyvy (1981), and Tishkoff *et al.* (1980) have used laser absorption to measure various hydrocarbons. Kline and Penner (1982) have used a He–Xe laser to detect formaldehyde in a shock tube. Hill and Majkowski (1980) have monitored sulfate in automobile emissions. Related fluorescence measurements are discussed in Section V.

To close this discussion on laser absorption spectroscopy, two novel laser absorption techniques are described. The first technique is saturated absorption (Goldsmith, 1981; Kychakoff *et al.*, 1982b), in which an intense laser beam is used to pump a large fraction of the molecules out of a level. The resulting decrease in absorption is monitored with a second laser beam which crosses the first beam. Only the molecules in the region intersected by the two beams contribute to the signal, thus the spatial resolution can be very good. As discussed in Section V on fluorescence, interpretation of saturation effects can be complicated due to collisional redistribution of populations.

The second novel technique is called intracavity dye laser spectroscopy. The absorbing species is placed between the mirrors of a CW (continuous wave) broadband dye laser, and the laser output is analyzed with a spectrometer. Even very small absorptions produce large dips in the laser spectral profile, thus the sensitivity is enhanced by many orders of magnitude. Harris and Weiner (1981a) have used the technique to measure oxygen atoms in a flame. This is one of the few techniques capable of detecting oxygen atoms. Harris and Weiner have also used intracavity absorption to measure C_2 in a flame as part of a study of soot formation (Harris and Weiner, 1981b). Unfortunately, stable operation of the laser is hard to achieve, and the theory for the enhancement is still somewhat controversial (Harris, 1982; Stoeckel et al., 1982).

V. Laser-Induced Fluorescence

Laser-induced fluorescence spectroscopy (LIF) is a sensitive technique for detecting molecules in flames. Several excellent reviews have been written (Bechtel and Chraplyvy, 1982; Crosley, 1981; Eckbreth, 1981a; Daily, 1980; Eckbreth et al., 1979; Schofield and Steinberg, 1981). In a typical experiment, a tunable laser is used to excite a transition of a molecule of interest. The fluorescence emitted by the excited molecules can be sensitively detected by photomultiplier tubes or optical multichannel detectors. By observing fluorescence from only a short length along the laser beam, good spatial resolution can be achieved. The signals can be large enough such that single-shot (nanosecond) measurements can be made. Two limitations of LIF are that many molecules do not absorb and fluoresce at convenient wavelengths and that systematic errors make quantitative measurements difficult. Many molecules, including N_2, O_2, H_2O, CO, CO_2, and most hydrocarbons, do not absorb in the visible or near-UV region. Although they do absorb in the vacuum UV (wavelength less than 200 nm), tunable laser sources in this region are not well developed, and strong, broad absorption by many species in this region makes spectroscopy very difficult. Some of these species absorb in the infrared, but fluorescence in this region is difficult due to the poor sensitivity of available detectors and the large background of flame radiation in the infrared. Fortunately, many of the interesting chemically reactive species in flames (the radicals) do absorb and emit in the visible or near-UV. The most commonly used laser is the pulsed dye laser which can be tuned to any region in the visible or near-UV using suitable dyes and harmonic generation crystals. With the high peak power of these lasers, detection limits in the parts per million to parts per billion range or better can be achieved.

There are several factors which make quantitative LIF difficult. These include collisional quenching of the fluorescence, uncertainties in detec-

tion of solid angle and observation volume, radiation trapping, and attenuation of the laser beam. Collisional effects will be discussed in detail here.
The other effects will be discussed later. Collisional effects are depicted in
Fig. 5, which shows some of the OH levels of interest for a LIF measurement (see Section IV for references on OH spectroscopy). The laser
excites molecules to a specific rotational–vibrational level of the $^2\Sigma^+$
electronic state. The number excited depends on the number in the lower
level of the transition, the intensity of the laser, and the strength of the

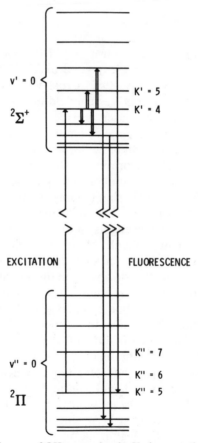

FIG. 5.   Schematic diagram of OH energy levels. Various rotational levels of the lowest
vibrational state in both the $^2\Pi$ and the $^2\Sigma^+$ electronic states are shown. Some of the
collision-induced energy-transfer processes are denoted by double-line arrows. Both spin
doubling and lambda doubling have been suppressed for clarity. [From Bechtel (1979) with
permission.]

absorption coefficient. The excited molecules spontaneously emit photons on a time scale determined by the radiative lifetime (about 700 nsec for OH). On a faster time scale, collisions will cause transitions to nearby rotational states (about 0.1 nsec at 1 atm) or to other vibrational or electronic states (several nanoseconds). Because collisional deactivation is typically much faster than the radiative rate, only a small fraction of the molecules emit photons. Also, unless the detection system has good spectral resolution, fluorescence from many collisionally excited rotational–vibrational levels will be observed. The signal is thus very dependent on the collisional processes which in turn depend on the pressure, temperature, and composition of the gas, as well as the quantum numbers of the excited states.

There have been many studies of collisional effects, particularly for the OH molecule. Stepowski and Cottereau (1981a,c) and Cottereau and Stepowski (1980) have used time-resolved OH fluorescence in low-pressure flames to measure the electronic de-excitation rate. They have also spectrally resolved the fluorescence to determine transition rates between rotational states. Lengel and Crosley (1977) and Crosley and Smith (1980b) have also studied OH collisions and have developed a detailed model of collisional energy transfer. Useful quenching information is also given by Fairchild *et al.* (1982) and Morley (1982). There are numerous other studies of collisions, although many of them were done at room temperature.

Several different approaches have been used to deal with collisional effects on fluorescence. In low-pressure flames, time-resolved fluorescence gives the quenching rate directly (Stepowski and Cottereau, 1979). If the temperature, the concentrations of the major species, and the quenching rates due to these species are known, the effects of quenching can be readily calculated (Bechtel and Teets, 1979; Morley, 1981). This is feasible for the well-studied OH molecule, but quenching cross sections for most other molecules under flame conditions are not known.

Schofield and Steinberg (1981) have used an alternate approach which does not require measured quenching cross sections (see also Muller *et al.*, 1979, 1980b). They used a spectrometer to detect the fluorescence only from the level excited by the laser, blocking the signal from nearby collisionally excited rotational levels. Although this can decrease the detected signal by several orders of magnitude, it does simplify the quenching considerations. When the signal from all rotational levels is detected, the vibrational and electronic collisional transitions cause the quenching. These transitions are often much less than the gas kinetic rate and can have strong dependencies on temperature and collision partner. When the resolved fluorescence from a single level is detected, all collisions, including rotational transitions, cause the quenching. Schofield and Steinberg argue that when all collisions are considered, the rate is likely to be close to the gas kinetic rate. Thus, they scale the quenching rate with $T^{-1/2}$ independent of the composition of the gas. They have shown this works well for a number of molecules including OH, SH, SO_2, S_2, and SO. Figure 6 is an example of the many species profiles they were able to measure in a flame doped with H_2S. Although this work is quite encouraging, there may be cases where the rotational collision rate is dominant and differs significantly for different collisional partners.

Perhaps the most interesting way to deal with quenching effects is to saturate the transition. Briefly, as the laser intensity increases, the rate of stimulated emission can become much larger than the quenching rate. The population of the upper level approaches that of the lower level and the signal saturates. In a two-level model, the population of the upper level is then half the original population of the lower level, and the signal depends

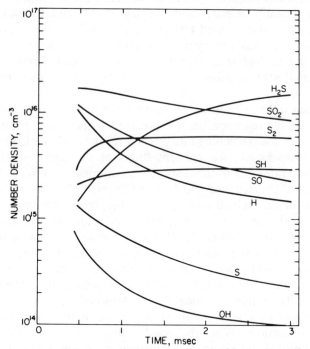

FIG. 6.   Concentration profiles in an $H_2/O_2/N_2$ flame with 1% H_2S added. The profiles were obtained using a spectrally resolved fluorescence method which allows correction for quenching in a straightforward manner (see text). [From Muller et al. (1979) with permission of the Combustion Institute.]

only on that original population, the radiative lifetime, and the fraction of the emitted light collected by the optics. Some atoms are approximately two-level systems, but most molecules are not. For molecules, collisional effects are still important. At high laser intensity, the populations of the upper and lower levels are still nearly equal, but collisions transfer molecules from the upper level to nearby rovibrational levels. Similarly, collisions repopulate the lower level, thus there is no simple relation between the population in the excited state and the original population in the lower level. Berg and Shackleford (1979) have shown that use of the two-level model is inadequate and have formulated a four-level model which gives insight into molecular saturation effects. Verdieck and Bonczyk (1981) used a simplified version of this model to analyze their measurements of CH and CN. Numerous researchers have formulated rate equations relating the populations of the various levels and have solved the equations under plausible assumptions (Kotlar *et al.*, 1980; Lucht and Laurendeau, 1979; Campbell, 1982; Chan and Daily, 1980a).

One successful approach is the balanced cross rate model of Lucht *et al.* (1980, 1982a,b,c, 1983). They postulate that the collision rate out of the upper level is approximately equal to the collision rate which repopulates the lower level. Thus, on short time scales, the sum of the populations of the upper and lower level is constant and is equal to the original population of the lower level. For many molecules in atmospheric pressure flames, Nd : YAG pumped dye lasers have sufficient intensity to equalize these upper and lower level populations on subnanosecond time scales. Thus observing the (spectrally resolved) fluorescence from the laser-populated upper level on short time scales gives a signal which is approximately independent of collision effects. Lucht *et al.* (1980, 1982a,b,c, 1983) use a spectrometer to isolate the fluorescence from the laser-pumped upper level, and gate the detector to observe the fluorescence only during the peak of the laser pulse, a few nanoseconds after the start of the pulse. Comparisons with absorption were used to verify the accuracy of the method. Figure 7 (from Lucht *et al.*, 1982a, 1983) shows that the method is relatively insensitive to flame pressure and composition. The figure shows the ratio of OH concentration measured by absorption and by saturated fluorescence for various $H_2/O_2/N_2$ flames at different pressures. Lucht and colleagues note that the quenching rate for the 240-Torr flame was comparable to the expected rate for an atmospheric hydrocarbon flame. Salmon *et al.* (1982) have used this method for NH measurements. Although this work is quite encouraging, it is not clear how well the approach will work for other molecules or for higher pressures.

Several other factors must be considered for saturated fluorescence.

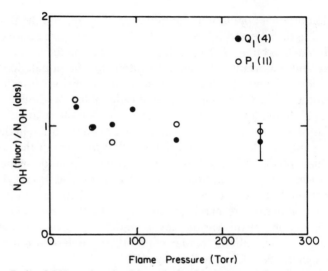

Fig. 7.   Ratio of OH number densities measured by saturated fluorescence and optical absorption versus flame pressure. The ratios are normalized to the value at a flame pressure of 72 Torr. The results show that in these $H_2/O_2/N_2$ flames, the saturated fluorescence measurements are insensitive to quenching effects. [From Lucht *et al.* (1982a, 1983) with permission.]

The transverse spatial profile of the laser intensity is likely to be roughly Gaussian, thus even if the center of the beam strongly saturates the transition, the edges of the beam will not. Daily (1978a) has shown that this significantly affects the saturation curve if fluorescence from the whole transverse profile is collected. This problem can be avoided if apertures are used to clip the edges of the beam, although this is difficult with high-power lasers. An alternate approach is to aperture the collection optics to observe mainly the signal from the center of the beam. Similarly, the signal should be gated to reject nonsaturated signals due to the leading and trailing edges of the pulse. A final complication is the possibility of laser-induced chemistry. Thus far, only energy-changing collisions have been discussed; however, reactive collisions can occur on submicrosecond time scales, and these collisions will alter the saturation rate equations. Muller *et al.* (1980a), Iino *et al.* (1981), and Daily and Chan (1978) have observed laser-induced chemistry for sodium excited by a microsecond pulse from a flashlamp dye laser. For saturation on the nanosecond time scale, this is less likely to be a problem. Other saturated LIF measurements are found in Pasternack *et al.* (1978), Baronavski and McDonald (1977a,b), Mailander (1978), Smith *et al.* (1977), and Ahn *et al.* (1982).

In addition to collisional effects, there are a few other considerations for accurate LIF measurements. The observation volume and detector efficiency must somehow be determined. This is usually done by calibrating either with a known signal from Rayleigh scattering (Bonczyk and Shirley, 1979; Verdieck and Bonczyk, 1981), or with fluorescence from a known concentration determined by absorption (Bechtel and Teets, 1979) or equilibrium considerations (Morley, 1981; Muller *et al.*, 1979, 1980b). For high concentrations it is necessary to correct for absorption of the laser prior to the observation volume and for absorption of the emitted fluorescence (radiation trapping, Daily, 1978b; Chan and Daily, 1979; Holstein, 1947). Also, the absorption lineshape must be considered. In most studies, relatively broadband lasers have been used which excite the whole absorption profile. In others, investigators have scanned narrowband lasers and have integrated the fluorescence versus frequency. If the absorption is not small, however, the lineshape must be considered when correcting for the absorption of the laser prior to the observation volume. Finally, Lengel and Crosley (1977) and Lucht *et al.* (1982c) have discussed polarization effects. A polarized laser will excite a coherent superposition of Zeeman sublevels (m levels). Ordinarily, this would cause the emitted spontaneous emission to be polarized, anisotropic, and to depend on the change in J of the emission transition. Rapid collisions tend to wash out these effects, but they can be observed using narrowband detection.

There have been many applications of LIF to flames in addition to those just given. Morley (1981), Dean *et al.* (1983), Anderson *et al.* (1980, 1981, 1982a,b), and Sullivan *et al.* (1981) have used LIF to study the chemistry of nitrogen in flames. Muller *et al.* (1979, 1980b) have studied sulfur chemistry in flames (see Fig. 6). Bechtel and Teets (1979) used LIF to help test a reaction mechanism for methane flames (see Fig. 8), and Cattolica and Schefer (1983a,b) have measured OH concentration in a combustion boundary layer. DiLorenzo *et al.* (1981), Cincotti *et al.* (1981), Coe and Steinfeld (1980a,b), Coe *et al.* (1981), and Miller *et al.* (1982) have observed fluorescence of polycyclic aromatic hydrocarbons in sooting flames. Other species studied include C_2 (Baronavski and McDonald, 1977a,b; Jones and Mackie, 1976), CH (Verdieck and Bonczyk, 1981; Bonczyk and Shirley, 1979; Cattolica *et al.*, 1982a; Filseth *et al.*, 1978), NO_2 (Barnes and Kircher, 1978; Agrawal *et al.*, 1977), and NO (Grieser and Barnes, 1980a,b).

Other fluorescence studies deserve comment. Several researchers have used a sheet of laser light to excite OH and have imaged the fluorescence onto an intensified vidicon (Dyer and Crosley, 1982) or reticon (Kychakoff *et al.*, 1982a,c) in order to measure simultaneously the distribution of

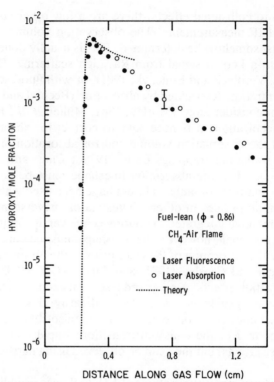

FIG. 8. OH concentration profiles in a methane–air flame. The fluorescence was calibrated by comparison with absorption measurements well above the flame (near 1.2 cm on the figure). The measurements support a theoretical calculation of Bechtel *et al.* (1981). [From Bechtel and Teets (1979) with permission.]

OH in a cross section of the flame. As shown in Fig. 9, this provides a graphic illustration of the flame structure and may be particularly useful for studies of turbulence (see also Aldén *et al.*, 1982a,b). Cole and Swords (1980) seeded air with NO_2 and used fluorescence to observe density fluctuations in a motored engine. Kychakoff and Hanson (1981) and Kimball-Linne *et al.* (1982) have developed a fiber-optic fluorescence probe which can be used with combustors where optical access is difficult.

Fluorescence has also been used to measure temperature. For example, using spectra such as those shown in Fig. 10, Bechtel (1979) measured the distribution of populations among the rotational states of OH. The observed Boltzmann distribution gave a rotational temperature which was equal, within experimental error, to the nitrogen vibrational temperature determined by Raman scattering (see Fig. 11). Anderson *et al.* (1982a) and Anderson (1979) have made similar measurements of rotational tempera-

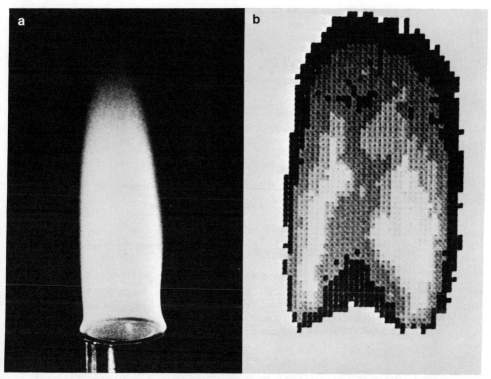

FIG. 9. (a) Photograph of a turbulent, premixed methane–air flame. (b) Digital picture of OH concentration in the flame, obtained by fluorescence excited by a single laser pulse. Each pixel represents the level of fluorescence from a $0.4 \times 0.4 \times 0.2$-mm region of the flame. [From Kychakoff *et al.* (1982a) with permission.]

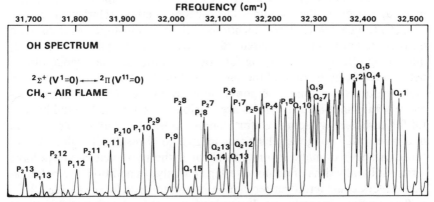

FIG. 10. Excitation scan of laser-induced fluorescence of OH. Some of the resolved rotational lines are identified using the notation of Dieke and Crosswhite (1961).

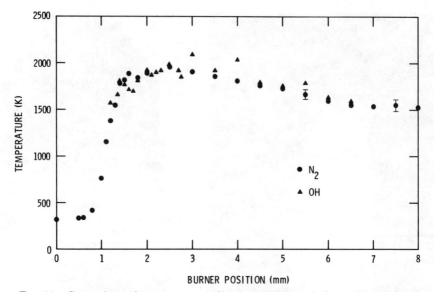

FIG. 11. Comparison of temperature profiles in a methane–air flame. The circles are nitrogen vibrational temperatures determined by Raman spectroscopy. The triangles are OH rotational temperatures determined from fluorescence spectra similar to those shown in Fig. 10. [From Bechtel (1979) with permission.]

ture. Crosley and Smith (1982) have observed that such rotational temperature measurements can be affected by collisions and by the spectral bandpass of the detection system. If, for example, only fluorescence from excited levels with small J is observed, then the measured distribution will be biased toward small J. Detection near a bandhead and with a large bandwidth can minimize these errors. An alternate temperature measurement approach is the two-line fluorescence method (Haraguchi et al., 1977; Cattolica, 1981). Two transitions with the same upper level but with different lower levels are excited. Because a common upper level is used, quenching effects are the same for both excitations. The temperature is determined from the ratio of the signals and the known relative absorption cross sections, assuming the populations are in thermal equilibrium. Because single-shot measurements are feasible, this approach could be useful for turbulent systems. Unfortunately, most atoms and molecules do not have wide enough level spacing to give good results at high temperatures. Also, systematic errors can occur if the two exciting lasers have different bandwidths or do not overlap spatially. In addition, the polarization effects considered earlier can be important (Lucht et al., 1982c).

Temperatures can also be determined from the relative populations of rotational levels of the excited electronic state. These populations can be determined from the spectrum of the emitted light. Calculation of the temperature requires a model of the upper state collisional processes. For example, if rotational relaxation is much faster than other processes, then the populations will have a Boltzmann distribution. For OH this is not quite true, but Zizak *et al.* (1981a,b) have observed that levels far from the excited level have a near-Boltzmann distribution which gives adequate temperatures. Chan and Daily (1980b) have used a model of the collision rates to calculate emission spectra as a function of temperature. Fitting these spectra to observed spectra gave temperatures in close agreement with thermocoupling measurements. Zizak and Winefordner (1982) and Crosley and Smith (1980a,b) have also used collision-assisted fluorescence to measure temperature. Whereas this work has shown that quantitative measurements are possible, the dependence of the techniques on poorly known collision rates is a definite disadvantage.

VI. Spontaneous Raman Spectroscopy

Spontaneous Raman spectroscopy is a simple, direct spectroscopy which has most of the attributes desired for species or temperature measurements (Lapp *et al.*, 1972, 1973; Lapp, 1973; Drake and Rosenblatt, 1976). It is limited only by weak signals. The Raman effect is the inelastic scattering of light by a molecule or atom. The incident photon induces a field in the molecule via its polarizability, and the molecule undergoes a transition to a new state, simultaneously emitting a photon $h\nu_i$ such that energy is conserved overall. The inelastically scattered photons are analyzed. In a typical spontaneous Raman experiment, as seen in Fig. 12, an incident laser of power I_0 is brought to a focus in the scattering region. The light scattered at right angles is collected and dispersed by a monochromator. The power of the scattered light arising from a single Raman transition i is

$$I_i = I_0[\partial\sigma_i(\nu_0)/\partial\Omega]n_i\delta\Omega\delta s\eta \tag{6}$$

where $\partial\sigma(\nu_0)/\partial\Omega$ is the cross section per molecule per solid angle at laser frequency ν_0, n_i is the number density in the lower state of the transition, $\delta\Omega$ is the collection angle, δs is the scattering length determined by the collection optics, and η is the transmission efficiency of the collection optics.

The spontaneous Raman signal is easily analyzed because it is a linear function of the number density and of the laser intensity. The collection factors $\delta\Omega$, δs, and η are determined by Raman scattering from a calibra-

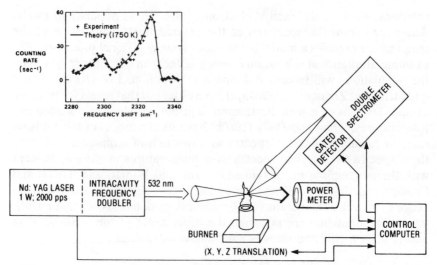

FIG. 12. A typical laser Raman apparatus for concentration and temperature measurements using a pulsed laser and photon-counting electronics. Insert: a Raman spectrum of the N_2 Q-branch from a flame at 2000 K.

tion gas. There are no quenching or time-delay effects. Optical-induced breakdown of the sample occurs long before the transition can be saturated because the effect is very weak. Of equal importance for combustion applications, the signal arises from a point, giving three-dimensional spatial resolution. Whereas most Raman measurements have been from a point, both one-dimensional (Smith and Giedt, 1977; Black and Chang, 1978; Sochet *et al.*, 1979) and two-dimensional (Hartley, 1974; Webber *et al.*, 1979) Raman visualizations have been performed.

Despite these attributes, spontaneous Raman spectroscopy is largely limited to major species concentrations (>1 mol%) and to low-noise environments because the cross sections are small. Typical spontaneous Raman cross sections for rotational and vibrational transitions are 10^{-29} and 10^{-31} cm^2/sr, respectively. These are 3 orders of magnitude smaller than Rayleigh (elastic) scattering and 10 orders of magnitude smaller than molecular absorptions.

For transitions within the ground electronic state the total cross section can be expressed following Placzek (1934; Schrötter and Klöckner, 1979) in terms of the wavenumber of the scattered photon ν_i and the matrix elements of the polarizability tensor $\alpha_{jk}(r_\ell)$ where r_ℓ are the internal vibrational coordinates

$$\sigma(\nu_i) = (2^7\pi^5/3^2)\nu_i^4(2J + 1)^{-1} \sum_{jk} |\langle 0|\alpha_{jk}|f\rangle|^2 \tag{7}$$

The cross sections for many transitions of many molecules have been measured at low temperatures (Penney *et al.*, 1974; Schrötter and Klöckner, 1979). Most of these have been analyzed assuming the polarizability tensor is linear in the r_ℓ and the molecule is a harmonic oscillator–rigid rotator. Neither of these assumptions is rigorously true and can lead to difficulties in calculating cross sections for some states of light molecules populated at high temperatures. James and Klemperer (1959) derived perturbation expressions for the effect of centrifugal distortion in diatomic molecules. Except for H_2, these corrections are adequate for temperatures below 2000 K (Drake and Rosenblatt, 1978; Drake *et al.*, 1980), but anharmonic terms must be considered near 3000 K (Drake, 1982). Anharmonic and nonlinear perturbation corrections have been calculated (Buckingham and Szabo, 1978), as well as highly accurate matrix elements for H_2 (Hamaguchi *et al.*, 1981; Cheung *et al.*, 1981). These high-order corrections are analogous to the corrections introduced in infrared absorption spectroscopy many years ago (Herman *et al.*, 1958).

The Raman spectra are usually the superposition of many lines which are partially resolved by the spectrometer of typical 1–2 cm^{-1} resolution. The spectral convolution is

$$I(\nu) = \sum_i g(\nu - \nu_i)I(\nu_i) \tag{8}$$

where g is the spectrometer slit function. Spectra are least-squares fit for species concentrations and temperatures (Hill *et al.*, 1979; Stephenson and Blint, 1979).

The pure vibrational transitions (Q-branches) of nitrogen have been frequently used to determine temperature. A typical spectrum is seen in Fig. 12. A succession of vibrational bands correspond to $v = 0 \to 1, 1 \to 2$, etc. The widths of each band are primarily determined by the rotational distribution. Both the bandwidth and the successive band ratios are temperature-sensitive features. The temperature distributions through many laminar flames have been determined: premixed C_3H_8–air (Stricker, 1976; Stephenson and Aiman, 1978; Stephenson, 1979; Bechtel *et al.*, 1981), premixed CH_4–air (Boiarski *et al.*, 1978; Bechtel *et al.*, 1981), premixed NH_3–O_2 (Setchell and Miller, 1978; Dasch and Blint, 1982), diffusion H_2–air (Aeschliman *et al.*, 1979), and premixed CH_4–N_2O (Beyer *et al.*, 1982). The N_2 vibrational temperatures have been compared and agree with CO_2 vibrational (Blint *et al.*, 1980; Stephenson, 1981) and OH rotational (Bechtel, 1979) temperatures in atmospheric pressure CH_4–air flames.

Similarly, the species profiles have been determined in several cases. Bechtel *et al.* (1981) performed an extended investigation of atmospheric pressure CH_4–air flames, measuring CH_4, O_2, H_2, CO, H_2O, and CO_2. As seen in Fig. 13, these profiles agree very well with detailed flame calcula-

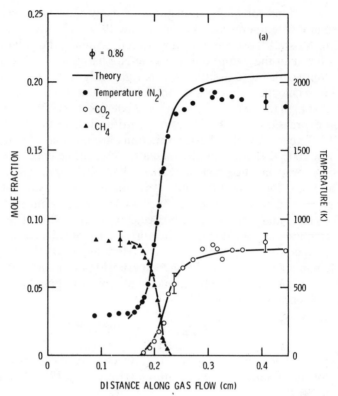

FIG. 13. Concentration and temperature profiles from spontaneous Raman spectroscopy of an atmospheric pressure CH_4–air flame. The distance has an arbitrary zero for this essentially free flame. The theoretical profile is based on a set of 24 chemical reactions. [From Bechtel *et al.* (1981) with permission.]

tions based on Peeters and Mahnen's (1973) low-pressure CH_4–air mechanism. This agreement demonstrates that the general transport, heat release, and reaction rate features have been accurately described. In NH_3–O_2–N_2 flames, Dasch and Blint (1982) found that current mechanisms are only partially successful in describing the NH_3, O_2, H_2, and H_2O profiles (see also the data of Setchell and Miller, 1978).

Although spontaneous Raman spectroscopy has been primarily applied to major stable species, Dasch and Bechtel (1981) demonstrated that atomic oxygen can be detected. They observed the Raman-allowed $J = 2 \rightarrow 1$ and $2 \rightarrow 0$ transitions in the ground ^3P electronic state. Teets and Bechtel (1981b) also observed these transitions with high spectral resolution using CARS. The very high oxygen atom concentrations in these experiments were generated in a hydrogen–oxygen flame. Setchel and Miller (1978) have detected modest amounts (2 mol%) of NO in NH_3 flames.

Several different laser and detector schemes have been used in these steady, atmospheric pressure flames. Usually, a visible laser is used in order to avoid laser-induced fluorescence, despite the increase of Raman cross sections as ν^4. Typically, 1–10 W of laser power will generate 10–1000 photons per second for major species (>1 mol%). In nonluminous flames, phase-sensitive detection has been used with chopped, continuous-wave lasers. It is possible to enhance the laser power by working within the laser cavity. To decrease the noise, pulsed lasers and gated photon counters have been used. Because the laser's duty cycle (ratio of pulse length to pulse separation) is usually less than 10^{-4}, the signal-to-noise ratio can be increased by 10^2. Maximum laser irradiance is limited to approximately 10^{11} W/cm^2 by the breakdown of the gas. Pulsed solid-state, dye, copper vapor, and excimer lasers have been used.

Spontaneous Raman spectroscopy as well as other laser techniques has the spatial resolution to probe within the flame front of premixed, high-pressure flames. Whereas atmospheric pressure flames are typically a few hundred micrometers thick, the probe laser can usually be focused to a few tens of micrometers. Raman spectroscopy improves at higher pressures such as in internal combustion engines because of the higher molecular densities.

The unambiguous analysis of spontaneous Raman spectra has led to increasing numbers of combustion applications with complicated transport phenomena. Bechtel and Blint (1980) probed the thin, nonluminous wall boundary layer of a premixed CH_4–air flame. This so-called quench zone was believed to be a major source of unburned hydrocarbons in small combustors. As seen in Fig. 14, reactant decay and product appearance were found within this zone. A two-dimensional flame calculation helped demonstrate that this nonquench behavior is a consequence of molecular diffusion to and from the boundary layer by both radicals and major species.

Recently, Raman spectroscopy has made important contributions in the understanding of turbulent diffusion flames. Drake et al. (1983) and Dibble et al. (1982) have measured mean and fluctuating concentrations and temperatures. These measurements were performed simultaneously with laser Doppler velocimetry, allowing fluxes to be calculated. The results demonstrate the dramatic ability of the conserved scalar approach (Bilger, 1976) to correlate the spatial distribution of time-averaged observables in both laminar and turbulent diffusion flames.

In order to obtain single laser pulse concentration and temperature

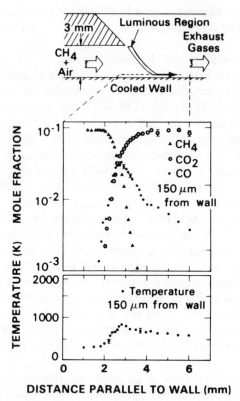

FIG. 14. Concentration and temperature profiles near a cool wall determined by spontaneous Raman spectroscopy. The profiles show fuel decay in this region in spite of the lowered temperature due to heat loss to the adjacent wall. [From Blint and Bechtel (1982a) with permission.]

measurements in turbulent or intermittent environments, single-detector monochromators have been replaced with multidetector polychromators and with intensified multielement detectors (see Fig. 15). In a polychromator arrangement the temperature can be determined from the intensity ratio of the N_2 anti-Stokes Q-branch ($\Delta v = -1$) to the Stokes Q-branch ($\Delta v = +1$).

Raman spectroscopy has been performed in motored and in fired internal combustion engines. Setchell (1978) measured cycle-averaged, preignition fuel/air ratios in a homogeneous-charge engine with a CW laser and gated detection. Johnston (1979) similarly determined air/fuel ratios in a direct-injection engine, showing large variations in the ratios and long

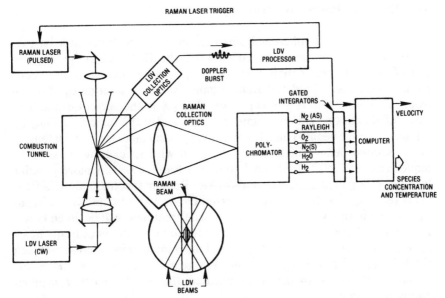

FIG. 15. A combined laser Doppler velocimetry/spontaneous Raman spectroscopy apparatus with polychromator. [From Dibble *et al.* (1982) with permission.]

mixing times. Johnston (1980) also observed the air/fuel ratio within the spark plug gap. Using single laser pulses and a polychromator, Smith (1980) observed temperature and nitrogen density cycle-to-cycle variations. The probability distribution functions (pdfs) for both temperature and density were Gaussian except near the time of flame arrival when the pdfs were bimodal. Smith (1982a) has shown that the bimodality is due to fluctuations in the flame arrival time.

To date, spontaneous Raman spectroscopy has not been successfully applied to highly luminous flames such as rich hydrocarbon flames. In addition to the high noise of the background radiation in these flames, the signals are overwhelmed by laser-induced fluorescence from small hydrocarbon fragments (Aeschliman and Setchell, 1975; Bailly *et al.*, 1976; Setchell and Aeschliman, 1977) and by laser-induced incandescence from soot particles (Eckbreth, 1977) and dust particles (Pealat *et al.*, 1977). Laser vaporization of soot particles was also observed and has been quantified by Dasch (1983). It may prove possible under some circumstances to destroy the soot with the leading edge of a pulse and to perform spectroscopy with the trailing edge, or to use two separate, closely spaced pulses (Pealat *et al.*, 1977).

VII. Coherent Raman Spectroscopy

There are several coherent Raman techniques which use at least two lasers, one of which can be tuned such that the difference in the frequencies of the two lasers equals a Raman frequency of interest. In one technique, known as stimulated Raman gain, Raman scattering of photons from one laser beam is stimulated by photons in the second beam. The stimulated emission is in the same direction and at the same frequency as the second laser beam. Thus, there is a very small increase in the intensity of the second beam which can be detected with special low-noise techniques (Eesley, 1979, 1981; Owyoung and Rahn, 1979). A more popular technique is called coherent anti-Stokes Raman spectroscopy (CARS). (This technique is thoroughly discussed by Hall and Eckbreth in a separate article in this volume, thus only a brief treatment will be given here.) In a typical CARS experiment, three intense laser beams at frequencies ν_1, ν_1, and ν_2 are crossed in a flame, and the nonlinear interaction between the light and the molecules produces a signal beam at $\nu_3 = 2\nu_1 - \nu_2$. Whereas the general theory is complicated, the signal intensity I_3 is given approximately by

$$I_3 \sim I_1^2 I_2 |b + n_i \sigma_i g(\nu_1 - \nu_2, \nu_i)|^2 \qquad (9)$$

Here σ_i is the Raman cross section for a transition with frequency ν_i, and n_i is the difference in number densities of the upper and lower levels of the Raman transition. There is a small constant background term b, and g is the collision-broadened Raman lineshape function. (The function g has both real and imaginary parts.)

Equation (9) shows many of the important features of CARS. The signal depends on the product of three intensities, thus large signals require high-peak-power, pulsed lasers. For large n_i, the signal is proportional to n_i^2 and can be quite large. The CARS signal from nitrogen in room air can be easily seen. In addition to strong signals, CARS has several other advantages over spontaneous Raman scattering (SRS). Because the CARS signal is a laser-like beam, spatial filtering can be used to discriminate against incandescent and chemiluminescent flame emission. This is particularly important in sooting flames where SRS is very difficult because of the bright soot incandescence. The small solid angle required by the CARS signal also makes optical access to practical combustors less difficult. On the other hand, the high peak intensities used for CARS can make window damage a significant problem. The large signals for CARS measurements of major flame species can be used for accurate thermometry (e.g., Eckbreth, 1981a).

Because of the n_i^2 signal dependence, the CARS signal decreases rapidly as the number density decreases. At high temperatures, many rotational and vibrational states are occupied and this also reduces n_i. For example, the peak CARS signal from 5% CO in a flame at 2000 K is five orders of magnitude less than the peak signal from pure CO at room temperature. At the 5% level in a flame, the signal from the CO is comparable to the background signal [from b in Eq. (9)]. Noise on this background due to laser-intensity fluctuations is an important limitation on CARS sensitivity. The background signal can be eliminated using polarizers, which also reduce the resonant signal by about a factor of 10 (Rahn et al., 1979; Eckbreth and Hall, 1981). These and other experiments indicate that practical CARS measurements in flames will be limited to species with concentrations greater than 0.1–1%. Improvements in lasers and experimental techniques may lower this limit slightly, but some of the limitations may be fundamental. For example, the laser intensities are limited by laser-induced breakdown of the gas and by saturation of the CARS signal due to stimulated Raman gain processes which deplete the population in the lower level of the transition. Because SRS does not suffer from these same limitations, there will be cases where it is more sensitive than CARS. Useful guidelines for assessing the sensitivities of the various Raman techniques are given by Rahn et al. (1981). Eesley (1979, 1981) analyzes in detail the signal-to-noise ratio for these techniques.

One additional advantage of CARS is that the spectral resolution can be several orders of magnitude better than for SRS. The resolution for SRS is limited by the spectrometer, whereas for CARS it can be determined by the laser linewidths. In cases where spectra of different species overlap, this can be a significant advantage (Teets and Bechtel, 1981b; Rahn et al., 1979).

The analysis of CARS spectra is considerably more complicated than that for SRS. In SRS, the observed spectrum is a convolution of the lineshapes of the laser and spectrometer with the Raman lineshape. Because the spectrometer linewidth is typically much larger than the Raman linewidth, the signal is essentially independent of the Raman lineshape. For CARS, the signal depends nonlinearly on the Raman lineshape, and even with broadband lasers the signal still depends on this lineshape (for details, see the accompanying article by Hall and Eckbreth in this volume). The dependence of these Raman linewidths on temperature and composition is not well known. For large molecules such as hexane, detailed modeling of the spectra is very difficult. With SRS, the number density from the area under the spectrum can be approximately determined using an averaged, temperature-dependent cross section. This is not possible for CARS because of the nonlinear dependence on the number density. Because of the experimental and theoretical complexity of CARS, spontaneous Raman scattering should be chosen whenever possible. However, there will be many cases where CARS will allow measurements which cannot be made in other ways.

VIII. Other Spectroscopies

As combustion diagnostics, other spectroscopies have been developed either because of the directness of the measurement (e.g., Rayleigh-scattering number density measurements) or for greater sensitivity for specific species. Many of the techniques for achieving the latter will prove increasingly important for detecting reactive intermediates which are important in pollutant formation and in gross flame properties such as flame speeds.

A. RAYLEIGH SCATTERING

Rayleigh scattering is the elastic scattering of light by a gas or by particles much smaller than the wavelength of the light. Only gases are dealt with here; particle scattering is considered in Section X.

The single Rayleigh line is the combined scattering of all species in the sample. The total scattered power I is given by

$$I = \sum I_0 \sigma_i x_i n \delta s = I_0 \sigma n \delta s \qquad (10)$$

where σ_i and x_i are the cross section and mole fraction, respectively, for species i, n is the total number density, δs is the scattering length, and σ is the average cross section. The cross sections depend upon molecular polarizabilities and can be calculated directly from refractive index data which are known with high precision. These cross sections are three orders of magnitude stronger than those of Raman scattering and allow a wider variety of conditions to be studied. For a spectrally narrow light source, the scattered light will be Doppler broadened, and, perhaps, Doppler shifted if the gas moves perpendicular to the so-called "scattering vector," which is the difference of the incident momentum vector and the momentum vector from the scattering point to the detector. Typically, only the light scattered at right angles is measured.

Rayleigh scattering has been used in combustion applications to measure number density or temperature. The scattering power is primarily dependent on the gas number density which is determined by temperature via the equation of state when there are no pressure gradients. In many systems, the measurement of number density or temperature is only semi-quantitative because the average cross section σ depends upon gas composition, which of course changes through a flame sheet. For a stoichiometric CH_4–air flame, σ decreases 3% from the unburned to the burned side. These effects can be minimized by using high concentrations of inert diluents or by selecting the gas composition to give a constant cross section, e.g., a 22% Ar in H_2–air diffusion flame (Dibble et al., 1982). Even when the cross section is not constant, Rayleigh scattering can provide significant insight into combustion behavior, especially flame shape. Difficulties have also been encountered with Rayleigh scattering from entrained dust particles. Usually these scattering spikes can be easily identified and discarded.

Despite the ease of measurement, there have been few applications of Rayleigh scattering to simple laminar flows. Pitz et al. (1976) spectrally analyzed the Rayleigh signals in a premixed hydrogen–air flame. Densities were determined from the intensity, and temperatures from the Doppler broadening. Spectrally shifted components (Brillouin scattering) were observed below the flame. Smith (1978) measured one-dimensional temperature profiles in a hydrogen–air diffusion flame using an independent flame calculation to estimate the average Rayleigh cross section. Muller-Dethlefs and Weinberg (1979) have measured small temperature changes as a function of flow rates above a porous plug burner. Schefer et al. (1980) have measured temperature profiles in H_2–air mixtures near a catalytic flat plate.

Many point measurements have been made in turbulent flames both in premixed systems (Bill et al., 1981, 1982; Gouldin and Dandekar, 1982; Namazian et al., 1983) and in nonpremixed systems (Dibble and Hollenbach, 1981). These measurements yield probability distribution functions for the fluctuating density or temperature. Simultaneous measurements of velocity and Rayleigh scattering in nonpremixed turbulent flames have rigorously tested $\overline{\rho'v'}$ correlations which are important for closure models of the conservation equations (Driscoll et al., 1983). These experiments demonstrate that transverse turbulent mass fluxes are diffusional whereas axial fluxes are not. These results suggest that second-order closure theories will be necessary to describe the nonlocal nature of turbulence. Point measurements have also been made in the preflame zone of a turbulent diffusion flame (Graham et al., 1974) and a turbulent heated boundary layer (Cheng et al., 1981).

Rajan et al. (1982) and Gouldin and Dandekar (1982) have also measured instantaneous, one-dimensional temperature profiles in turbulent premixed flames by imaging the Rayleigh scattering along a short length of the laser onto a multielement detector array. They observed instantaneous flame widths much smaller than the mean (time-averaged) flame widths. These results indicate that the average flame widths in these flames are dominated by flame flutter rather than microscale structure. The instantaneous flame widths were very close to laminar flame widths (Bechtel *et al.*, 1981). They also observed some examples of flame folding and breaking (see Fig. 16) (see also the results of Smith, 1978).

Because of its relatively high data rates, Rayleigh scattering has already demonstrated great utility. There will certainly be many more practical

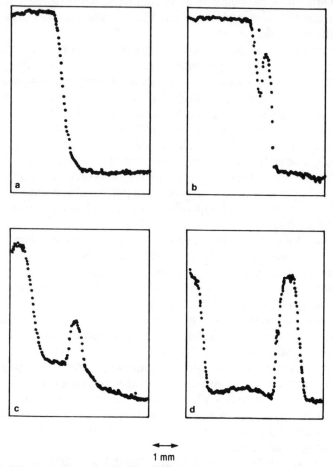

1 mm

FIG. 16.   Rayleigh scattering intensity profiles along a pulsed laser beam and in a turbulent premixed flame. Profile (a) is illustrative of a simple flame front, (b) and (c) illustrate hot and cold gas mixing near the flame front, and (d) illustrates flame breakup or air entrainment. [From Rajan *et al.* (1982) with permission.]

applications to the measurement of temperature profiles, including two-dimensional distributions.

B. MULTIPHOTON FLUORESCENCE SPECTROSCOPY

Multiphoton fluorescence uses two or more photons in order to excite a resonant state which in turn fluoresces. Two photons allow excitation of species where either (1) no suitable one-photon laser sources exist or (2) the single photon is absorbed by background gases. The powers from currently available pulsed, tunable lasers ($>10^4$ W) easily allow excitation of two-photon transitions. The fluorescence photons typically have energies different from other possibly interfering photons. These properties allow sensitivities comparable to single-photon fluorescence spectroscopy as discussed earlier. Despite these advantages, the process is nonlinear in the laser intensities such that the relationship between signal and laser power depends upon beam focusing. Also, for most species, the two-photon cross sections are in doubt, and it is necessary to have a reliable calibration source of the species.

Multiphoton fluorescence and ionization spectroscopies allow detection of atoms much more sensitively than other techniques. Using two-photon fluorescence, Bischel et al. (1981, 1982) have measured O and N atoms at densities of 10^{14} cm^{-3} in flowing discharges. A tunable, pulsed dye laser which was frequency doubled and Raman shifted (226 nm) excited the 2p $^3P_2 \rightarrow$ 3p ^3P two-photon transition, and 3p $^3P \rightarrow$ 3s $^3S_1$ fluorescence at 845 nm was observed. Aldén et al. (1982c) subsequently observed O atoms in a C_2H_2–O_2 flame using the same transitions. In both cases, the signals were two orders of magnitude smaller than expected using Pindzola's (1978) calculated two-photon cross section. Goldsmith (1982b) has detected H and O atoms by multiphoton ionization in a flame as discussed in the following.

C. OPTOGALVANIC SPECTROSCOPY

The optogalvanic effect has been defined as any change in the electrical properties of a plasma due to photoexcitation (Goldsmith and Lawler, 1981). A species is laser excited by one or more photons to a resonant state close to the ionization limit. The excited species subsequently undergoes either collisional ionization or photoionization. The term "optogalvanic" is sometimes restricted to cases where collisional ionization predominates, and "resonant multiphoton ionization" is used where photoionization predominates. Flames are weak plasmas with charge densities on the order of 10^{10}–10^{13} cm^{-3} and are good media for optogalvanic spectroscopy.

The effect can be sensitively measured by a change in conductivity between two biased electrodes (see Fig. 17). Essentially all of the generated charges can be collected. Many metal species which were doped into flames have been detected (Turk et al., 1979; Schenk and Hastie, 1981). Goldsmith (1982a) has detected the nascent hydrogen atoms in an atmospheric pressure flame. By frequency multiplying and mixing, a tunable beam of 224-nm photons and a nontunable beam of 266-nm photons were generated. These can excite the $n = 1 \to 2$ two-photon transitions which occur at 122 nm. The detection limit corresponds to densities as low as 10^{13} cm^{-3}. The relationship between signal and concentration is very complicated, thus a reliable calibration source is needed.

Goldsmith also observed interfering signals from NO which arise from two-photon (1 + 1) resonantly enhanced ionization. Mallard et al. (1982) studied these transitions in an H_2–air–N_2O flame and estimated a detection limit for NO of 10^{13} cm^{-3}. Rockney et al. (1982) observed similar amounts of NO by four-photon (2 + 2) resonantly enhanced ionization in a CH_4–air flame.

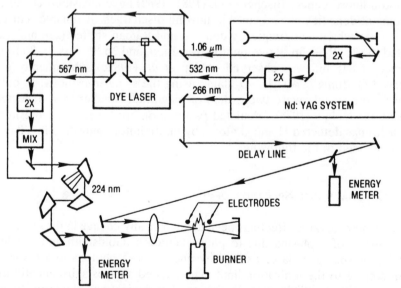

FIG. 17. Optogalvanic spectroscopy apparatus illustrating nonlinear optical mixing to generate 266-nm and tunable 224-nm wavelengths. These wavelengths can excite the $n = 1$ to $n = 2$ transition in atomic hydrogen. [From Goldsmith (1982a) with permission.]

D. Photoacoustic and Photothermal Deflection Spectroscopy

Photoacoustic and photothermal deflection spectroscopy are both based on detecting the energy which a sample has absorbed from an optical probe beam. In a sense, these spectroscopies are the physical complement of absorption spectroscopy in which the attenuation of the probe beam is measured. Because photoacoustic and photothermal techniques depend on the deposition of energy, very small absorptions can be measured and signals improve with more powerful lasers.

Photoacoustic spectroscopy detects the acoustic pressure wave which is generated when the absorbed energy is converted into translational energy. The pressure wave is sensitively measured with a microphone. Detectivities of 10^{-12} have been obtained for static gases in specially enhanced cells (Patel and Tam, 1981). Transient, low exhaust soot concentrations have been measured in such systems using both visible and infrared lasers (Roessler and Faxvog, 1979; Roessler, 1982; Faxvog and Roessler, 1979).

In flames, Allen et al. (1977) first observed signals from added Na. The microphone was located several centimeters outside the flame. NO_2 has similarly been measured in a CH_4–air flame (Tennal et al., 1982). Although the technique is very sensitive, it has limited utility in flames because the signal is generated along the entire length of the optical beam, and the pressure wave must traverse an acoustically long, nonuniform path to reach the detector. The microphone signals are very complicated because of multiple, acoustic reflections.

Photothermal deflection spectroscopy (Boccara et al., 1980) offers solutions to these problems by detecting density changes close to the pump laser beam. As seen in Fig. 18, a second probe laser is introduced near the pump laser and at a small crossing angle. The thermal wave with its slightly smaller refractive index propagates from the pump beam through the probe beam causing the probe to be deflected. Using a position-sensitive detector which is not sensitive to intensity fluctuations, very small deflections of the probe laser can be measured (Jackson et al., 1981). There will be severe difficulties for this method in combustion environments with refractive index fluctuations, as in turbulent flames. This technique is physically related to earlier thermal lensing experiments (Long et al., 1976; Swofford and Morrell, 1978).

Using thermal deflection spectroscopy, Rose et al. (1982) have detected NO_2 in a methylamine-doped flame. These dye-laser-excited signals could not be quantitatively analyzed for concentrations because the visible spectrum of NO_2 is very complicated at high temperatures. Zapka et al. (1982) have measured temperatures from the time delay between the

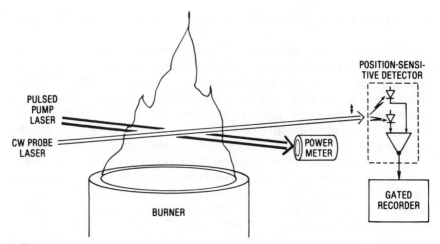

FIG. 18. Photothermal deflection experiment showing a crossed beam arrangement which gives quasi-point spatial resolution.

pump laser pulse, which induced a spark, and the probe deflection (speed of sound measurement).

E. Laser-Induced Breakdown Spectroscopy

Sparks can be generated by tightly focusing a high-power pulsed laser. R. W. Schmieder (unpublished) has found that the emission spectra from these sparks are primarily atomic and ionic lines. He has shown that with the proper selection of lines, it is possible to perform elemental analyses for H, C, N, and O. The technique has also been used to observe qualitatively the mixing behavior in laminar and turbulent CH_4–air systems (Schmieder and Kerstein, 1980).

IX. Optical Refraction, Holography, and Tomography

In addition to quasi-point spectroscopic measurements in flames, other methods based on index of refraction gradients have been used. These methods are often directed at obtaining a whole-field image of a combustion process. Many of these imaging techniques predate lasers, and the techniques were developed for gas dynamic research. Examples include shadow and schlieren photography, interferometry, and deflection-mapping experiments. In these procedures, a laser may or may not be used as a light source for the experiment. The use of optical techniques in conjunction with a high-speed camera or a pulsed light source to capture the time dependence of a dynamic event such as an explosion is of interest in many of these experiments. The optical techniques are based on the changing index of refraction associated with combustion or other gas dynamic events. The index of refraction is determined by both temperature and composition. This dependence is known as the Lorenz–Lorentz relation. As light passes through a region of changing index of refraction, both the direction and phase of the light are changed.

The shadow and schlieren methods have been used for many years to better understand sparks, explosions, air flows, shock waves, and sound waves. These methods have helped in visualizing turbulence effects, convection at hot surfaces, air flow around projectiles, jets, diffusion boundaries, and flame propagation. Extensive reviews of shadow, schlieren, and deflection-mapping experiments have been presented by Beams (1954), Weinberg (1960, 1963), and Davies (1981). Many arrangements of mirrors, lenses, and stops are possible for both shadow and schlieren photography. These have been reviewed by Weinberg (1963), and only a few of the most common arrangements will be mentioned here.

The advantage of shadow and schlieren photography compared with simply photographing the flame's luminosity is that the flame front (where the fuel oxidation occurs) is often much more easily visualized by either shadow or schlieren method. For either the shadow or schlieren method, a continuous source of illumination may be used, e.g., a lamp or a CW laser in conjunction with a high-speed camera to record the gas dynamic event; a pulsed light source appropriately timed to record on film a single record of an event may also be used. Because lasers have a large brightness, they are now often used for either shadow or schlieren photography. Lasers can be made with pulse durations much shorter than incoherent stroboscopic light sources; thus they can produce "stop-action" photographs of supersonic events. The disadvantage of lasers is the coherent noise (speckle) that can be produced in the image by imperfections in the optical elements of the imaging system.

Density gradients cause light to deflect toward regions of higher density and a larger index of refraction. In a shadow system, these gradients cause a change in the intensity of light on a viewing screen or film. Because a uniform deflection caused by a constant index gradient would displace all rays uniformly and would not redistribute the light intensity, the formation of a shadow depends on gradients in deflection. Thus it can be said that the shadowgraph responds to the second derivative of the index of refraction with respect to distance. Because of this, the shadowgraph is sometimes well suited for visualization of shock waves or other structures that have very rapid changes in index gradients. Examples of shadowgraph systems are schematically illustrated in Fig. 19.

There have been numerous applications of lasers as light sources for shadowgraph photography. For example, Johnston et al. (1979) have used an argon-ion laser and a high-speed camera for shadowgraph recording of density fluctuations and fuel injection in a stratified-charge engine. Additional experiments (Johnston, 1979) on flow visualization of both liquid and gas-phase fuel injection show the differences in the fuel mixing process with different types of injectors and injection pressures. Other laser shadowgraph experiments by Johnston (1980) show more rapid combustion in a direct-injection stratified-charge engine when the mixture was ignited early relative to fuel injection, rather than later when mixing was more complete.

Shadowgraph experiments have been used to visualize the effect of intake air motion and swirl on combustion in a constant-volume combustion bomb (Dyer, 1979), in a direct-injection stratified-charge engine (Witze, 1980b), and in a homogeneous-charge spark-ignition engine (Witze and Vilchis, 1981). Shadow methods have been used to visualize flame propagation changes when spark plug location is varied in a homogeneous-charge spark-ignition engine (Witze, 1982). Laser shadowgraphs

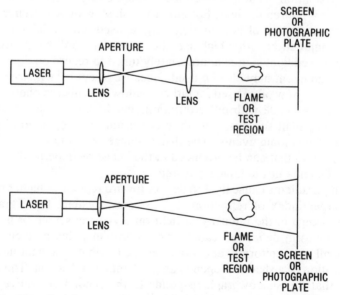

FIG. 19. Examples of two possible optical configurations for laser shadow methods of flow visualization.

can display the degree of flame wrinkling as engine speed is changed (Smith, 1982a), and the results for a homogeneous-charge spark-ignition engine show an increase in flame wrinkling as engine speed is increased.

Schlieren systems have a knife-edged, graded-filter, slit, circular opening or an obstacle or other type of stop to block either the deflected or undeflected rays from the test region. Because the schlieren system depends on deflection, it can be said to respond to the first derivative of the index of refraction in the test region. For schlieren photography there are many possible choices for optical arrangements that are based on either mirrors, lenses, or combinations of these. Lens systems can provide a large field of view but have the disadvantage of being more expensive than similarly sized mirror systems. Typical schlieren systems are shown in Fig. 20. Here the laser is used as a light source, although, as mentioned previously, incoherent sources may also be used.

There are several features of schlieren systems that distinguish them from shadow systems. Shadow systems do not require the use of some type of stop whereas for schlieren systems this is necessary. Placing a circular opening at the focus of the focusing lens or schlieren lens deflects rays such that they do not pass through the opening and thus do not reach the viewing screen or film. The regions of large index gradient (such as a

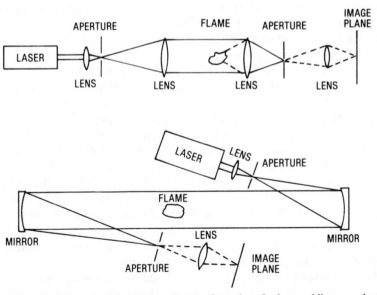

FIG. 20. Examples of two possible optical configurations for laser schlieren methods of flow visualization.

flame front) are seen as dark regions on the bright background of the viewing screen. If a circular obstacle is at the focus of the schlieren lens, then only the deflected rays are seen, bright regions on a dark background of the viewing screen. Because the focusing lens or schlieren lens images the test section onto the viewing screen (the two regions are optically conjugate), there is no shadow superimposed on the schlieren image.

There have been numerous examples of schlieren systems that use lasers as light sources and which have been employed in combustion experiments. One example is the study of different air flow patterns that are obtained by changing the intake valve geometry on a rapid-compression machine (Matekunas, 1979). This type of combustion device simulates the combustion in a spark-ignition engine but allows independent control of operating variables that are not independent in a spark-ignition engine. Experiments by Groff (1982) have employed schlieren photography to show the development of instabilities in a flame front of a confined, near-spherical flame and to measure the burning velocity. In an internal combustion engine the influence of turbulence on flame structure was observed as well as the flame geometry at times just after spark ignition (Smith, 1982b). The data show that the turbulent Reynolds number decreases with increase in engine speed because the Taylor microscale (the size of a turbulent eddy) decreases faster than the turbulence intensity increases. In still other applications, schlieren photography has been used to visualize the trajectory of a fuel spray in a direct-injection stratified-charge engine (Sinnamon et al., 1980).

Both laser shadow and schlieren photography have been used to measure both flame propagation and development using a plasma jet igniter that had different gases within the plasma jet cavity (Orrin et al., 1981). The results showed that either hydrogen- or methane-filled cavities can produce faster flame growth than cavities filled with argon, and much faster flame growth than conventional spark ignition.

Laser schlieren photography has also been used to visualize both intake air motion and fuel–air interactions in a simulation of diesel combustion in a rapid-compression machine (Meintjes and Alkidas, 1982; Peterson and Alkidas, 1982). The results showed that the fuel injected into the chamber burns as both evaporated fuel and as droplet combustion. The visualization also allows the determination of the influence of fuel-injector orientation on the combustion process. Another application of laser schlieren photography is shock velocity measurements in a combustible mixture (Oppenheim et al., 1975; Meyer and Oppenheim, 1971).

Most schlieren photography has been used for the qualitative visualization of flow patterns in combustion or other aerodynamic phenomena. However, it is possible to obtain quantitative information about the index of refraction from either quasi-one-dimensional or axial symmetric index distributions (Kogelschatz and Schneider, 1972).

An alternative method to the schlieren technique of obtaining index of refraction information is based on Moiré deflectometry. A typical system is shown in Fig. 21. This system consists of a collimated laser beam and two Ronchi rulings (these devices are transmitting gratings with alternating opaque and transmitting stripes) with parallel planes separated by a distance L and with rulings which are rotated with respect to one another by an angle θ. If a ray of light is deflected by an index gradient before it reaches the first Ronchi ruling, it will disturb the Moiré fringe pattern that is produced after the second Ronchi ruling. The fringe pattern may be directly related to the index distribution by an integral transform inversion for flames that have cylindrical symmetry. This technique has been used to determine temperature distributions in flames from the measured index of refraction distribution (Kafri, 1980; Keren et al., 1981; Bar-Ziv et al., 1983).

The shadow, schlieren, and Moiré methods utilize rate deflections due to index of refraction gradients. An alternate approach of recovering index of refraction information is based on interferometry, which depends on wavefront phase changes. A Mach–Zehnder interferometer is shown in Fig. 22. The interference between the wavefronts from the different paths of the interferometer can occur if the optical path is changed in the test beam. Such a change, for example, could occur because of a temperature change in a part of the test region. Each of the fringes produced represents a region of constant phase, and the phase shift is proportional to $2\pi[\int_{\text{path}} (n - n_0)ds]/\lambda_0$, where n is the index of refraction in the test region, n_0 the index of refraction in the reference beam, and λ_0 is the reference beam wavelength. If the two recombined beams are parallel, the result is an infinite fringe interferogram; if the beams intersect at some small angle, a finite fringe interferogram is obtained.

Because the phase shift is an integral effect of the index of refraction along the entire optical path, the index of refraction may be determined

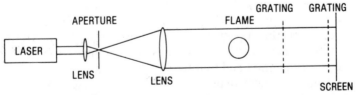

FIG. 21. Moiré deflectometry system for measuring beam deflections which may be used to determine the index of refraction for a flame with axial symmetry.

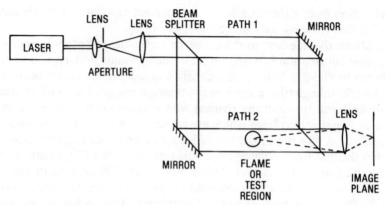

FIG. 22.   Mach–Zehnder interferometer that may be used to visualize flames. The temperature change in the flame produces an index of refraction change which leads to fringe shifts.

from the fringe distribution only if assumptions are made about the symmetry of the index distribution. The problem may be solved by the methods of inverse-scattering theory (Baltes, 1978). Special cases may be solved if there is radial symmetry by noting that the optical pathlength of a nonrefracted ray may be related to the index distribution by a form of the Abel integral equation. If this equation is inverted, the index distribution is found. The theory of the Abel integral equation applied to a flame or other axisymmetric index of refraction distribution neglects beam-bending effects (refraction) by the flame. The effects may be accounted for if the object is imaged onto the film used in the recording of the interferogram (Vest, 1974). The image fringe pattern may be related to the index of refraction by an inversion scheme (Horglotz–Weichert method) that was originally developed for seismology data.

An example of the use of a Mach–Zehnder interferometer for combustion research is the investigation of electrical sparks and flame growth in combustible and noncombustible mixtures (Herden and Maly, 1977). The results show that the flame radius or plasma radius may be obtained as a function of time when a pulsed N_2 laser is used as the light source for the interferometer. A time resolution of <20 nsec was reported. Additional experiments (Maly and Vogel, 1979) have shown that the interferograms may be used to measure the spatial profile of temperature at different times after spark ignition and that the breakdown phase of ignition gives more rapid flame growth than either arc or glow discharges that have equal energy. Other examples of laser interferometry include flow visualization near an ignition source (Lee et al., 1975), temperature measure-

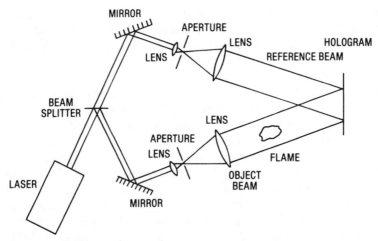

FIG. 23.   Schematic diagram of a holographic interferometer that may be used to record an interferogram of a flame.

ments during flame spread over polymethylmethacrylate surfaces (Fernández-Pello and Williams, 1975), temperature measurements during flame spread over polyethylene, polystyrene, and cellulose surfaces (Fernández-Pello and Williams, 1976), temperature measurements and fuel concentration measurements by two-color interferometry for laser ignition of polymethylmethacrylate (Mutoh *et al.*, 1979), and the visualization of laser ignition of titanium (Clark *et al.*, 1975). Other applications of laser interferometry include temperature measurements in conical-shaped diffusion flames (South and Hayward, 1976) and unburned-gas density measurements in a combustion bomb (Garforth, 1976).

Holography has also been used in combustion research, and has been applied to the determination of particle size (Thompson *et al.*, 1967; Thompson, 1974; Trolinger and Heap, 1979), to flow visualization (Trolinger, 1974), and to the recording of interferograms holographically. One of the advantages of holographic interferometry is that it is not restricted to symmetric temperature or index of refraction fields. A typical holographic interferometer experiment is schematically illustrated in Fig. 23. Two sequential exposures are used to form the holographic interferogram. The film plate is first exposed to the reference beam and to a second beam that has traversed the test region without a flame present. The flame is then ignited and a second exposure (on the same plate or film) is made. The use of diffuse illumination (by placing a diffuser before the test section) allows the recording of fringe patterns corresponding to many different viewing directions through the flame. This is equivalent to many different Mach–Zehnder interferograms, each recording a different orientation of the flame. The disadvantage of double-pulse holographic interferometry is that it requires that the optics positions must be stable to a fraction of a wavelength, otherwise the fringe pattern will be disturbed (Trolinger, 1975).

In addition to double-exposure holographic interferometry, real-time holographic interferometry is possible. For this method a hologram of the test region is made and the developed plate is mounted in its original position. If the test and reference beams illuminate the hologram, an interference will be observed between the test beam and a reconstruction of the test object that was recorded on the hologram. If the first exposure is made prior to the presence of a flame, any changes produced by a flame in the test section may be photographed as they occur. The major disadvantage of real-time holography is that the developed hologram must be repositioned to its original position very accurately (Trolinger, 1975).

Holographic interferometry has been used to measure air–fuel ratios in internal combustion engines (Dent, 1980) by the double-exposure method. The data were inverted by assuming axial symmetry for the fuel jet which was injected into a combustion bomb or engine cylinder. In other experiments, Varde (1975) measured flame speeds after spark ignition for different flame positions in a constant-volume combustion bomb. Temperatures near the spark gap were estimated from the holographic interferograms. Temperature has also been measured in premixed laminar flames by double-pulse holographic interferometry (Reuss, 1983). The double-pulse method has an advantage over conventional Mach–Zehnder interferograms because fringes from optical windows or fringes from precombustion gas are cancelled by the double-pulse technique. Temperature data by holographic interferometry were compared to thermocouple measurements in the flame, and the results agreed to within ± 50 K.

The effects of beam refraction on holographic interferometry have been considered in the measurement of temperature in axisymmetric laminar flames (Montgomery and Reuss, 1982). The accuracy of performing only an Abel inversion was numerically tested by ray tracing through an assumed temperature distribution to form a computer-generated holographic interferogram. This interferogram was then inverted by solving the Abel equation to reconstruct the temperature distribution. The reconstructed temperature distribution was compared to the original assumed temperature distribution to determine the error produced by the Abel inversion. Results were obtained both for different flames and for different regions in a given flame. Minimum errors were obtained for temperature if the test section was imaged onto the recording film. Other applications of holographic interferometry to combustion research include the measurement of flame spread along two parallel sheets of paper in air. The holographic interferometer was used to measure the temperature distribution at various times during the flame propagation and for various separations of the paper sheets (Kurosaki *et al.*, 1979).

Similar inversion techniques to convert absorption "line-of-sight" measurements to "point" measurements are also possible. Several absorption measurements can be combined along different paths; if these data can be uniquely inverted, a concentration distribution can be obtained. For example, if the species concentration distribution is axisymmetric, then either an Abel inversion or an "onion-peeling" method (Chen and Goulard, 1976) may be used to infer the radial distribution of species concentration from the absorption measurements. The method has also been used for intracavity absorption and applied to the determination of C_2 concentration in flames (Harris and Weiner, 1981b).

The line-of-sight measurements may be used for nonaxisymmetric concentration distributions (Chen and Goulard, 1976; J. Y .Wang, 1976) by dividing the region to be mapped into zones of uniform concentration. The absorption measurements may be inverted to yield a concentration distribution.

The advances in computerized tomography for X-ray medical diagnostics (Swindell and Barrett, 1977; Barrett and Swindell, 1977) have also been applied to flow diagnostics. For example, multiangular absorption along M equally spaced parallel beams at N equally spaced angles forms an $M \times N$ data set from which it is possible to reconstruct a concentration distribution. This method has been applied to the determination of the concentration of methane in both methane–argon and methane–air jets (Santoro *et al.*, 1980, 1981; Emmerman *et al.*, 1980). Temperature and concentration simulations have been performed on the reconstruction of data from an assumed OH concentration distribution; laser experiments are in progress (Ray and Semerjian, 1982).

Another application of laser absorption is to obtain information about time-averaged vapor concentrations in sprays. The effects of droplets on the vapor-probing laser may be calculated from Mie theory if the size distribution of droplets is known. If the droplet attenuation is factored out of the total attenuation, any remaining attenuation is due to vapor-phase absorption. A method for determining the effect of droplets on total extinction has been proposed by Chraplyvy (1981). This method employs a two-wavelength absorption/scattering technique. The droplet extinction is determined by a laser wavelength that is not absorbed by the vapor. This method has been applied to the determination of vapor concentration in *n*-heptane fuel sprays (Tishkoff *et al.*, 1980; Chraplyvy, 1981). These experiments used an axisymmetric spray and an "onion-peeling" inversion procedure. Other experiments have also examined axisymmetric fuel sprays, but these were restricted to droplet extinction and did not attempt to explicitly measure vapor concentration (Yule *et al.*, 1981, 1982).

X. Particle Size Measurements

One of the most important aspects of the application of lasers to combustion research is the determination of particle size. This includes not only particles formed by combustion, e.g., soot, but also fuel droplets in spray combustion. Drop size distributions in spray combustion affects radiant heat transfer, flame length, smoke formation, and oxides of nitrogen formation (McCreath and Beer, 1976). Combustion in diesel engines and gas turbine engines is also greatly influenced by fuel droplet size distributions and the interaction between drops and evaporated fuel.

The measurement of droplet size has been accomplished by both optical and nonoptical techniques. Nonoptical methods include droplet capture, freezing, and cascade impaction. These methods, however, have great limitations because they disturb the flow field as well as the droplet and vapor concentrations that are being measured. These methods in general lack the capability of time resolution and have uncertainties about both agglomeration effects and the measurement of very small particle sizes. In contrast, optical methods can yield time-resolved *in situ* measurements. Optical methods, however, require good optical access to the region of interest, and no single optical technique is adequate for measuring particle sizes that span the region from much smaller than the optical wavelength to much larger than the optical wavelength.

Although liquid drops are usually spherical in shape due to surface tension forces, solid particles can be very irregular in shape if there has been particle agglomeration. For this reason, optical scattering methods of particle size measurement must be applied to solid particles with caution because these methods usually assume (1) spherical shape, (2) a known index of refraction, (3) homogeneous particle material, and (4) a known size distribution if the particles are not measured one at a time. Details of these limitations will be discussed later in this section; the following discussion summarizes some of the laser methods that have been applied to the determination of particle sizes. Additional details may be obtained from extensive reviews by Stevenson (1978), Azzopardi (1979), Penner and Chang (1981), and Kerker (1969).

One of the oldest methods of determining the properties of droplets is by flash photography. Early experiments (York and Stubbs, 1952) did not use lasers, but lasers can replace electric-spark optical sources. In effect, a shadow of a drop is measured at high magnification. If the drops are moving rapidly, a short-duration pulse is very desirable for minimizing blurring effects. The laser can provide such short pulses and give a uniform background lighting as well. With laser illumination, particle sizes as small as 2 μm in diameter may be measured (Chigier, 1977).

Both single-flash and double-flash photographic methods can be used. The double-flash methods allow not only the determination of the size distribution, but also the particle's velocity component in the plane of the exposed photograph. The obvious attractions of photography are that it can detect and analyze both spherical and nonspherical particles, it is not sensitive to the optical properties of the particle, and it is not very sensitive to multiple scattering effects. Photography's big disadvantages are that it requires large particles and large amounts of data processing. Part of the data-processing problem has now been simplified by the development of a TV analyzer (Simmons and Lapera, 1969) whose output is recorded on video tape. Numerous applications of photography to particle size analysis may be found in the literature (DeCorso, 1960; Briffa and Dombrowski, 1966; Chigier *et al.*, 1973, 1974; Mullinger and Chigier, 1974; Chigier and McCreath, 1974; McCreath *et al.*, 1972; Mellor *et al.*, 1971; Chigier, 1974).

An alternative imaging method to photography is laser holography. In addition to recording the light intensity as in photography, holography also records the phase of the light signal that is scattered by the particles. There are several advantages to holography. A greater depth of field and a more uniform magnification of the particles can be obtained compared to conventional photography. The result is that a larger volume of a spray or group of particles may be probed. Droplets and particles with sizes of 5–500 μm have been measured with holographic methods (Thompson, 1974; Thompson *et al.*, 1967; Trolinger and Heap, 1979; Hickling, 1969; Belz and Dougherty, 1972; Fourney *et al.*, 1969; Briones and Wuerker, 1977; Seeker *et al.*, 1981; Samuelsen *et al.*, 1981; Polymeropoulos and Sernas, 1977; Reynolds, 1976). Holography as well as photography requires extensive data analysis, and multiple scattering effects in dense particle fields degrade the image.

Laser scattering methods are convenient ways of obtaining information about particle sizes. If a parallel beam of light is scattered by a particle or group of particles, the intensity of the scattered light depends on the shape and size of the particle or group of particles, the number density, and the real and imaginary parts of the complex index of refraction. The relation of scattered intensity to incident intensity also depends on the scattering angle, the wavelength of the incident light, and the polarization of both the scattered and the incident radiation. The inversion of the light-scattering data to obtain information on particle size thus requires knowledge or assumptions about the shape of the particles, the homogeneity of the particles, and the complex index of refraction. Given these caveats, particle size measurements may be obtained by several different types of light-scattering methods. These include scattered intensity versus scattering angle for a fixed wavelength, or scattered intensity versus incident wavelength for a fixed scattering angle. Other methods include the ratio of polarization in the scattering plane to the polarization perpendicular to the scattering plane as a function of either scattering angle or incident wavelength. If the particles are spheres, a mathematical theory of the electromagnetic scattering of light (Mie theory) is used to find the particle size from the light-scattering data. The Mie theory has been extended to nonspherical shapes such as cylinders and ellipsoids and to inhomogeneous particles as well (Jones, 1979; Druger et al., 1979; Kerker, 1969; Purcell and Pennypacker, 1973).

One method of determining particle sizes of spheres of a known index of refraction is based on the ratio of light intensities at two different scattering angles near the forward direction. This method was proposed by Hodkinson (1966) and an instrument based on this technique was developed by Gravett (1973). This method is useful only over a limited size range, i.e., $1 < \pi d/\lambda < 20$, but the ratio of scattered intensities is relatively insensitive to the particle's index of refraction (Hodkinson, 1966). Here d is the diameter of the sphere and λ is the wavelength. Because this method is based on the ratio of intensities from a single particle, the distribution of sizes may be found after many samples, and it does not matter where the particle crosses the laser beam. A disadvantage of this method is that the ratio of scattered intensities at two forward angles may exhibit secondary maxima as a function of sphere diameter; consequently, a single ratio measurement can result in large errors for a broad size distribution. For this reason this method has been extended to several scattering-angle ratios by Hirleman and Wittig (1977) which allow for determination of when large particles are giving misleading results. Hirleman and Wittig (1977) also note that for nonspherical particles, the properties of the forward-scattering lobe are not sensitive to the particle shape, and the scattering is representative of a projected area onto the forward direction. Additional details of nonspherical effects on forward scattering are given by Holland and Gagne (1970), Hirleman (1980), and Latimer et al. (1978).

Another type of particle sizing technique also counts and sizes individual particles one at a time, but this method measures the total scattered intensity from a focused Gaussian laser beam at a small scattering angle

(Holve and Self, 1979a,b). Because this method uses near-forward scattering, it also is relatively insensitive to both the index of refraction and particle shape. This technique requires that the system response be calibrated with particles of known size and is sensitive to alignment errors in the optics that image the focused laser beam waist onto the detector. In addition, the intensity of the scattered light depends not only on the particle size and the complex index of refraction, but also on the trajectory of the particle through the laser beam waist and the intensity distribution of the laser at its waist. Inversion algorithms which account for different trajectories are known. These allow particle size distributions to be determined from the intensity of the scattered light (Holve et al., 1981; Holve, 1980, 1982a; Holve and Davis, 1982; Holve and Self, 1979c).

Another method of droplet sizing in the range 2–500 μm is based on Fraunhofer diffraction. Hodkinson and Greenleaves (1963) demonstrated that for drops with a size parameter $\pi d/\lambda > 3(n - 1)$, the main contribution to far-field scattering in the forward direction is Fraunhofer diffraction. Here d is the drop diameter, λ is the wavelength, and n is the index of refraction. A parallel beam of monochromatic light interacts with a particle or group of particles, and the forward-scattered diffraction pattern is collected and transformed to the far field by a lens. This method measures the properties of all particles in the beam. Dobbins et al. (1963) applied this method to measure the volume-to-surface ratio of a spray (Sauter mean diameter) from the Fraunhofer diffraction intensity profile. Roberts and Webb (1964) determined the sensitivity of this technique to the functional form of the particle size distribution and found that it was applicable to nearly any unimodal, polydisperse distribution. Sauter mean diameters have been determined by this method for numerous sprays (Rao and Lefebvre, 1975; Rizkalla and Lefebvre, 1975; Lorenzetto and Lefebvre, 1977; Dieck and Roberts, 1970).

This method has been extended by Swithenbank et al. (1977) to measure not just the Sauter mean diameter, but also the size distribution by collecting light in several annular rings in the focal plane of the collecting lens. A commercial instrument to evaluate drop size based on this method has been produced by Malvern Instruments, Ltd. The big advantage of laser diffraction methods over imaging methods is that an on-line minicomputer may derive the drop distribution in times ranging from a few minutes to a few seconds. There is no limitation on the velocity of the particles, but drop velocities cannot be determined by this method. Also, multiple scattering effects in a very dense spray or refraction effects in hot flows may produce large errors in the interpretation of data.

Various types of detectors have been used to measure the shape and the intensity of the diffraction. These include rotating masks (Cornillault, 1972; Wertheimer and Wilcock, 1976) and optical fiber arrays (McSweeney and Rivers, 1972; Swithenbank et al., 1977).

Additional details of the applications of laser diffraction to particle size analysis are given by Anderson and Beissner (1970), Caroon and Borman (1979), and Hirleman (1982). The droplet sizes measured by laser diffraction are line-of-sight averages. If the spray exhibits nonuniformities in particle sizes, the local size distribution can be determined by deconvolution methods (Hammond, 1981; Yule et al., 1981).

The laser diffraction method of particle sizing has been used by Styles and Chigier (1977) to measure droplet size distributions in air-blast kerosene spray flames. The results were used to show the competition between atomization in regions of the spray near the nozzle (where a decrease in the effective drop size is produced) and fast vaporization of small drops at larger distances from the nozzle (where an increase in the effective drop size is produced).

Laser Doppler velocimetry (LDV) has been frequently used to measure velocities in flowing media; a detailed discussion of this method and its applications to combustion research are given in the next section. Laser Doppler velocimetry may also be applied to particle size analysis (Farmer, 1972, 1974, 1976; Adrian and Orloff, 1977; Jones, 1974; Hong and Jones, 1976; Farmer et al., 1979; Fristrom et al., 1973).

In LDV experiments, the sampled volume is formed by two intersecting laser beams. Because the laser light is coherent, an interference effect causes fringes to be formed in the sample volume. Particles that pass through this region scatter light in proportion to the local light intensity, and the velocity of the particle may be determined by the frequency of modulation. The depth of modulation is a function of the particle size. If the particle diameter is small relative to the fringe spacing, full modulation is observed. If the particle is larger than the fringe spacing, the fringe visibility is reduced. Thus if the fringe spacing may be changed, particle sizes may be derived. This restricts the LDV method to particle sizes with diameters greater than the wavelength of the light. This type of measurement is most useful if the fringe spacing is at least 25% larger than the particle diameter, but not so much larger that the visibility is near unity. Unfortunately, the fringe visibility depends on a number of other factors, including the position of the particle when it crosses the probe volume, the collection optics, the relative intensity of the crossed beams, and particle index of refraction. Because of these complexities, accurate measurement of particle size by fringe visibility has proven to be difficult (Roberds, 1977; Chu and Robinson, 1977; Adrian and Orloff, 1977).

A derivative of the fringe visibility method has been proposed by dos

Santos and Stevenson (1976) in which the droplets fluoresce rather than scatter light. This makes the interpretation of the visibility somewhat simpler because the angular distribution of the fluorescence does not exhibit the strong angular anisotropy of Mie scattering. Limitations of the fluorescence method include the frequency response of the dye fluorescence for small fringe spacing and large flow velocities. This method may also be employed in the back-scatter mode which is desirable if the particles are accessible only through one window.

An alternate procedure to derive particle size is also based on LDV. Because the light-scattering signals are also proportional to the particle size, the measurement of the total light scattered from the crossed beams may be related to the particle size (Yule *et al.*, 1977; Chigier *et al.*, 1979a,b). This type of particle sizing system also has the disadvantage that the signal depends not only on the particle size but also on how the particle crosses the measurement volume. This measurement method is applicable to particles much greater in diameter than the fringe spacing, but because it requires an absolute intensity measurement it is sensitive to laser stability and detector drifts.

For some measurements a combination of several particle sizing methods have been used simultaneously. An example is the experiment of Bachalo (1979) in which the fringe visibility of a laser Doppler interferometer was measured at one laser wavelength and forward-scattering intensity ratios were measured at a second laser wavelength. These two wavelengths were obtained from the blue (0.488-μm) and green (0.5145-μm) lines of an argon-ion laser. The optical system provided focusing of the two laser wavelengths into a common volume. The ratio method permitted sizes of 0.25–3 μm to be measured while the LDV visibility system permitted particles with diameters of 3–25 μm to be measured.

Faxvog has employed a crossed-beam geometry similar to LDV to obtain particle sizes from the measured extinction in the focus of two beams. A coincidence circuit (Faxvog, 1974, 1977) that measured the extinction in both beams ensured that the particles crossed the beams near the beams' waists. The transverse profile of both of the laser beams was a doughnut mode which facilitated the measuring of only particles passing through the center of the focus. This type of instrument has been used for the measurement of particle sizes from diesel engines in the range of 0.05–5 μm.

Light-scattering measurements that are based on the spectral broadening of scattered laser light are nearly independent of the index of refraction and may be used to infer particle sizes for very small particles ($\pi d/\lambda \ll 1$). The laser light that is scattered by the particles is detected at some angle with respect to the incident laser beam. The power spectrum of the scattered light is measured by beating the scattered optical signal with a reference beam or with itself to produce either a heterodyne or homodyne signal in the audio or radio frequency range. Because the particles in the laser beam have a Brownian motion due to their thermal energy, the scattered light is spectrally broadened. The broadening is directly related to the diffusion coefficient of the particles that scatter the light beam. The particle size may be determined from the power spectrum linewidth if expressions such as the Stokes–Einstein relation or Epstein relation are used to relate the particle diameter to the particle diffusion coefficient, gas viscosity, and temperature. Although some of the first examples of diffusion-broadening spectroscopy were applied to particles in liquids (Cummins et al., 1964; Foord, 1970), experiments may now be performed on aerosols (Hinds and Reist, 1972a,b).

Initial flame experiments were done on particles of soot that were formed in an ethylene–oxygen flame. The data were interpreted on the basis of spherical soot particles of uniform size (Penner et al., 1976a). Later experiments and calculations extended the results to polydisperse particle size distributions (Penner et al., 1976b; Penner and Chang, 1978, 1981). The effects of turbulence on the power spectrum of a signal have also been calculated (Chang and Penner, 1981a).

Driscoll et al. (1979) have determined particle sizes in acetylene–oxygen flames by the diffusion-broadening spectroscopy method. Their results show that the effective diameter of the particles increases as the particles move from the flame front into the exhaust gas region of a flat-flame burner. Penner and Chang (1981) give similar results for methane–oxygen flames; Flower (1982) has also obtained these trends in both methane–oxygen and propane–oxygen flames.

Other experiments which detected the polarization of scattered argon-ion laser light from a CH_4/O_2 flame (Chang and Penner, 1981b) have shown that the particles have light-scattering properties of isotropic spheres and not of the chain-like structures often observed in electron micrographs. Particle sizes that are determined by the angular dependence of the scattered radiation of assumed spherical carbon particles are compared with particle sizes determined by diffusion-broadening spectroscopy. The results agree to within 10% at several positions in a rich methane–oxygen flame that is stabilized on a flat-flame burner.

The scattering/extinction method of D'Alessio et al. (1975) has been applied to the measurement of particle sizes of soot in flames. In this technique both the light-scattering strength and the extinction are measured. These two measured quantities allow the number density and the diameter to be measured. The method also requires knowledge of the complex index of refraction of the particles (Lee and Tien, 1981; Roessler and Faxvog, 1980, 1981; Roessler et al., 1981; Roessler, 1983). The method also depends on particle sphericity and on the size distribution that is assumed for the sample (Graham and Robinson, 1976). The results of these types of experiments (D'Alessio et al., 1975) show particle sizes ranging from approximately 5 nm immediately after the reaction zone to approximately 150 nm about 1 cm above the primary reaction zone in a methane–oxygen premixed flame at atmospheric pressure. Similar particle size measurements have also been performed in other methane–oxygen flames (D'Alessio et al., 1977) and in propane–oxygen flames (Prado et al., 1981).

The method of scattering and extinction has also been used to determine the effect of alkali metal additives and alkaline earth additives on soot formation and coagulation in premixed ethylene–oxygen–nitrogen flames (Haynes et al., 1979). The measured particle sizes were similar to those reported by D'Alessio et al. (1975, 1977). This method also allows particle number density to be measured, and this density decreases at increasingly larger distances above the flame front. This decrease in density is attributed to a coagulation process being greater than nucleation processes. The addition of the metal additives tends to produce a small decrease in the extinction coefficient which indicates less soot is being formed. However, these metal additives also tend to produce large decreases in the scattering with increasing metal additive concentration. This reduction in scattering intensity is due to a reduction in the average particle size and increase in particle number density caused by metal additive. Consequently, the major effect of the metal additive is to impede the coagulation process.

The scattering extinction method has also been applied to laminar diffusion flames (Kent et al., 1981; Kent and Wagner, 1982) to determine particle size and number density. Other applications include particle sizing in diesel sprays that have both small soot particles and much larger fuel droplets (Beretta et al., 1981). The sizes of both the fuel droplets and the soot particles were determined by including measurements of the polarization ratio. Additional details on the various scattering and extinction methods for particle size determination have been given by Bonczyk (1979).

XI. Laser Doppler Velocimetry

Optical imaging methods of flow visualization were discussed in Section VIII. An alternate method of obtaining flow velocities is based on the scattering of light by particles that are in a combustion system and are sufficiently small to follow the gas-flow velocity. Several different optical configurations have been designed to allow measurement of the particle velocity, but one of these configurations is the most widely used techniques. In this configuration (see Fig. 24) two equal intensity beams simultaneously illuminate a particle. Light is scattered from both beams which have propagation vectors $\mathbf{K}_1$ and $\mathbf{K}_2$. Because different Doppler shifts arise from each beam for any direction of observation, a beat signal will be produced which depends on the particle velocity. This type of system is called a differential Doppler system and the beat frequency ν_d is independent of observation direction.

A way of visualizing the scattering process is to remember that the intersecting laser beams form an interference pattern in the football-shaped region where they intersect. This pattern consists of alternating bright and dark planes or fringes. Since the fringe spacing depends only on the optics, the Doppler frequency provides a measure of the component of velocity perpendicular to the fringes. The fringe spacing $\Delta X = \lambda/[2 \sin(\Theta/2)]$. Here λ is the laser wavelength, and Θ is the angle of beam intersection. The observed Doppler frequency is just $f_D = V/\Delta X$. If the laser beams have the same frequency, the Doppler signal is the same for a particle going in either a positive or negative direction through the fringes. Directional sensitivity is achieved by frequency shifting one or both beams such that a net frequency difference exists between them in the probed volume. If the beams have a net frequency difference between them, the fringe pattern can be thought of as moving rather than being stationary. The frequency shifting is usually achieved with an acoustooptic modulator. Often it is desired to measure more than one component of the particle's velocity. Because this can be achieved by adding a second set of fringes whose planes are mutually perpendicular to the first set of fringes, a means must be obtained to separate the signals from the two sets of fringes. This is usually achieved by using two different laser wave-

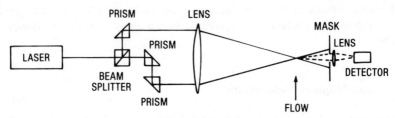

FIG. 24. Schematic diagram of an experimental apparatus for the determination of flow velocities by laser Doppler velocimetry. The flow is seeded with particles that scatter light from the interference pattern formed at the crossing of the two laser beams.

lengths such as the 514.5- and 488-nm lines of an argon-ion laser. Orthogonal polarizations may also be used. There are a variety of electronic schemes for processing the signal; these include frequency trackers which are based on phase-lock loops that follow the frequency of the Doppler signal, and counters that effectively count the number of fringes through which a particle passes for each particle passing through the measurement volume. Photon correlation has also been used as a means of obtaining LDV measurements at low-scattering light levels (Pike, 1979). Still another method of signal processing is to use a high-speed transient digitizer with computer processing of the recorded signal. Any of these methods of signal processing may be used for laminar flows to give accurate measurements of fluid velocities if the particles follow the flow. For measurements in turbulent flows more caution is required. For example, if a counter is used to obtain velocities from discrete samples of particles, and an ensemble average is used to obtain a mean velocity, errors can occur. These velocity bias errors (McLaughlin and Tiederman, 1973) result from high-velocity samples outnumbering low-velocity samples in a given time period.

Although the differential Doppler velocimeter is the most widely used configuration for laser velocimetry measurement, it is possible to measure flow velocities by non-Doppler methods. A common method of achieving this is through a time-of-flight configuration of one or more laser beams. A simple example would be two parallel laser beams with electronics to detect the time difference in the scattering signals as a particle passes through each beam (Lading, 1979). Similar two-beam arrangements have been used with tunable dye lasers to measure the velocity of individual atoms as they pass through the beams (She *et al.,* 1978). In this case, the laser-induced fluorescence instead of the particle scattering was measured as a function of time.

Transit-time velocimetry may also be used in conjunction with a single, focused Gaussian beam (Hirleman, 1978). The intensity distribution in any plane perpendicular to the beam propagation vector is given by $I(r) = I_0 \exp(-2r^2/w^2)$ where $I(r)$ is the intensity at radial distance r from the optic axis, I_0 is the peak intensity, and w is the beam radius. Consider a particle traveling in the plane normal to the laser beam direction with velocity V_1 and in the y direction at some distance x from the optic axis. It will see geometrically similar Gaussian intensity distributions because $I(r) = I(x,y) = I_0 \exp(-2x^2/w^2)\exp(-2y^2/w^2)$. Here for convenience the y coordinate was chosen along the particle trajectory. Consequently, a measure of the transit time t_1 between the maximum intensity and the $1/e^2$ intensity uniquely determines the magnitude of the particle's velocity V_1 in the plane normal to the beam axis. Thus $V_1 = w/t_1$ regardless of the transverse position x. This analysis assumes that the particles have diameters that are small compared to w. This type of velocity measurement has been applied by Holve (1982b) to measure flow velocities in coal combustion burners.

There have been many applications of laser velocimetry and much of the early work has been reviewed by Self and Whitelaw (1976). Subsequent measurements have been applied to understand the role of turbulence in both premixed and diffusion flames. For example, Moreau and Boutier (1977) used LDV to measure velocity and velocity fluctuations in methane–air flames confined to a constant-cross-section chamber. A variety of preflame mixing conditions were employed and the measurements showed maximum velocity fluctuations in the flame reaction zone. Other measurements of the structure of premixed turbulent flames include those of Boyer et al. (1981) and Yoshida (1981). Laser velocimetry measurements have been applied to the determination of correlations between temperature fluctuations and velocity fluctuations in premixed turbulent flames (Yanagi and Mimura, 1981; Dandekar and Gouldin, 1981). These measurements have been applied to improve the understanding of turbulent burning velocities and the role of turbulent diffusion for heat and mass transfer in flames. Measurements of temperature and velocity in an unconfined, premixed turbulent flame by Yoshida and Tsuji (1979) have also demonstrated a tendency for turbulence to increase within the flame zone. Other experiments by Moss (1980) and Bray et al. (1981) have applied LDV to determining the relative role of buoyancy and Reynolds stresses in flame-generated turbulence.

In addition to measurements of the role of velocity fluctuations on turbulence in premixed flames, LDV has been applied to laboratory flames as well as industrial burner flames. For example, Blint and Bechtel (1982a) applied laser Doppler velocimetry to the measurement of flow velocities adjacent to a metal wall that was part of a laboratory burner (see Fig. 25). The velocity measurements were used in conjunction with laser Raman spectroscopy measurement to determine how rapidly the fuel would decay as it was burned along the wall. Baker et al. (1973) have reported LDV measurements in an industrial burner with an LDV apparatus that had also been used to characterize laboratory flames (Durst et al., 1972). These results were extended in subsequent reviews (Baker et al., 1974, 1975). These measurements were used to quantify the turbulence levels in recirculation zones in these burners and to measure the effect of swirl on the flow velocities. Other swirl-burner measurements (Gupta et al., 1977) show the importance of matching the directions and concentrations of reactants such that high fuel concentrations overlap

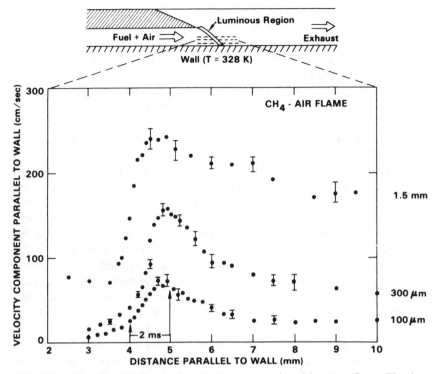

FIG. 25. Example of laser velocimetry measurement in a laboratory flame. The data show a velocity component that is parallel to the wall of the laboratory burner. The flame was a laminar CH₄–air flame at atmospheric pressure. Data profiles are exhibited for three different distances from the burner wall.

regions of large shear stress in the flow. If this is achieved, the heat release rate of the multiannular swirl burner may be increased.

Laser Doppler velocimetry has also been used to measure fluid motion in internal combustion engines (Cole and Swords, 1979a,b; Hutchinson *et al.*, 1978; Rask, 1979a,b; Witze, 1979; Asanuma and Obokata, 1979; Johnston *et al.*, 1979). Because the piston motion and intake manifold geometry produce complicated gas motion inside of the cylinder, LDV is a very important tool for understanding the fluid motion within internal combustion engines. In general, the mean velocity and the turbulence intensity are measured at various positions within the cylinder and for different positions of the crankshaft during the engine cycle.

Laser velocimetry has been used to assess the validity of hot-wire anemometry measurements of air motion in internal combustion engines

(Witze, 1980a). The results show that for conditions of known flow direction, low turbulence level, and low compression ratio, the hot-wire anemometer can provide useful results, but the hot-wire method is in general very sensitive to gas temperature variations and is not as useful for velocity measurement as laser velocimetry methods.

Laser velocimetry measurements have also been used to study the effect of valve geometry on spark-ignition engine performance (Witze and Vilchis, 1981), and recent LDV measurements of turbulence have been used for input to theoretical predictions of combustion in spark-ignition engines (Borgnakke et al., 1982).

Laser velocimetry has found applications to measuring flame spread over solid fuels. For example, Fernández-Pello and Santoro (1979) have measured gas velocities produced by the flame-induced natural convection near burning polymer fuels of various thicknesses. When these results were combined with temperature measurements, the relative importance of heat transfer through the gas and heat transfer through the fuel could be inferred. Laser velocimetry has been applied to forced flows along burning solid fuels. Fernández-Pello (1979) measured the dependence of flame spread rate on the component of gas velocity in the direction of flame propagation and compared these results to theoretical calculations. Similar measurements have also been performed (Fernández-Pello et al., 1981) on flame speed with the forced flow opposing the direction of flame propagation. For both flow cases the increase or decrease in the flame spread rate depends not only on the flow velocity but also on the solid fuel thickness and on the oxygen concentration in the flow. Other examples of the application of laser velocimetry to boundary-layer combustion include hydrogen–air combustion in a heated, turbulent boundary layer (Cheng et al., 1981) and the ignition of a combustible liquid near a heated vertical surface (Chen and Faeth, 1982).

It is important to obtain time-resolved turbulence levels in turbulent diffusion flames in order to test the predictions of models of this type of combustion. Some recent experiments with laser velocimetry have addressed this problem by measuring the velocity in turbulent diffusion flames (Chigier and Dvorak, 1975; Ballantyne and Bray, 1977; Glass and Bilger, 1978; Starner and Bilger, 1980).

Correlations between velocity fluctuations and temperature have also been reported for turbulent diffusion flames (Warshaw et al., 1980; Dibble et al., 1981, 1982). These measurements are obtained by combining laser velocimetry with laser Raman scattering or laser Rayleigh scattering to obtain velocity and temperature measurement with both space and time resolution at the same point in a flame. Similar experiments have been reported for combinations of laser velocimetry and Mie scattering from particles injected into a turbulent diffusion flame (Kennedy and Kent, 1979, 1981; Starner and Bilger, 1981), and laser velocimetry has been applied to velocity measurements in a variety of experiments on spray combustion (Khalil and Whitelaw, 1977; Styles and Chigier, 1977; Yule *et al.*, 1982).

Finally, there have been measurements (Marko and Rimai, 1982a,b) indicating that individual seed particles do not have a constant velocity as they transit the LDV fringe volume. This means that conventional processing techniques yield only velocity distribution averages and neglect trajectory information from individual signals.

XII. Conclusions

This article has presented the wealth of laser techniques which have been brought to bear on combustion problems. These methods include not only spectroscopies, which reveal the chemical kinetic aspects of combustion, but also include refractive index and light-scattering experiments which allow the thermal and hydrodynamic properties to be measured. It is the complicated and competitive nature of combustion which requires the measurement of so many properties in order to understand most combustion behavior.

It has been shown how major species and temperature profiles can be measured by either spontaneous or coherent Raman spectroscopies under conditions of high luminosity, turbulence, and small-scale structure. These techniques are complemented by more sensitive but specialized techniques such as laser-induced fluorescence, optogalvanic, and optoacoustic spectroscopy which have permitted the detection of OH, O, H, NH, NH_2, NO, SH, SO_2, and other species under combustion conditions. This list will grow to include more complicated radicals and will place on a more secure basis the growing list of extended combustion kinetic mechanisms.

It has also been shown how Rayleigh scattering, holography, shadowgraph, and schlieren photography allow whole-field measurements of various derivatives of the refractive index and of particle and droplet properties. Whole-field or, at least, one-dimensional methods are essential to measure the spatial structure of flames. The spatial structure and transport are certainly among the most important properties of practical combustors.

The outpouring and proliferation of techniques would be of limited consequence if similar advances had not been made in calculations of combustion phenomena. As discussed at various points in this article, available computational techniques can now treat either the very detailed chemistry and diffusion of one-dimensional flames or the very complicated hydrodynamics of turbulent, two-dimensional flames. Together, laser diagnostics and computer simulations have given rise to a renaissance in combustion science.

References

Adrian, R. J., and Orloff, K. L. (1977). Laser anemometry signals: Visibility characteristics and applications to particle sizing. *Appl. Opt.* **16**, 677.

Aeschliman, D. P., and Setchell, R. E. (1975). Fluorescence limitations to combustion studies using Raman spectroscopy. *Appl. Spectrosc.* **29**, 426.

Aeschliman, D. P., Cummings, J. C., and Hill, R. A. (1979). Raman spectroscopic study of a laminar hydrogen diffusion flame in air. *J. Quant. Spectrosc. Radiat. Transfer* **21**, 293.

Agrawal, Y., Hadeishi, T., and Robben, F. (1977). Laser fluorescence detection of nitrogen dioxide in combustion. *Prog. Astronaut. Aeronaut.* **53**, 279.

Ahn, B. T., Bastiaans, G. J., and Albahadily, F. (1982). Practical determination of flame species via laser saturation fluorescence spectroscopy. *Appl. Spectrosc.* **36**, 106.

Aldén, M., Edner, H., Holmstedt, G., Svanberg, S., and Hoegberg, T. (1982a). Single-pulse laser induced hydroxyl fluorescence in an atmospheric flame, spatially resolved with a diode array. *Appl. Opt.* **21**, 1236.

Aldén, M., Edner, H., and Svanberg, S. (1982b). Simultaneous spatially resolved monitoring of C_2 and OH in a C_2H_2/O_2 flame using a diode array detector. *Appl. Phys.* **B29**, 93.

Aldén, M., Edner, H., Grafstrom, P., and Svanberg, S. (1982c). Two-photon excitation of atomic oxygen in a flame. *Opt. Commun.* **42**, 244.

Allen, J. E., Jr., Anderson, W. R., and Crosley, D. A. (1977). Optoacoustic pulses in a flame. *Opt. Lett.* **1**, 118.

Amann, C. A. (1982). Seeing is believing—A look at engine diagnostics. *Res. Publ.—Gen. Mot. Corp., Res. Lab.* **GMR-4068**.

Anderson, W. L., and Beissner, R. E. (1970). Counting and classifying small objects by far-field light scattering. *Appl. Opt.* **10**, 1503.

Anderson, W. R. (1979). Laser excited fluorescence measurement of hydroxyl rotational temperature in a methane/nitrous oxide flame. *East. Sect. Combust. Inst., Pap.* **3**.

Anderson, W. R., Decker, L. J., and Kotlar, A. J. (1980). Measurement of hydroxyl and imido concentration profiles in stoichiometric methane/nitrous oxide flames by laser excited fluorescence. *East. Sect. Combust. Inst., Pap.* **67**.

Anderson, W. R., Vanderhoff, J. A., Kotlar, A. J., Decker, L. J., and Beyer, R. A. (1981). Laser excitation of NCO A-X system fluorescence in a methane/nitrous oxide flame using an argon ion laser. *East. Sect. Combust. Inst. Pap.* **47**.

Anderson, W. R., Decker, L. J., and Kotlar, A. J. (1982a). Temperature profile of a stoichiometric CH_4/N_2O flame from laser excited fluorescence measurements on OH. *Combust. Flame* **48**, 163.

Anderson, W. R., Decker, L. J., and Kotlar, A. J. (1982b). Concentration profiles on NH and OH in a stoichiometric CH_4/N_2O flame by laser excited fluorescence and absorption. *Combust. Flame* **48**, 179.

Asanuma, T., and Obokata, T. (1979). Gas velocity measurements in a motored and firing engine by laser anemometry. *SAE Tech. Pap. Ser.* **790096**.

Attal, B., Pealat, M., and Taran, J.-P. (1980). CARS diagnostics of combustion. *J. Energy* **4**, 135.

Azzopardi, B. J. (1979). Measurement of drop sizes. *Int. J. Heat Mass. Transfer* **22**, 1245.

Bachalo, W. D. (1979). On-line particle diagnostics systems for application in hostile environments. *In* "Laser Velocimetry and Particle Sizing" (H. D. Thompson and W. H. Stevenson, eds.), p. 506. Hemisphere Publ. Corp., Washington, D.C.

Bailly, R., Pealat, M., and Taran, J.-P. E. (1976). Raman investigation of a subsonic jet. *Opt. Commun.* **17**, 68.

Baker, R. J., Bourke, P. J., and Whitelaw, J. H. (1973). Applications of laser anemometry to the measurement of flow properties of industrial burner flames. *Symp. (Int.) Combust.* [*Proc.*] **14**, 699.

Baker, R. J., Hutchinson, P., and Whitelaw, J. H. (1974). Velocity measurements in the recirculation region of an industrial burner flame by laser anemometry with light frequency shifting. *Combust. Flame* **23**, 57.

Baker, R. J., Hutchinson, P., Khalil, E. E., and Whitelaw, J. H. (1975). Measurement of three velocity components in a model furnace with and without combustion. *Symp. (Int.) Combust.* [*Proc.*] **15**, 553.

Ballantyne, A., and Bray, K. N. C. (1977). Investigations into the structure of jet diffusion flames using time resolved optical measuring techniques. *Symp. (Int.) Combust.* [*Proc.*] **16**, 777.

Baltes, H. P. (1978). "Inverse Source Problems in Optics." Springer-Verlag, Berlin and New York.

Barnes, R. H., and Kircher, J. F. (1978). Laser NO_2 fluorescence measurements in flames. *Appl. Opt.* **17**, 1099.

Baronavski, A. P., and McDonald, J. R. (1977a). Measurement of C_2 concentrations in an oxygen-acetylene flame: An application of saturation spectroscopy. *J. Chem. Phys.* **66**, 3300.

Baronavski, A. P., and McDonald, J. R. (1977b). Application of saturation spectroscopy to the measurement of C_2, $^3\Pi_\mu$ concentrations in oxy-acetylene flames. *Appl. Opt.* **16**, 1897.

Barrett, H. H., and Swindell, W. (1977). Analog reconstruction methods for transaxial tomography. *Proc. IEEE* **65**, 89.

Bar-Ziv, E., Sgulim, S., Kafri, O., and Keren, E. (1983). Temperature mapping in flames by Moiré deflectometry. *Appl. Opt.* **22**, 698.

Beams, J. W. (1954). Shadow and Schlieren methods. *In* "Physical Measurements in Gas Dynamics and Combustion" (R. W. Ladenburg ed.), p. 24. Princeton Univ. Press, Princeton, New Jersey.

Bechtel, J. H. (1979). Temperature measurements of the hydroxyl radical and molecular nitrogen in premixed, laminar flames by laser techniques. *Appl. Opt.* **18**, 2100.

Bechtel, J. H. (1980). Laser probes of premixed laminar methane-air flames and comparison with theory. *In* "Laser Probes for Combustion Chemistry" (D. R. Crosley, ed.), p. 85. Am. Chem. Soc., Washington, D.C.

Bechtel, J. H. (1981). Laser spectroscopy in combustion. *In* "Physics in the Automotive Industry" (F. E. Jamerson, ed.), p. 127. Am. Inst. Phys., New York.

Bechtel, J. H., and Blint, R. J. (1980). Structure of a laminar flame-wall interface by laser Raman spectroscopy. *Appl. Phys. Lett.* **37**, 576.

Bechtel, J. H., and Chraplyvy, A. R. (1982). Laser diagnostics of flames, combustion products and sprays. *Proc. IEEE* **70**, 658.

Bechtel, J. H., and Teets, R. E. (1979). Hydroxyl and its concentration profile in methane-air flames. *Appl. Opt.* **18**, 4138.

Bechtel, J. H., Blint, R. J., Dasch, C. J., and Weinberger, D. A. (1981). Atmospheric pressure, premixed hydrocarbon-air flames: Theory and experiment. *Combust. Flame* **42,** 197.

Belz, R. A., and Dougherty, N. S. (1972). In-line holography of reacting liquid sprays. *SPIE Symp. Eng. Appl. Hologr., 1972,* p. 209.

Beretta, F., Cavaliere, A., Ciajolo, A., D'Alessio, A., Langella, C., and DiLorenzo, A. (1981). Laser light scattering emission/extinction spectroscopy and thermogravimetry analysis in the study of soot behavior in oil spray flames. *Symp. (Int.) Combust. [Proc.]* **18,** 1091.

Berg, J. O., and Shackleford, W. L. (1979). Rotational redistribution effects in saturated laser-induced fluorescence. *Appl. Opt.* **18,** 2093.

Beyer, R. A., Vanderhoff, J. A., Kotlar, A. J., and DeWilde, M. A. (1982). Temperature and species profiles in premixed flames. *West. Sect., Combust. Inst., Pap.* **WSS/CI-82-55.**

Bilger, R. W. (1976). Turbulent jet diffusion flames. *Prog. Energy Combust. Sci.* **1,** 87.

Bill, R. G., Namer, I., Talbor, L., Cheng, R. K., and Robben, F. (1981). Flame propagation in grid-induced turbulence. *Combust. Flame* **43,** 229.

Bill, R. G., Namer, I., Talbot, L., Cheng, R. K., and Robben, F. (1982). Density fluctuations of flames in grid-induced turbulence. *Combust. Flame* **44,** 277.

Bischel, W. K., Perry, B. E., and Crosley, D. R. (1981). Two-photon laser-induced fluorescence in oxygen and nitrogen atoms. *Chem. Phys. Lett.* **82,** 85.

Bischel, W. K., Perry, B. E., and Crosley, D. R. (1982). Detection of fluorescence from oxygen and nitrogen atoms induced by two-photon absorption. *Appl. Opt.* **21,** 1419.

Black, P.C., and Chang, R. K. (1978). Laser-Raman optical multichannel analyzer for transient gas concentration profile and temperature determination. *AIAA J.* **16,** 295.

Blint, R. J., and Bechtel, J. H. (1982a). Flame-wall interface: Theory and experiment. *Combust. Sci. Technol.* **27,** 87.

Blint, R. J., and Bechtel, J. H. (1982b). Hydrocarbon combustion near a cooled wall. *SAE Tech. Pap. Ser.* **820063.**

Blint, R. J., Bechtel, J. H., and Stephenson, D. A. (1980). Concentration and temperature in flames by Raman spectroscopy. *J. Quant. Spectrosc. Radiat. Transfer* **23,** 89.

Boccara, A. C., Fournier, D., and Badoz, J. (1980). Thermo-optical spectroscopy: Detection by the mirage effect. *Appl. Opt.* **36,** 130.

Boiarski, A. A., Barnes, R. H., and Kircher, J. F. (1978). Flame measurements utilizing Raman scattering. *Combust. Flame* **32,** 111.

Bonczyk, P. A. (1979). Measurement of particulate size by in situ laser-optical methods: A critical evaluation applied to fuel pyrolyzed carbon. *Combust. Flame* **35,** 191.

Bonczyk, P. A., and Shirley, J. A. (1979). Measurement of CH and CN concentration in flames by laser-induced saturated fluorescence. *Combust. Flame* **34,** 253.

Borgnakke, C., Martin, J. K., and Witze, P. O. (1982). Turbulent combustion rate in a spark ignition engine—some comparisons between model predictions and experiments. *West. Sect., Combust. Inst., Pap.* **WSS/CI-82-69.**

Boyer, L., Clavin, P., and Sabathier, F. (1981). Dynamic behavior of a premixed turbulent flame front. *Symp. (Int.) Combust. [Proc.]* **18,** 1041.

Bray, K. N. C., Libby, P. A., Masuya, G., and Moss, J. B. (1981). Turbulence production in premixed turbulent flames. *Combust. Sci. Technol.* **25,** 127.

Briffa, F. E. J., and Dombrowski, N. (1966). Entrainment of air into a liquid spray. *AIChE J.* **12,** 708.

Briones, R. A., and Wuerker, R. F. (1977). Holography of solid propellent combustion. *Proc. Soc. Photo-Opt. Instrum. Eng.* **125,** 90.

Bryant, L. (1967). The origin of the automobile engine. *Sci. Am.* **216,** 102.

Buckingham, A. D., and Szabo, A. (1978). Determination of the polarizability anisotropy in a diatomic molecule from relative Raman intensities. *J. Raman Spectrosc.* **7**, 46.

Campbell, D. H. (1982). Vibrational level relaxation effects on laser-induced fluorescence measurements of hydroxide number density in a methane-air flame. *Appl. Opt.* **21**, 2912.

Caroon, T. A., and Borman, G. (1979). Comments on utilizing the Fraunhofer diffraction method for droplet size distribution measurements. *Combust. Sci. Technol.* **19**, 255.

Carrington, T. (1959). Line shape and f value in the OH $^2\Sigma^+ - ^2\Pi$ transition. *J. Chem. Phys.* **31**, 1243.

Cattolica, R. J. (1981). OH rotational temperature form two-line laser-excited fluorescence. *Appl. Opt.* **20**, 1156.

Cattolica, R. J. (1982). Nonequilibrium OH in flames. *Combust. Flame* **44**, 43.

Cattolica, R. J., and Schefer, R. W. (1983a). The effect of surface chemistry on the development of the [OH] in a combustion boundary layer. *Symp. (Int.) Combust. [Proc.]* **19**, 311.

Cattolica, R. J., and Schefer, R. W. (1983b). Laser fluorescence measurements of the OH concentration in a combustion boundary layer. *Combust. Sci. Technol.* **30**, 205.

Cattolica, R. J., Stepowski, D., Puechberty, D., and Cottereau, M. (1982a). Laser fluorescence measurements of the CH radical in a low-pressure flame. *Sandia Lab. [Tech. Rep.] SAND* **SAND82-8615**.

Cattolica, R. J., Yoon, S., and Knuth, E. L. (1982b). OH concentration in an atmospheric pressure methane-air flame from molecular beam mass spectrometry and laser absorption spectroscopy. *Combust. Sci. Technol.* **28**, 225.

Chan, C., and Daily, J. W. (1979). Near-resonant Rayleigh scattering and atomic flame fluorescence spectroscopy. *J. Quant. Spectrosc. Radiat. Transfer* **21**, 527.

Chan, C., and Daily, J. W. (1980a). Laser excitation dynamics of hydroxyl radical in flames. *Appl. Opt.* **19**, 1357.

Chan, C., and Daily, J. W. (1980b). Measurement of temperature in flames using laser induced fluorescence spectroscopy of OH. *Appl. Opt.* **19**, 1963.

Chang, P. H. P., and Penner, S. S. (1981a). Determination of turbulent velocity fluctuations and mean particle radii in flames using scattered laser power spectra. *J. Quant. Spectrosc. Radiat. Transfer* **25**, 97.

Chang, P. H. P., and Penner, S. S. (1981b). Particle-size measurements in flames using light scattering; comparison with diffusion broadening spectroscopy. *J. Quant. Spectrosc. Radiat. Transfer* **25**, 105.

Chen, F. P., and Goulard, R. (1976). Retrieval of arbitrary concentration and temperature fields by multiangular scanning techniques. *J. Quant. Spectrosc. Radiat. Transfer* **16**, 819.

Chen, L.-D., and Faeth, G. M. (1982). Ignition of supercritical fluids during natural convection from a vertical surface. *Combust. Flame* **44**, 169.

Cheng, R. K., Bill, R. G., Jr., and Robben, F. (1981). Experimental study of combustion in a turbulent boundary layer. *Symp. (Int.) Combust. [Proc.]* **18**, 1021.

Cheung, L. M., Bishop, D. M., Drapcho, D. L., and Rosenblatt, G. M. (1981). Relative Raman intensities for H_2 and D_2, correction factors for nonrigidity. *Chem. Phys. Lett.* **80**, 445.

Chidsey, I. L., and Crosley, D. R. (1980). Calculated rotational transition probabilities for the A-X system of OH. *J. Quant. Spectrosc. Radiat. Transfer* **23**, 187.

Chigier, N. A. (1974). Velocity measurement of particles in sprays. *Flow: Its Meas. Control Sci. Ind.* **1**, 823.

Chigier, N. A. (1977). Instrumentation techniques for studying heterogeneous combustion. *Prog. Energy Combust. Sci.* **3,** 175.

Chigier, N. A., and Dvorak, K. (1975). Laser anemometry measurements in flames with swirl. *Symp. (Int.) Combust. [Proc.]* **15,** 573.

Chigier, N. A., and McCreath, C. G. (1974). Combustion of droplets in sprays. *Acta Astronaut.* **1,** 687.

Chigier, N. A., Makepeace, R. W., and McCreath, C. G. (1973). Aerodynamic interaction between burning sprays and recirculation zones. *In* "Combustion Institute European Symposium 1973" (F. J. Weinberg, ed.), p. 577. Academic Press, New York.

Chigier, N. A., McCreath, C. G., and Makepeace, R. W. (1974). Dynamics of droplets in burning and isothermal kerosene sprays. *Combust. Flame* **23,** 11.

Chigier, N. A., Ungut, A., and Yule, A. J. (1979a). Particle sizing in flames with laser velocimeters. *In* "Laser Velocimetry and Particle Sizing" (H. D. Thompson and W. H. Stevenson, eds.), p. 416. Hemisphere Publ. Corp., Washington, D.C.

Chigier, N. A., Ungut, A., and Yule, A. J. (1979b). Particle size and velocity measurement in flames by laser anemometry. *Symp. (Int.) Combust. [Proc.]* **17,** 315.

Chraplyvy, A. R. (1981). Nonintrusive measurements of vapor concentrations inside sprays. *Appl. Opt.* **20,** 2620.

Chu, W. P., and Robinson, D. M. (1977). Scattering from a moving spherical particle by two crossed coherent plane waves. *Appl. Opt.* **16,** 619.

Cincotti, V., D'Alessio, A., Menna, P., and Venitozzi, C. (1981). U.V. absorption spectroscopy and laser excited fluorescence in the study of formation of high molecular mass compounds in the rich combustion of methane. *Riv. Combust.* **35,** 59.

Clark, A. F., Moulder, J. C., and Runyan, C. C. (1975). Combustion of bulk titanium in oxygen. *Symp. (Int.) Combust. [Proc.]* **15,** 489.

Coe, D. S., and Steinfeld, J. I. (1980a). Laser-induced fluorescence of polycyclic aromatic hydrocarbons in a flame. *In* "Laser Probes for Combustion Chemistry" (D. R. Crosley, ed.), p. 159. Am. Chem. Soc., Washington, D.C.

Coe, D. S., and Steinfeld, J. I. (1980b). Fluorescence excitation and emission spectra of polycyclic aromatic hydrocarbons at flame temperatures. *Chem. Phys. Lett.* **76,** 485.

Coe, D. S., Haynes, B. S., and Steinfeld, J. I. (1981). Identification of a source of argon-ion-laser excited fluorescence in sooting flame. *Combust. Flame* **43,** 211.

Cole, J. B., and Swords, M. D. (1979a). Optical studies of turbulence in an internal combustion engine. *Symp. (Int.) Combust. [Proc.]* **17,** 1295.

Cole, J. B., and Swords, M. D. (1979b). Laser Doppler anemometry measurements in an engine. *Appl. Opt.* **18,** 1539.

Cole, J. B., and Swords, M. D. (1980). Measurement of concentration fluctuations in an internal combustion engine. *J. Phys. D* **13,** 733.

Cornillault, J. (1972). Particle size analyzer. *Appl. Opt.* **11,** 265.

Cottereau, M. J., and Stepowski, D. (1980). Laser-induced fluorescence spectroscopy applied to the hydroxyl radical in flame. *In* "Laser-Probes for Combustion Chemistry" (D. R. Crosley, ed.), p. 131. Am. Chem. Soc., Washington, D.C.

Crosley, D. R., ed. (1980a). "Laser Probes for Combustion Chemistry." Am. Chem. Soc., Washington, D.C.

Crosley, D. R. (1980b). Lasers, chemistry, and combustion. *In* "Laser Probes for Combustion Chemistry" (D. R. Crosley, ed.), p. 3. Am. Chem. Soc., Washington, D.C.

Crosley, D. R. (1981). Collisional effects on laser-induced fluorescence flame measurements. *Opt. Eng.* **20,** 511.

Crosley, D. R. (1982). Laser-induced fluorescence in spectroscopy, dynamics, and diagnostics. *J. Chem. Educ.* **59,** 446.

Crosley, D. R., and Smith, G. P. (1980a). Vibrational energy transfer in hydroxyl in flames. *East. Sect. Combust. Inst., Pap.* **69.**

Crosley, D. R., and Smith, G. P. (1980b). Vibrational energy transfer in laser-excited OH as a flame thermometer. *Appl. Opt.* **19,** 517.

Crosley, D. R., and Smith, G. P. (1982). Rotational energy transfer and LIF temperature measurements. *Combust. Flame* **44,** 27.

Cummins, H. Z., Knable, N., and Yeh, Y. (1964). Observations of diffusion broadening of Rayleigh scattered light. *Phys. Rev. Lett.* **12,** 150.

Cuzillo, B. C., Metcalf, J. T., and Daily, J. W. (1982). Laser absorption measurement of hydrocarbons in a spark ignition square piston I. C. engine. *West. Sect., Combust. Inst., Pap.* **WSS/CI-82-59.**

Daily, J. W. (1978a). Saturation of fluorescence in flames with a Gaussian laser beam. *Appl. Opt.* **17,** 225.

Daily, J. W. (1978b). Detectability limit and uncertainty considerations for laser induced fluorescence spectroscopy in flames. *Appl. Opt.* **17,** 1610.

Daily, J. W. (1980). Laser-induced fluorescence spectroscopy in flames. *In* "Laser Probes for Combustion Chemistry" (D. R. Crosley, ed.), p. 61. Am. Chem. Soc., Washington, D.C.

Daily, J. W., and Chan, C. (1978). Laser-induced fluorescence measurements of sodium in flames. *Combust. Flame* **33,** 47.

D'Alessio, A., DiLorenzo, A., Sarofim, A. F., Beretta, F., Masi, S., and Venitozzi, C. (1975). Soot formation in methane-oxygen flames. *Symp. (Int.) Combust. [Proc.]* **15,** 1427.

D'Alessio, A., DiLorenzo, A., Borghese, A., Berretta, F., and Masi, S. (1977). Study of the soot nucleation zone of rich methane-oxygen flames. *Symp. (Int.) Combust. [Proc.]* **16,** 695.

Dandekar, K. V., and Gouldin, F. C. (1981). Temperature and velocity measurements in premixed turbulent flames. *AIAA Pap.* **81-0179.**

Dasch, C. J. (1983). Laser vaporization of soot particles. *Opt. Lett.* (submitted for publication).

Dasch, C. J., and Bechtel, J. H. (1981). Spontaneous Raman scattering by ground-state oxygen atoms. *Opt. Lett.* **6,** 36.

Dasch, C. J., and Blint, R. J. (1982). A mechanistic and experimental study of ammonia flames. *East. Sect. Combust. Inst., Pap.* **71.**

Davies, T. P. (1981). Schlieren photography—short bibliography and review. *Opt. Laser Technol.* **13,** 37.

Dean, A. M., Chou, M. S., and Stern, M. (1982). Laser absorption measurements on OH, NH, and NH_2 in ammonia-oxygen flames: Determination of an oscillator strength for NH_2. *J. Chem. Phys.* **76,** 5334.

Dean, A. M., Chou, M. S., and Stern, M. (1983). Laser induced fluorescence and absorption measurements of NO in ammonia-oxygen and methane-air flames. *J. Chem. Phys.* (submitted for publication).

DeCorso, S. M. (1960). Effect of ambient and fuel pressure on spray drop size, *J. Eng. Power* **82,** 10.

Dent, J. C. (1980). Potential applications of holographic interferometry to engine combustion research. *In* "Combustion Modeling in Reciprocating Engines" (J. N. Mattavi and C. A. Amann, eds.), p. 265. Plenum, New York.

Dibble, R. W., and Hollenbach, R. E. (1981). Laser Rayleigh thermometry in turbulent flames. *Symp. (Int.) Combust. [Proc.]* **18,** 1489.

Dibble, R. W., Rambach, G. D., Hollenbach, R. E., and Ringland J. T. (1981). Simultaneous measurement of velocity and temperature in flames using LDV and CW laser Rayleigh thermometry. *Symp. Turbul., 7th, 1981* (unpublished).

Dibble, R. W., Kollmann, W., and Schefer, R. W. (1982). Conserved scaler fluxes measured in a turbulent nonpremixed flame by combined laser Doppler velocimetry and laser Raman scattering. *West. Sect. Combust. Inst., Pap.* **WSS/CI-82-52.**

Dicke, R. H. (1953). The effect of collisions on the Doppler width of spectral lines. *Phys. Rev.* **89,** 472.

Dieck, R. H., and Roberts, R. L. (1970). The determination of Sauter mean droplet diameter in fuel nozzle sprays. *Appl. Opt.* **9,** 2007.

Dieke, G. H., and Crosswhite, H. M. (1961). The ultraviolet bands of OH. *J. Quant. Spectrosc. Radiat. Transfer* **2,** 97.

DiLorenzo, A., D'Alessio, A., Cincotti, V., Masi, S., Menna, P., and Venitozzi, C. (1981). UV absorption, laser excited fluorescence and direct sampling in the study of the formation of polycyclic aromatic hydrocarbons in rich CH_4-O_2 flames. *Symp. (Int.) Combust. [Proc.]* **18,** 485.

Dixon-Lewis, G. (1979). Kinetic mechanism, structure and properties of premixed flames in hydrogen-oxygen-nitrogen mixture. *Philos. Trans. R. Soc. London, Ser. A* **292,** 45.

Dobbins, R., Crocco, L., and Glassman, I. (1963). Measurement of mean particle sizes by sprays from diffractively scattered light. *AIAA J.* **1,** 1882.

dos Santos, R., and Stevenson, W. H. (1976). Aerosol sizing by means of laser-induced fluorescence. *Appl. Phys. Lett.* **30,** 236.

Drake, M. C. (1982). Rotational Raman intensity correction factors due to vibrational anharmonicity. *Opt. Lett.* **7,** 440.

Drake, M. C., and Rosenblatt, G. M. (1976). Flame temperatures from Raman scattering. *Chem. Phys. Lett.* **44,** 313.

Drake, M. C., and Rosenblatt, G. M. (1978). Rotational Raman scattering from premixed diffusion flames. *Combust. Flame* **33,** 179.

Drake, M. C., Asawaroengchai, C., and Rosenblatt, G. M. (1980). Temperature from rotational and vibrational Raman scattering: Effects of vibrational-rotational interactions and other corrections. *In* "Laser Probes for Combustion Chemistry" (D. R. Crosley, ed.), p. 231. Am. Chem. Soc., Washington, D.C.

Drake, M. C., Bilger, R. W., and Starner, S. H. (1983). Raman measurements and conserved scalar modelling in turbulent diffusion flames. *Symp. (Int.) Combust. [Proc.]* **19,** 459.

Driscoll, J. F., Mann, D. M., and McGregor, W. K. (1979). Submicron particle size measurements in an acetylene-oxygen flame. *Combust. Sci. Technol.* **20,** 41.

Driscoll, J. F., Schefer, R. W., and Dibble, R. W. (1983). Mass fluxes $\overline{\rho'\mu'}$ and $\overline{\rho'\nu'}$ measured in a turbulent, nonpremixed flame. *Symp. (Int.) Combust. [Proc.]* **19** (to be published).

Druger, S. D., Kerker, M., Wang, D. S., and Cooke, D. D. (1979). Light scattering by inhomogeneous particles. *Appl. Opt.* **18,** 3888.

Durst, F., Melling, A., and Whitelaw, J. H. (1972). The application of optical anemometry to measurement in combustion systems. *Combust. Flame* **18,** 197.

Dyer, M. J., and Crosley, D. R. (1982). Two-dimensional imaging of OH laser-induced fluorescence in a flame. *Opt. Lett.* **7,** 382.

Dyer, T. M. (1979). Characterization of one and two dimensional homogeneous combustion phenomena in a constant volume bomb. *SAE Tech. Pap. Ser.* **790353.**

Eckbreth, A. C. (1977). Effects of laser modulated particulate incandescence on Raman scattering diagnostics. *J. Appl. Phys.* **48,** 4473.

Eckbreth, A. C. (1981a). Recent advances in laser diagnostics for temperature and species concentration in combustion. *Symp. (Int.) Combust. [Proc.]* **18,** 1471.

Eckbreth, A. C. (1981b). Spatially precise laser diagnostics for combustion. *Proc. Int. Congr. Instrum. Aerosp. Simul. Facil., 1981,* p. 71.

Eckbreth, A. C., and Hall, R. J. (1981). CARS concentration sensitivity with and without nonresonant background suppression. *Combust. Sci. Technol.* **25,** 175.

Eckbreth, A. C., Bonczyk, P. A., and Verdieck, J. R. (1979). Combustion diagnostics by laser Raman and fluorescence techniques. *Prog. Energy Combust. Sci.* **5,** 253.

Eesley, G. L. (1979). Coherent Raman Spectroscopy. *J. Quant. Spectrosc. Radiat. Transfer* **22,** 507.

Eesley, G. L. (1981). "Coherent Raman Spectroscopy." Pergamon, Oxford.

Emmerman, P. J., Goulard, R., Santoro, R. J., and Semerjian, H. G. (1980). Multiangular absorption diagnostics of a turbulent argon-methane jet. *J. Energy* **4,** 70.

Fairchild, P. W., Smith, G. P., and Crosley, D. R. (1982). Bimolecular quenching rate constants for OH at high temperatures. *West. Sect., Combust. Inst., Pap.* **WSS/CI-82-47.**

Farmer, W. M. (1972). Measurement of particle size, number density, and velocity using a laser interferometer. *Appl. Opt.* **11,** 2603.

Farmer, W. M. (1974). Observation of large particles with a laser interferometer. *Appl. Opt.* **13,** 610.

Farmer, W. M. (1976). Sample space for particle size and velocity measuring interferometers. *Appl. Opt.* **15,** 1984.

Farmer, W. M., Harwell, K. E., Hornkohl, J. O., and Schwartz, F. A. (1979). Particle sizing interferometer measurements in rocket exhausts. *In* "Laser Velocimetry and Particle Sizing" (H. D. Thompson and W. H. Stevenson, eds.), p. 518. Hemisphere Publ. Corp., Washington, D.C.

Faxvog, F. R. (1974). Detection of airborne particles using optical extinction measurements. *Appl. Opt.* **13,** 1913.

Faxvog, F. R. (1977). New laser particle sizing instrument. *SAE Tech. Pap. Ser.* **770140.**

Faxvog, F. R., and Roessler, D. M. (1979). Optoacoustic measurements of diesel particulate emissions. *J. Appl. Phys.* **50,** 7880.

Fernández-Pello, A. C. (1979). Flame spread in a forward forced flow. *Combust. Flame* **36,** 63.

Fernández-Pello, A. C., and Santoro, R. J. (1979). On the dominant mode of heat transfer in downward flame spread. *Symp. (Int.) Combust. [Proc.]* **17,** 1201.

Fernández-Pello, A. C., and Williams, F. A. (1975). Laminar flame spread over PMMA surfaces. *Symp. (Int.) Combust. [Proc.]* **15,** 217.

Fernández-Pello, A. C., and Williams, F. A. (1976). Experimental techniques in the study of laminar flame spread over solid combustibles. *Combust. Sci. Technol.* **14,** 155.

Fernández-Pello, A. C., Ray, S. R., and Glassman, I. (1981). Flame spread in an opposed forced flow: The effect of ambient oxygen concentration. *Symp. (Int.) Combust. [Proc.]* **18,** 579.

Filseth, S. V., Zacharias, H., Danon, J., Wallenstein, R., and Welge, K. H. (1978). Laser excited fluorescence of CH in a low pressure flame. *Chem. Phys. Lett.* **58,** 140.

Flower, W. L. (1982). Optical measurement of soot formation in premixed flames. *West. Sect. Combust. Inst., Pap.* **WSS/CI-82-67.**

Foord, R. (1970). Determination of diffusion coefficients of halmocyamin at low concentrations by intensity fluctuations spectroscopy of scattered laser light. *Nature (London)* **227,** 242.

Fourney, M. E., Matkin, J. H., and Waggoner, A. P. (1969). Aerosol size and velocity determination via holography. *Rev. Sci. Instrum.* **40,** 205.

Fristrom, R. M., and Westenberg, A. A. (1965). "Flame Structure." McGraw-Hill, New York.

Fristrom, R. M., Jones, A. R., Schwar, M. J. R., and Weinberg, F. J. (1973). Particle sizing by interference fringes and signal coherence in Doppler velocimetry. *Symp. Faraday Soc.* **7**, 183.

Galatry, L. (1961). Simultaneous effect of Doppler and foreign gas broadening on spectral lines. *Phys. Rev.* **122**, 1218.

Garforth, A. M. (1976). Unburnt gas density measurement in spherical combustion bomb by infinite-fringe laser interferometry. *Combust. Flame* **26**, 343.

Gaydon, A. G. (1974). "Spectroscopy of Flames," 2nd ed. Chapman & Hall, London.

Gaydon, A. G., and Wolfhard, H. G. (1979). "Flames: Their Structure, Radiation, and Temperature," 4th ed. Chapman & Hall, London.

Glass, M., and Bilger, R. W. (1978). The turbulent jet diffusion flame in a co-flowing stream—some velocity measurements. *Combust. Sci. Technol.* **18**, 165.

Glassman, I. (1977). "Combustion." Academic Press, New York.

Goldman, A., and Gillis, J. R. (1981). Spectral line parameters for the A $^2\Sigma$-X^2 II (0,0) band of OH for atmosphereic and high temperatures. *J. Quant. Spectrosc. Radiat. Transfer* **25**, 111.

Goldsmith, J. E. M. (1981). Spatially resolved saturated absorption spectroscopy in flames. *Opt. Lett.* **6**, 525.

Goldsmith, J. E. M. (1982a). Resonant multiphoton optogalvanic detection of hydrogen in flames. *Opt. Lett.* **7**, 437.

Goldsmith, J. E. M. (1982b). Resonant multiphoton optogalvanic detection of atomic hydrogen and oxygen in flames. *West. Sect. Combust. Inst. Pap.* **WSS/CI-82-44.**

Goldsmith, J. E. M., and Lawler, J. E. (1981). Optogalvanic spectroscopy. *Contemp. Phys.* **22**, 255.

Goulard, R. (1976). "Combustion Measurements," Hemisphere Publ. Corp., Washington, D.C.

Gouldin, F. C., and Dandekar, R. V. (1982). Time resolved density measurements in premixed turbulent flames. *AIAA Pap.* **82-0036.**

Grabner, L., and Hastie, J. W. (1982). Flame boundary layer effects in line-of-sight optical measurements. *Combust. Flame* **44**, 15.

Graham, S. C., and Robinson, A. (1976). A comparison of numerical solutions to the self-preserving size distribution for aerosol coagulation in the free-molecular regime. *J. Aerosol Sci.* **7**, 261.

Graham, S. C., Grant, A. J., and Jones, J. M. (1974). Transient molecular concentration measurements in turbulent flows using Rayleigh light scattering. *AIAA J.* **12**, 1140.

Gravett, C. (1973). Real-time measurement of the size distribution of particulate matter by a light scattering method. *J. Air Pollut. Control Assoc.* **23**, 1035.

Green, R. M., and Miller, J. A. (1981). The measurement of relative concentration profiles of NH$_2$ using laser absorption spectroscopy. *J. Quant. Spectrosc. Radiat. Transfer* **26**, 313.

Grieser, D. R., and Barnes, R. H. (1980a). Nitric oxide measurements in a flame by laser fluorescence. *Appl. Opt.* **19**, 741.

Grieser, D. R., and Barnes, R. H. (1980b). Nitric oxide detection in flames by laser fluorescence. *In* "Laser Probes for Combustion Chemistry" (D. R. Crosley, ed.), p. 153. Am. Chem. Soc., Washington, D.C.

Groff, E. G. (1982). The cellular nature of confined spherical propane-air flames. *Combust. Flame* **48**, 51.

Gupta, A. K., Beér, J. M., and Swithenbank, J. (1977). Concentric multi-annular swirl burner: Stability limits and emission characteristics. *Symp. (Int.) Combust. [Proc.]* **16**, 79.

Hall, R. J., and Eckbreth, A. C. (1981). Combustion diagnostics by coherent anti-Stokes Raman spectroscopy (CARS). *Opt. Eng.* **20**, 494.

Hamaguchi, H., Buckingham, A. D., and Jones, W. J. (1981). Determination of derivatives of polarizability anisotropy in diatomic molecules. II. The hydrogen and nitrogen molecules. *Mol. Phys.* **43**, 1311.

Hammond, D. C., Jr. (1981). Deconvolution technique for line-of-sight optical scattering measurements in axisymmetric sprays. *Appl. Opt.* **20**, 493.

Hanson, R. K. (1977). Shock tube spectroscopy: Advanced instrumentation with a tunable diode laser. *Appl. Opt.* **16**, 1479.

Hanson, R. K. (1980). Absorption spectroscopy in sooting flames using a tunable diode laser. *Appl. Opt.* **19**, 482.

Hanson, R. K., and Falcone, P. K. (1978). Temperature measurement technique for high temperature gases using a tunable diode laser. *Appl. Opt.* **17**, 2477.

Hanson, R. K., Kuntz, P. A., and Kruger, C. H. (1977). High resolution spectroscopy of combustion gases using a tunable IR diode laser. *Appl. Opt.* **16**, 2045.

Hanson, R. K., Varghese, P. L., Schoenung, S. M., and Falcone, P. K. (1980). Absorption spectroscopy of combustion gases using a tunable IR diode laser. *In* "Laser Probes for Combustion Chemistry" (D. R. Crosley, ed.), p. 413. Am. Chem. Soc., Washington, D.C.

Haraguchi, H., Smith, B., Weeks, S., Johnson, D. J., and Winefordner, J. D. (1977). Measurement of small volume flame temperature by the two-line atomic fluorescence method. *Appl. Spectrosc.* **31**, 156.

Harris, S. J. (1982). Power dependence of continuous-wave intracavity spectroscopy. *Opt. Lett.* **7**, 497.

Harris, S. J., and Weiner, A. M. (1981a). "Detection of atomic oxygen by intracavity spectroscopy. *Opt. Lett.* **6**, 142.

Harris, S. J., and Weiner, A. M. (1981b). Intracavity laser tomography of C_2 in an oxyacetylene flame. *Opt. Lett.* **6**, 434.

Hartley, D. L. (1974). Raman gas mixing measurements and Ramanography. *In* "Laser Raman Gas Diagnostics" (M. Lapp and C. M. Penney, eds.), p. 311. Plenum, New York.

Haynes, B. S., Jander, H., and Wagner, H. Gg. (1979). The effect of metal additives on the formation of soot in premixed flames. *Symp. (Int.) Combust. [Proc.]* **17**, 1365.

Herden, W., and Maly, R. (1977). Interferometric study of nonstationary combustion using a N_2 laser (1 bar). *Combust. Flame* **30**, 207.

Herman, R., Rothery, R. W., and Rubin, R. J. (1958). Line intensities in vibrational rotational bands of diatomic molecules. *J. Mol. Spectrosc.* **2**, 369.

Herzberg, G. (1945). "Infrared and Raman Spectra of Polyatomic Molecules." Van Nostrand-Reinhold, Princeton, New Jersey.

Herzberg, G. (1950). "Spectra of Diatomic Molecules." Van Nostrand-Reinhold, Princeton, New Jersey.

Herzberg, G. (1966). "Electronic Structure and Electronic Spectra of Polyatomic Molecules." Van Nostrand-Reinhold, Princeton, New Jersey.

Hickling, R. (1969). Holography of liquid droplets. *J. Opt. Soc. Am.* **59**, 1334.

Hill, J. C., and Majkowski, R. F. (1980). Time resolved measurement of vehicle sulfate and methane emissions with tunable diode lasers. *SAE Tech. Pap. Ser.* **800510**.

Hill, R. M. (1979). Application of curve of growth to absorption spectral lines. *J. Quant. Spectrosc. Radiat. Transfer* **21**, 19.

Hill, R. A., Mulac, A. J., Aeschliman, D. P., and Flower, W. L. (1979). Temperatures from rotational-vibrational Raman Q-branches. *J. Quant. Spectrosc. Radiat. Transfer* **21**, 213.

Hinds, W., and Reist, P. C. (1972a). Aerosol measurement by laser Doppler spectroscopy. I. Theory and experimental results for aerosols homogeneous. *Aerosol Sci.* **3**, 501.

Hinds, W., and Reist, P. C. (1972b). Aerosol measurements by laser Doppler spectroscopy. II. Operational limits, effects of polydispersity, and applications. *Aerosol Sci.* **3,** 515.

Hirleman, E. D. (1978). Laser technique for simultaneous particle size and velocity measurements. *Opt. Lett.* **3,** 19.

Hirleman, E. D. (1980). Laser-based single particle counters for *in situ* particulate diagnostics. *Opt. Eng.* **19,** 854.

Hirleman, E. D. (1982). Calibration studies of laser diffraction droplet sizing instruments. *West. Sect. Combust. Inst., Pap.* **WSS/CI-82-54.**

Hirleman, E. D., and Wittig, S. L. K. (1977). *In situ* optical measurement of automobile exhaust gas particulate size distributions: Regular fuel and methanol mixtures. *Symp. (Int.) Combust. [Proc.]* **16,** 245.

Hirschfelder, J. O., Curtis, C. F., and Bird, R. B. (1964). "Molecular Theory of Gases and Liquids." Wiley, New York.

Hodkinson, J. R. (1966). Particle sizing by means of the forward scattering lobe. *Appl. Opt.* **5,** 839.

Hodkinson, J. R., and Greenleaves, I. (1963). Computations of light scattering and extinction by spheres according to diffraction and geometrical optics and comparisons with Mie theory. *J. Opt. Soc. Am.* **53,** 577.

Holland, A. C., and Gagne, G. (1970). The scattering of polarized light by polydisperse systems of irregular particles. *Appl. Opt.* **9,** 1113.

Holstein, T. (1947). Imprisonment of resonance radiation in gases. *Phys. Rev.* **72,** 1212.

Holve, D. J. (1980). *In situ* optical particle sizing technique. *J. Energy* **4,** 176.

Holve, D. J. (1982a). *In situ* optical particle counter for submicron particles. *AIAA Pap.* **82-0237.**

Holve, D. J. (1982b). Transit timing velocimetry (TTV) for two-phase reacting flows. *Combust. Flame* **48,** 105.

Holve, D. J., and Davis, G. W. (1982). Analysis of a single-particle counter for application to general aerosol measurements, *West. Sect. Combust. Insti., Pap.* **WSS/CI-82-56.**

Holve, D. J., and Self, S. (1979a). Optical particle sizing for *in situ* measurements. Part I. *Appl. Opt.* **18,** 1632.

Holve, D. J., and Self, S. (1979b). Optical particle sizing for *in situ* measurements. Part II. *Appl. Opt.* **18,** 1646.

Holve, D. J., and Self, S. (1979c). An optical particle-sizing counter for *in situ* measurements. *In* "Laser Velocimetry and Particle Sizing" (H. D. Thompson and W. H. Stevenson, eds.), p. 397. Hemisphere Publ. Corp., Washington, D.C.

Holve, D. J., Tichenor, D., Wang, J. C. F., and Hardesty, D. R. (1981). Design criteria and recent developments of optical single particle counters for fossil fuel systems. *Opt. Eng.* **20,** 529.

Hong, N. S., and Jones, A. R. (1976). A light scattering technique for particle sizing based on laser fringe anemometry. *J. Phys. D* **9,** 1839.

Huber, K. P., and Herzberg, G. (1979). "Constants of Diatomic Molecules." Van Nostrand-Reinhold, Princeton, New Jersey.

Hui, A. K., Armstrong, B. H., and Wray, A. A. (1978). Rapid computation of the Voigt and complex error functions. *J. Quant. Spectrosc. Radiat. Transfer* **19,** 509.

Humlicek, J. (1982). Optimized computation of the Voigt and complex probability functions. *J. Quant. Spectrosc. Radiat. Transfer* **27,** 437.

Hutchinson, P., Morse, A. P., and Whitelaw, J. H. (1978). Velocity measurements in motored engines: Experience and prognosis. *SAE Tech. Pap. Ser.* **780061.**

Iino, M., Yano, H., Takubo, Y., and Shimazu, M. (1981). Saturation characteristics of laser-induced Na fluorescence in a propane air flame: The role of chemical reaction. *J. Appl. Phys.* **52,** 6025.

Jackson, W. B., Ames, N. M., Boccara, A. C., and Fournier, D. (1981). Photothermal deflection spectroscopy and detection. *Appl. Opt.* **20**, 1344.

James, T. C., and Klemperer, W. (1959). Line intensities in the Raman effect of $^1\Sigma$ diatomic molecules. *J. Chem. Phys.* **31**, 130.

Johnston, S. C. (1979). Precombustion fuel/air distribution in a stratified charge engine using laser Raman spectroscopy. *SAE Tech. Pap. Ser.* **790433**.

Johnston, S. C. (1980). Raman spectroscopy and flow visualization study of stratified charge engine combustion. *SAE Tech. Pap. Ser.* **800136**.

Johnston, S. C., Robinson, C. W., Rorke, W. S., Smith, J. R., and Witze, P. O. (1979). Application of laser diagnostics to an injected engine. *SAE Tech. Pap. Ser.* **790092**.

Jones, A. R. (1974). Light scattering by a sphere situated in an interference pattern with reference to fringe anemometry and particle sizing. *J. Phys. D* **7**, 1369.

Jones, A. R. (1979). Scattering of electromagnetic radiation in particulate laden fluids. *Prog. Energy Combust. Sci.* **5**, 73.

Jones, D. G., and Mackie, J. C. (1976). Evaluation of C_2 resonance fluorescence as a technique for transient flame studies. *Combust. Flame* **27**, 143.

Kafri, O. (1980). Noncoherent method for mapping phase objects. *Opt. Lett.* **5**, 555.

Kennedy, I. M., and Kent, J. H. (1979). Measurements of a conserved scalar in turbulent jet diffusion flames. *Symp. (Int.) Combust. [Proc.]* **17**, 279.

Kennedy, I. M., and Kent, J. H. (1981). Scalar measurements in a CO-flowing turbulent diffusion flame. *Combust. Sci. Technol.* **25**, 109.

Kent, J. H., and Wagner, H. Gg. (1982). Soot measurement in laminar ethylene diffusion flames. *Combust. Flame* **47**, 53.

Kent, J. H., Jander, H., and Wagner, H. Gg. (1981). Soot formation in a laminar diffusion flame. *Symp. (Int.) Combust. [Proc.]* **18**, 1117.

Keren, E., Bar-Ziv, E., Glatt, I., and Kafri, O. (1981). Measurements of temperature distribution of flames by Moiré deflectometry. *Appl. Opt.* **20**, 4263.

Kerker, M. (1969). "The Scattering of Light and Other Electromagnetic Radiation." Academic Press, New York.

Khalil, E. E., and Whitelaw, J. H. (1977). Aerodynamic and thermodynamic characteristics of kerosene-spray flames. *Symp. (Int.) Combust. [Proc.]* **16**, 569.

Kimball-Linne, M. A., Kychakoff, G., Hanson, R. K., and Booman, R. A. (1982). Fiberoptic fluorescence probe for species measurements in combustors. *West. Sect. Combust. Inst., Pap.* **WSS/CI-82-50**.

Klim, A. (1981). A comparison of methods for the calculation of Voigt profiles. *J. Quant. Spectrosc. Radiat. Transfer* **26**, 537.

Kline, J. M., and Penner, S. S. (1982). Temperature dependence of the absorption coefficient of CH_2O at the 3.508 μm HeXe laser line. *J. Quant. Spectrosc. Radiat. Transfer* **27**, 127.

Kogelschatz, U., and Schneider, W. R. (1972). Quantitative schlieren techniques applied to high current arc investigations. *Appl. Opt.* **11**, 1822.

Kotlar, A. J., Gelb, A., and Crosley, D. R. (1980). A multilevel model of response to laser-fluorescence excitation in the hydroxyl radical. *In* "Laser Probes for Combustion Chemistry" (D. R. Crosley, ed.), p. 137. Am. Chem. Soc., Washington, D.C.

Kurosaki, Y., Ito, A., and Chiba, M. (1979). Downward flame spread along two vertical, parallel sheets of thin combustible solid. *Symp. (Int.) Combust. [Proc.]* **17**, 1211.

Kychakoff, G., and Hanson, R. K. (1981). Tunable laser fiberoptic probe for absorption-fluorescence combustion measurements. *West. Sect. Combust. Inst., Pap.* **WSS/CI-81-50**.

Kychakoff, G., Howe, R. D., Hanson, R. K., and McDonald, J. C. (1982a). Quantitative visualization of combustion species in a plane. *Appl. Opt.* **21**, 3225.

Kychakoff, G., Howe, R. D., and Hanson, R. K. (1982b). Spatially resolved combustion using crossed-beam saturated absorption spectroscopy. *Tech. Dig.—Conf. Lasers Electro-Opt.* (*CLEO '82*), p. 84.

Kychakoff, G., Knapp, K., Howe, R. D., and Hanson, R. K. (1982c). Quantitative flow visualization in combustion gases. *West. Sect. Combust. Inst., Pap.* **WSS/CI-82-60**.

Lading, L. (1979). The time-of-flight laser anemometer versus the laser Doppler anemometer. *In* "Laser Velocimetry and Particle Sizing" (H. D. Thompson and W. H. Stevenson, eds.), p. 26. Hemisphere Publ. Corp., Washington, D.C.

Lapp, M. (1973). Raman band shapes for flame gases. *Adv. Raman Spectrosc.* **1**, 256.

Lapp, M., and Hartley, D. L. (1976). Raman scattering studies of combustion. *Combust. Sci. Technol.* **13**, 199.

Lapp, M., and Penney, C. M. (1974). "Laser Raman Gas Diagnostics." Plenum, New York.

Lapp, M., Goldman, L. M., and Penney, C. M. (1972). Raman scattering from flames. *Science* **175**, 1112.

Lapp, M., Penney, C. M., and Goldman, L. M. (1973). Vibrational Raman scattering temperature measurements. *Opt. Commun.* **9**, 195.

Latimer, P., Brunsting, A., Pyle, B. E., and Moore, C. (1978). Effects of asphericity on single particle scattering. *Appl. Opt.* **17**, 3152.

Learner, R. C. M. (1962). The influence of vibrational-rotational interaction on intensities in the electronic spectra of diatomic molecules. I. The hydroxyl radical. *Proc. R. Soc. London, Ser. A* **269**, 311.

Lederman, S. (1977). The use of laser Raman diagnostics in flow fields and combustion. *Prog. Energy Combust. Sci.* **3**, 1.

Lee, J. H., Knystautas, R., and Guiraa, C. M. (1975). Critical power density for direct initiation of unconfined gaseous detonations. *Symp. (Int.) Combust. [Proc.]* **15**, 53.

Lee, S. C., and Tien, C. L. (1981). Optical constants of soot in hydrocarbon flames. *Symp. (Int.) Combust. [Proc.]* **18**, 1159.

Lengel, R. K., and Crosley, D. R. (1977). Energy transfer in A $^2\Sigma$ OH. I. Rotational. *J. Chem. Phys.* **67**, 2085.

Libby, P. A., and Williams, F. A. (1980). "Turbulent Reacting Flows." Springer-Verlag, Berlin and New York.

Long, M. E., Swofford, R. L., and Albrecht, A. C. (1976). Thermal lens technique: A new method of absorption spectroscopy. *Science* **191**, 183.

Lorenzetto, J., and Lefebvre, A. H. (1977). Measurement of drop size on a plain air-blast atomizer. *AIAA J.* **15**, 1006.

Lucht, R. P., and Laurendeau, N. M. (1979). Two-level model for near saturated fluorescence in diatomic molecules. *Appl. Opt.* **18**, 856.

Lucht, R. P., Sweeney, D. W., and Laurendeau, N. M. (1980). Balanced cross-rate model for saturated molecular fluorescence in flames using a nanosecond pulse length laser. *Appl. Opt.* **19**, 3295.

Lucht, R. P., Sweeney, D. W., and Laurendeau, N. M. (1982a). Laser saturated fluorescence measurements of OH concentration in flames. *West. Sect. Combust. Inst., Pap.* **WSS/CI-82-42**.

Lucht, R. P., Sweeney, D. W., and Laurendeau, N. M. (1982b). Laser-saturated fluorescence measurements of OH in a sooting, atmospheric pressure $CH_4/O_2/N_2$ flame. *West. Sect. Combust. Inst., Pap.* **WSS/CI-82-57**.

Lucht, R. P., Laurendeau, N. M., and Sweeney, D. W. (1982c). Temperature measurement by two-line laser saturated OH fluorescence in flames. *Appl. Opt.* **21**, 3729.

Lucht, R. P., Sweeney, D. W., and Laurendeau, N. M. (1983). Laser-saturated fluorescence measurements of OH concentration in flames. *Combust. Flame* **50**, 189.

Lück, K. C., and Muller, F. J. (1977). Simultaneous determination of temperature and OH-concentration in flames using high-resolution laser-absorption spectroscopy. *J. Quant. Spectrosc. Radiat. Transfer* **17**, 403.

Lück, K. C., and Thielen, W. (1978). Measurements of temperatures and OH-concentrations in a lean methane-air flame using high-resolution laser-absorption spectroscopy. *J. Quant. Spectrosc. Radiat. Transfer* **20**, 71.

Lück, K. C., and Tsatsaronis, G. (1979). A study of flat methane-air flames at various equivalence ratios. *Acta Astronaut.* **6**, 467.

Lyon, R. K., and Benn, D. (1979). Kinetics of the $NO-NH_3-O_2$ reaction. *Symp. (Int.) Combust. [Proc.]* **17**, 601.

McCreath, C. G., and Beer, J. M. (1976). A review of drop size measurements in fuel sprays. *Appl. Energy* **2**, 3.

McCreath, C. G., Roett, M. F., and Chigier, N. A. (1972). A technique for measurement of velocities and size of particles in flames. *J. Phys. E.* **5**, 601.

McDonald, J. R. (1980). Laser probes for combustion applications. *In* "Laser Probes for Combustion Chemistry" (D. R. Crosley, ed.), p. 19. Am. Chem. Soc., Washington, D.C.

McLaughlin, D. K., and Tiederman, W. G. (1973). Biasing correcting for individual realization of laser anemometer measurements in turbulent flows. *Phys. Fluids* **16**, 2082.

McSweeney, A., and Rivers, W. (1972). Optical fibre array for measuring radial distribution of light intensity for particle size analysis. *Appl. Opt.* **11**, 2101.

Mailander, M. (1978). Determination of absolute transition probabilities and particle densities by saturated fluorescence excitation. *J. Appl. Phys.* **49**, 1256.

Mallard, W. G., Miller, J. H., and Smyth, K. C. (1982). Resonantly enhanced two-photon photoionization of NO in an atmospheric flame. *J. Chem. Phys.* **76**, 3483.

Maly, R., and Vogel, M. (1979). Initiation and propagation of flame fronts in lean CH_4-air mixtures by the three modes of the ignition spark. *Symp. (Int.) Combust. [Proc.]* **17**, 821.

Marko, K. A., and Rimai, L. (1982a). Single-particle velocity and trajectory measurements of microscale inhomogeneities in unstable airflows. *Opt. Lett.* **7**, 130.

Marko, K. A., and Rimai, L. (1982b). Rapid digitized storage of seed-particle-track images in microscopically inhomogeneous flows. *Opt. Lett.* **7**, 328.

Matekunas, F. A. (1979). A Schlieren study of combustion in a rapid compression machine simulating the spark ignition engine. *Symp. (Int.) Combust. [Proc.]* **17**, 1283.

Meintjes, K., and Alkidas, A. C. (1982). An experimental and computational investigation of the flow in diesel prechambers. *SAE Tech. Pap. Ser.* **820275**.

Mellor, R., Chigier, N. A., and Beer, J. M. (1971). Hollow cone liquid spray in uniform air stream. *In* "Combustion and Heat Transfer in Gas Turbine Systems" (E. R. Norster, ed.), p. 291. Pergamon, Oxford.

Meyer, J. W., and Oppenheim, A. K. (1971). On the shock-induced ignition of explosive gases. *Symp. (Int.) Combust. [Proc.]* **13**, 1153.

Miller, J. H., Mallard, W. G., and Smyth, K. C. (1982). The observation of laser-induced visible fluorescence in sooting diffusion flames. *Combust. Flame* **47**, 205.

Montgomery, G. P., Jr., and Reuss, D. L. (1982). Effects of refraction on axisymmetric flame temperatures measured by holographic interferometry. *Appl. Opt.* **21**, 1373.

Moreau, P., and Boutier, A. (1977). Laser velocimetry measurements in a turbulent flame. *Symp. (Int.) Combust. [Proc.]* **16**, 1747.

Morley, C. (1981). Mechanism of NO formation from nitrogen compounds in hydrogen flames studied by laser fluorescence. *Symp. (Int.) Combust. [Proc.]* **18**, 23.

Morley, C. (1982). The application of laser fluorescence to detection of species in atmospheric pressure flames—relative quenching rates of OH by H_2O, O_2, H_2, and CO. *Combust. Flame* **47**, 67.

Moss, J. B. (1980). Simultaneous measurements of concentration and velocity in an open premixed turbulent flame. *Combust. Sci. Technol.* **22**, 119.

Muller, C. H., III, Schofield, K., Steinberg, M., and Broida, H. P. (1979). Sulfur chemistry in flames. *Symp. (Int.) Combust. [Proc.]* **17**, 867.

Muller, C. H., III, Schofield, K., and Steinberg, M. (1980a). Laser induced flame chemistry of Li and Na. Implications for other saturated mode measurements. *J. Chem. Phys.* **72**, 6620.

Muller, C. H., III, Schofield, K., and Steinberg, M. (1980b). Laser-induced fluorescence: A powerful tool for the study of flame chemistry. *In* "Laser Probes for Combustion Chemistry" (D. R. Crosley, ed.), p. 103. Am. Chem. Soc., Washington, D.C.

Muller-Dethlefs, K., and Weinberg, F. J. (1979). Burning velocity measurements based on laser Rayleigh scattering. *Symp. (Int.) Combust. [Proc.]* **17**, 985.

Mullinger, P. J., and Chigier, N. A. (1974). The design and performance of internal mixing multijet twin-fluid atomizers. *J. Inst. Fuel* **47**, 251.

Mutoh, N., Hirano, T., and Akita, K. (1979). Experimental study on radiative ignition of polymethylmethacrylate. *Symp. (Int.) Combust. [Proc.]* **17**, 1183.

Namazian, M., Talbot, L., and Robben, F. (1983). Two point Rayleigh scattering measurements in a V-shaped turbulent flame. *Symp. (Int.) Combust. [Proc.]* **19**, 489.

Nibler, J. W., and Knighten, G. W. (1979). Coherent anti-Stokes Raman spectroscopy. *In* "Raman Spectroscopy of Gases and Liquids" (A. Weber, ed.), p. 253. Springer-Verlag, Berlin and New York.

Olsen, D. B., Mallard, W. G., and Gardiner, W. C. (1978). High temperature absorption of the 3.39 μm He-Ne laser line by small hydrocarbons. *Appl. Spectrosc.* **32**, 489.

Oppenheim, A. K., Cohen, L. M., Short, J. M., Cheng, R. K., and Hom, K. (1975). Dynamics of the exothermic process in combustion. *Symp. (Int.) Combust. [Proc.]* **15**, 1503.

Orrin, J. E., Vince, I. M., and Weinberg, F. J. (1981). A study of plasma jet ignition mechanisms. *Symp. (Int.) Combust. [Proc.]* **18**, 1755.

Ottesen, D. K., and Stephenson, D. A. (1982). Fourier transform infrared (FT-IR) measurements in sooting flames. *Combust. Flame* **46**, 95.

Owyoung, A., and Rahn, L. A. (1979). High resolution inverse Raman spectroscopy in a methane-air flame. *IEEE J. Quantum Electron.* **QE-15**, 25.

Pasternack, L., Baronavski, A. P., and McDonald, J. R. (1978). Application of saturation spectroscopy for measurement of atomic Na and MgO in acetylene flames. *J. Chem. Phys.* **69**, 4830.

Patel, C. K. N., and Tam, A. C. (1981). Pulsed optoacoustic spectroscopy of condensed matter. *Rev. Mod. Phys.* **53**, 517.

Pealat, M., Brilly, R., and Taran, J.-P. E. (1977). Real time study of turbulence in flames. *Opt. Commun.* **22**, 91.

Pealat, M., Taran, J.-P., and Moya, F. (1980). CARS spectrometer for gases and flames. *Opt. Laser Technol.* **12**, 21.

Peeters, J., and Mahnen, G. (1973). Reaction mechanism and rate constants of elementary steps in methane-oxygen flames. *Symp. (Int.) Combust. [Proc.]* **14**, 133.

Penner, S. S. (1959). "Quantitative Molecular Spectroscopy and Gas Emissivities." Addison-Wesley, Reading, Massachusetts.

Penner, S. S., and Chang, P. (1978). On the determination of log-normal particle size distributions using half widths and detectabilities of scattered laser power spectra. *J. Quant. Spectrosc. Radiat. Transfer* **20**, 447.

Penner, S. S., and Chang, P. H. P. (1981). Particle sizing in flames. *In* "Combustion in Reactive Systems" (J. R. Bowen, N. Manson, A. K. Oppenheim, and R. I. Soloukhin, eds.), pp. 1–30. Am. Inst. Aeronaut. Astronaut., New York.

Penner, S. S., and Jerskey, T. (1973). Use of lasers for local measurement of velocity components, species densities, and temperatures. *Annu. Rev. Fluid Mech.* **5**, 9.

Penner, S. S., and Kavanagh, R. W. (1953). Radiation from isolated spectral lines with combined Doppler and Lorentz broadening. *J. Opt. Soc. Am.* **43**, 385.

Penner, S. S., Bernard, J. M., and Jerskey, T. (1976a). Power spectra observed in laser scattering from moving, polydisperse particle systems in flames. I. Theory. *Acta Astronaut.* **3**, 69.

Penner, S. S., Bernard, J. M., and Jerskey, T. (1976b). Laser scattering from moving, polydisperse particles in flames. II. Preliminary experiments. *Acta Astronaut.* **3**, 93.

Penney, C. M., St. Peters, R. L., and Lapp, M. (1974). Absolute rotational Raman cross sections for N_2, O_2 and CO_2. *J. Opt. Soc. Am.* **64**, 712.

Peterson, R. C., and Alkidas, A. C. (1982). A visual study of divided-chamber diesel combustion using a rapid compression machine. *West. Sect. Combust. Inst., Pap.* **WSS/CI-82-79**.

Pike, E. R. (1979). Burst correlation processing in LDV. *In* "Laser Velocimetry and Particle Sizing" (H. D. Thompson and W. H. Stevenson, eds.), p. 41. Hemisphere Publ. Corp., Washington, D.C.

Pindzola, M. S. (1978). Two-photon excitation of atomic oxygen. *Phys. Rev. A* **17**, 1021.

Pitz, R. W., Cattolica, R., Robben, F., and Talbot, L. (1976). Temperature and density in a hydrogen-air flame from Rayleigh scattering. *Combust. Flame* **27**, 313.

Placzek, G. (1934). Rayleigh Streung und Raman Effekt. *In* "Handbuch der Radiologie" (G. Marks, ed.), Vol. 6, Chap. 2, p. 205. Akad. Verlagsges., Leipzig.

Polymeropoulos, C. E., and Sernas, V. (1977). Measurement of droplet size and fuel-air ratio in sprays. *Combust. Flame* **29**, 123.

Prado, G., Jagoda, J., Neoh, K., and Lahaye, J. (1981). A study of soot formation in premixed propane/oxygen flames by *in-situ* optical techniques and sampling probes. *Symp. (Int.) Combust. [Proc.]* **18**, 1127.

Purcell, E. M., and Pennypacker, C. R. (1973). Scattering and absorption of light by nonspherical dielectric grains. *Astrophys. J.* **186**, 705.

Rahn, L. A., Zych, L. J., and Mattern, P. L. (1979). Background-free CARS studies of carbon monoxide in a flame. *Opt. Commun.* **39**, 249.

Rahn, L. A., Mattern, P. L., and Farrow, R. L. (1981). A comparison of coherent and spontaneous Raman combustion diagnostics. *Symp. (Int.) Combust. [Proc.]* **18**, 1533.

Rajan, S., Smith, J. R., and Rambach, G. D. (1982). Internal structure of a premixed turbulent flame. *West. Sect. Combust. Inst., Pap.* **WSS/CI-82-88**.

Rao, K. V. L., and Lefebvre, A. H. (1975). Fuel atomization in a flowing airstream. *AIAA J.* **13**, 1413.

Rask, R. B. (1979a). Laser Doppler anemometry measurements in an internal combustion engine. *SAE Tech. Pap. Ser.* **790094**.

Rask, R. B. (1979b). Velocity measurements inside the cylinder of a motored internal combustion engine. *In* "Laser Velocimetry and Particle Sizing" (H. D. Thompson and W. H. Stevenson, eds.), p. 251. Hemisphere Publ. Corp., Washington, D.C.

Ray, S. R., and Semerjian, H. G. (1982). Laser tomography for temperature and concentration measurement in reacting flows. *West. Sect. Combust. Inst., Pap.* **WSS/CI-82-64**.

Reuss, D. L. (1983). Temperature measurements in a radially symmetric flame using holographic interferometry. *Combust. Flame* **49**, 207.

Revet, J. M., Puechberty, D., and Cottereau, M. J. (1978). A direct comparison of hydroxyl concentration profiles measured in a low-pressure flame by molecular beam mass spectrometry and ultraviolet absorption spectroscopy. *Combust. Flame* **33**, 5.

Reynolds, G. O. (1976). Holography and holographic interferometry. *In* "Combustion Measurements" (R. Goulard, ed.), p. 253. Hemisphere Publ. Corp., Washington, D.C.

Rizkalla, A. A., and Lefebvre, A. H. (1975). The influence of air and liquid properties on airblast atomization. *J. Fluids Eng.* **97**, 316.

Roberds, D. W. (1977). Particle sizing using laser interferometry. *Appl. Opt.* **16**, 1861.

Roberts, J., and Webb, M. (1964). Measurement of droplet size for wide range particle distributions. *AIAA J.* **2**, 583.

Rockney, B. H., Cool, T. A., and Grant, E. R. (1982). Detection of nascent NO in a methane/air flame by multiphoton ionization. *Chem. Phys. Lett.* **87**, 141.

Roessler, D. M. (1982). Opacity and photoacoustic measurements of diesel particle mass emissions. *SAE Tech. Pap. Ser.* **820460**.

Roessler, D. M. (1983). Optical properties and dispersion exponent of carbon particles. *Res. Publ.—Gen. Mot. Corp., Res. Lab.* **GMR-4227**.

Roessler, D. M., and Faxvog, F. R. (1979). Optoacoustic measurements of optical absorption in acetylene smoke. *J. Opt. Soc. Am.* **69**, 1699.

Roessler, D. M., and Faxvog, F. R. (1980). Optical properties of agglomerated acetylene smoke particles at 0.5145 μm and 10.6 μm wavelengths. *J. Opt. Soc. Am.* **70**, 230.

Roessler, D. M., and Faxvog, F. R. (1981). Changes in diesel particulates with engine air/fuel ratio. *Combust. Sci. Technol.* **26**, 225.

Roessler, D. M., Faxvog, F. R., Stevenson, R., and Smith, G. W. (1981). Optical properties and morphology of particulate carbon: Variation with air/fuel ratio. *In* "Particulate Carbon Formation During Combustion" (D. C. Siegla and G. W. Smith, eds.), p. 57. Plenum, New York.

Roose, T. R., Hanson, R. K., and Kruger, C. H. (1981). A shock tube study of the decomposition of NO in the presence of NH_3. *Symp. (Int.) Combust. [Proc.]* **18**, 853.

Rose, A., Pyrum, J. D., Muzny, C., Salamo, G. J., and Gupta, R. (1982). Applications of the photothermal deflection technique to combustion diagnostics. *Appl. Opt.* **21**, 2663.

Salmon, J. T., Lucht, R. P., Laurendeau, N. M., and Sweeney, D. W. (1982). Laser-saturated fluorescence measurements of NH in a premixed subatmospheric $CH_4/N_2O/Ar$ flame. *West. Sect. Combust. Inst., Pap.* **WSS/CI-82-61**.

Samuelsen, G. S., Trolinger, J. D., Heap, M. P., and Seeker, W. R. (1981). Observation of the behavior of coal particles during thermal decomposition. *Combust. Flame* **40**, 7.

Santoro, R. J., Semerjian, H. G., Emmerman, P. J., and Goulard R. (1980). Optical tomography for flow field diagnostics. *AIAA Pap.* **80-1541**.

Santoro, R. J., Semerjian, H. G., Emmerman, P. J., and Goulard, R. (1981). Optical tomography for flow field diagnostics. *Int. J. Heat Mass Transfer* **24**, 1139.

Schefer, R. W., Robben, F., and Chong, R. K. (1980). Catalyzed combustion of H_2–air mixtures in a flat-plate boundary layer. I. Experimental results. *Combust. Flames* **38**, 51.

Schenk, P. K., and Hastie, J. W. (1981). Optogalvanic spectroscopy-application to combustion systems. *Opt. Eng.* **20**, 522.

Schmieder, R. W., and Kerstein, A. (1980). Imaging a conserved scalar in gas mixing by means of a linear spark. *Appl. Opt.* **19**, 4210.

Schoenung, S. M., and Hanson, R. K. (1981a). CO and temperature measurements in a flat flame by laser absorption spectroscopy and probe techniques. *Combust. Sci. Technol.* **24**, 227.

Schoenung, S. M., and Hanson, R. K. (1981b). Temporally and spatially resolved measurements of fuel mole fraction in a turbulent CO diffusion flame. *West. Sect. Combust. Inst., Pap.* **WSS/CI-81-33**.

Schofield, K., and Steinberg, M. (1981). Quantitative atomic and molecular laser fluorescence in the study of detailed combustion processes. *Opt. Eng.* **20**, 501.

Schrötter, H. W., and Klöckner, H. W. (1979). Raman scattering cross sections in gases and liquids. In "Raman Spectroscopy of Gases and Liquids" (A. Weber, ed.), p. 123. Springer-Verlag, Berlin and New York.

Seeker, W. R., Samuelsen, G. S., Heap, M. P., and Trolinger, J. D. (1981). The thermal decomposition of pulverized coal particles. *Symp. (Int.) Combust. [Proc.]* **18**, 1213.

Self, S. A., and Whitelaw, J. H. (1976). Laser anemometry for combustion research. *Combust. Sci. Technol.* **13**, 171.

Sell, J. A. (1981). Infrared diode laser spectroscopy of nitric oxide. *J. Quant. Spectrosc. Radiat. Transfer* **25**, 19.

Sell, J. A., Herz, R. K., and Monroe, D. R. (1980). Dynamic measurement of carbon monoxide concentrations in automotive exhaust using infrared diode laser spectroscopy. *SAE Tech. Pap. Ser.* **800463**.

Sell, J. A., Herz, R. H., and Perry, E. C. (1981). Time resolved measurements of carbon monoxide in the exhaust of a computer command controlled engine. *SAE Tech. Pap. Ser.* **810276**.

Setchell, R. E. (1978). Initial measurements within an internal combustion engine using Raman spectroscopy. *Sandia Lab. [Tech. Rep.]* **SAND78-1220**.

Setchell, R. E., and Aeschliman, D. P. (1977). Fluorescence interference in Raman scattering from combustion products. *Appl. Spectrosc.* **31**, 530.

Setchell, R. E., and Miller, J. A. (1978). Raman scattering measurements of nitric oxide in ammonia/oxygen flames. *Combust. Flame* **33**, 23.

She, C. Y., Fairbank, W. M., Jr., and Billman, K. W. (1978). Measuring the velocity of individual atoms in real time. *Opt. Lett.* **2**, 30.

Simmons, H. C., and Lapera, D. L. (1969). A high speed spray analyzer for gas turbine fuel nozzles. *Gas Turbine Combust. Fuels Technol. Conf., Am. Soc. Mech. Eng. 1969.*

Sinnamon, J. F., Lancaster, D. R., and Steiner, J. C. (1980). An experimental and analytical study of engine fuel spray trajectories. *SAE Tech. Pap. Ser.* **800135**.

Smith, B., Winefordner, J. D., and Omenetto, N. (1977). Atomic fluorescence of sodium under continuous-wave laser excitation. *J. Appl. Phys.* **48**, 2676.

Smith, G. P., and Crosley, D. R. (1981). Quantitative laser-induced fluorescence in OH: Transition probabilities and the influence of energy transfer. *Symp. (Int.) Combust. [Proc.]* **18**, 1511.

Smith, J. R. (1978). Rayleigh temperature profiles in a hydrogen diffusion flame. *Proc. Soc. Photo-Opt. Instrum. Eng.* **158**, 85.

Smith, J. R. (1980). Temperature and density measurements in an engine by pulsed Raman spectroscopy. *SAE Tech. Pap. Ser.* **800137**.

Smith, J. R. (1982a). Turbulent flame structure in a homogeneous-charge engine. *SAE Tech. Pap. Ser.* **820043**.

Smith, J. R. (1982b). The influence of turbulence in flame structure in an engine. *Sandia Lab. [Tech. Rep.]* **SAND82-8722** (unpublished).

Smith, J. R., and Giedt, W. H. (1977). Temperature and concentration profiles in transient gas flow by rotational Raman scattering. *Int. J. Heat Mass Transfer* **20**, 899.

Sochet, L.-R., Lucquin, M., Bridoux, M., Crunelle-Cras, M., Grase, F., and Delhaye, M. (1979). Use of multichannel pulsed Raman spectroscopy as a diagnostic technique in flames. *Combust. Flame* **36**, 109.

South, R., and Hayward, B. M. (1976). Temperature measurement in conical flames by laser interferometry. *Combust. Sci. Technol.* **12**, 183.

Starner, S. H., and Bilger, R. W. (1980). LDA measurements in a turbulent diffusion flame with axial pressure gradient. *Combust. Sci. Technol.* **21**, 259.

Starner, S. H., and Bilger, R. W. (1981). Measurement of scalar-velocity correlation in a turbulent diffusion flame. *Symp. (Int.) Combust. [Proc.]* **18**, 921.

Stephenson, D. A. (1979). Non-intrusive profiles of atmospheric premixed hydrocarbon-air flames. *Symp. (Int.) Combust. [Proc.]* **17**, 993.

Stephenson, D. A. (1981). High temperature Raman spectra of CO_2 and H_2O for combustion diagnostics. *Appl. Spectrosc.* **35**, 582.

Stephenson, D. A., and Aiman, W. R. (1978). A laser probe of a premixed laminar flame. *Combust. Flame* **31**, 85.

Stephenson, D. A., and Blint, R. J. (1979). Theoretical fitting of computer processed laser Raman spectra from methane- and propane-air flames. *Appl. Spectrosc.* **33**, 41.

Stepowski, D., and Cottereau, M. J. (1979). Direct measurement of OH local concentration in a flame from the fluorescence induced by a single laser pulse. *Appl. Opt.* **18**, 354.

Stepowski, D., and Cottereau, M. J. (1981a). Time resolved study of rotational energy transfer in OH in a flame by laser induced fluorescence. *J. Chem. Phys.* **74**, 6674.

Stepowski, D., and Cottereau, M. J. (1981b). Use of laser-induced fluorescence of hydroxyl to study the perturbation of a flame by a probe. *Symp. (Int.) Combust. [Proc.]* **18**, 1567.

Stepowski, D., and Cottereau, M. J. (1981c). Study of the collisional lifetime of hydroxyl radicals in flames by time-resolved laser-induced fluorescence. *Combust. Flame* **40**, 65.

Stevenson, W. H. (1978). Spray and particulate diagnostics in combustion systems: A review of optical methods. *Cent. Sect. Meet., Combust. Inst., 1978.*

Stevenson, W. H. (1982). Laser Doppler velocimetry: A status report. *Proc. IEEE* **70**, 652.

Stoeckel, F., Melieres, M., and Chenevier, M. (1982). Quantitative measurements of very weak H_2O absorption lines by time resolved intracavity laser spectroscopy. *J. Chem. Phys.* **76**, 2191.

Stricker, W. (1976). Local temperature measurements in flames by laser Raman spectroscopy. *Combust. Flame* **27**, 133.

Styles, A. C., and Chigier, N. A. (1977). Combustion of air blast atomized spray flames. *Symp. (Int.) Combust. [Proc.]* **16**, 619.

Sullivan, B. J., Smith, G. P., Crosley, D. R., and Black, G. (1981). Laser-induced fluorescence studies of the NCO molecule. *East. Sect. Combust. Inst., Pap.* **44**.

Swindell, W., and Barrett, H. H. (1977). Computerized tomography: Taking sectional x-rays. *Phys. Today* **30**, 32.

Swithenbank, J., Beer, J., Taylor, D. S., Abbot, D., and McCreath, C. G. (1977). A laser diagnostic technique for the measurement of droplet and particle size distribution. *In* "Experimental Diagnostics in Gas-Phase Combustion Systems" (B. T. Zinn, ed.), p. 421. Am. Inst. Aeron. Astron., New York.

Swofford, R. L., and Morrell, J. A. (1978). Analysis of the repetitively pulsed dual-beam thermo-optical absorption spectrometer. *J. Appl. Phys.* **49**, 3667.

Teets, R. E., and Bechtel, J. H. (1981a). Sensitivity analysis of a model for the radical recombination region of hydrocarbon air flames. *Symp. (Int.) Combust. [Proc.]* **18**, 425.

Teets, R. E., and Bechtel, J. H. (1981b). Coherent antistokes Raman spectra of oxygen atoms in flames. *Opt. Lett.* **6**, 458.

Tennal, K., Salamo, G. J., and Gupta, R. (1982). Minority species concentration measurements in flames by the photoacoustic technique. *Appl. Opt.* **21**, 2133.

Thompson, B. J. (1974). Holographic particle sizing techniques. *J. Phys. E* **7**, 781.

Thompson, B. J., Ward, J. H., and Zinky, W. R. (1967). Application of hologram techniques for particle size analysis. *Appl. Opt.* **6**, 519.

Thompson, H. D., and Stevenson, W. H., eds. (1979). "Laser Velocimetry and Particle Sizing." Hemisphere Publ. Corp., Washington, D.C.

Tishkoff, J. M., Hammond, D. C., Jr., and Chraplyvy, A. R. (1980). Diagnostic measurements of fuel spray dispersion. *Am. Soc. Mech. Eng.* [*Pap.*] **80-WA/HT-35**.

Tolles, W. M., Nibler, J. W., McDonald, J. R., and Harvey, A. B. (1977). A review of the theory and application of coherent anti-Stokes Raman spectroscopy. *Appl. Spectrosc.* **31**, 253.

Trolinger, J. D. (1974). Laser instrumentation for flow field diagnostics. *AGARDograph* **AGARD-AG-186**.

Trolinger, J. D. (1975). Holographic interferometry as a diagnostic tool for reactive flows. *Combust. Sci. Technol.* **13**, 229.

Trolinger, J. D., and Heap, M. P. (1979). Coal particle combustion studied by holography. *Appl. Opt.* **18**, 1757.

Turk, G. C., Travis, J. C., DeVoe, J. R., and O'Haver, T. C. (1979). Laser enhanced ionization spectroscopy in analytical flames. *Anal. Chem.* **51**, 1890.

Varde, K. S. (1975). A laser interferometer study of combustion near an ignition source in a static chamber. *SAE Tech. Pap. Ser.* **750887**.

Varghese, P. L., and Hanson, R. K. (1981). Collision width measurements of CO in combustion gases using a tunable diode laser. *J. Quant. Spectrosc. Radiat. Transfer* **26**, 339.

Verdieck, J. F., and Bonczyk, P. A. (1981). Laser-induced saturated fluorescence investigations of CH, CN, and NO in flames. *Symp.* (*Int.*) *Combust.* [*Proc.*] **18**, 1559.

Vest, C. M. (1974). Interferometry of strongly refracting axisymmetrix phase objects. *Appl. Opt.* **14**, 1601.

Wang, C. C., Myers, M. T., and Zhou, D. (1981). Observation of competition of rotational effects in the intensity of ultraviolet bands of OH. *Phys. Rev. Lett.* **47**, 490.

Wang, C. P. (1976). Laser applications to turbulent reactive flows; density measurements by resonance absorption and resonance scattering techniques. *Combust. Sci. Technol.* **13**, 211.

Wang, J. Y. (1976). Laser absorption methods for simultaneous determination of temperature and species concentration through a cross section of a radiating flow. *Appl. Opt.* **15**, 768.

Warnatz, J. (1979). Flame velocity and structure of laminar hydrocarbon-air flames. *Prog. Astronaut. Aeronaut.* **76**, 501.

Warnatz, J. (1981). The structure of laminar alkane, alkene, and acetylene flames. *Symp.* (*Int.*) *Combust.* [*Proc.*] **18**, 369.

Warshaw, S., Lapp, M., Penney, C. M., and Drake, M. (1980). Temperature-velocity correlation measurements for turbulent diffusion flames from vibrational Raman scattering data. *In* "Laser Probes for Combustion Chemistry" (D. R. Crosley, ed.), p. 239. Am. Chem. Soc., Washington, D.C.

Webber, B. F., Long, M. B., and Chang, R. K. (1979). Two dimensional average concentration in a jet flow by Raman scattering. *Appl. Phys. Lett.* **35**, 119.

Weinberg, F. J. (1960). Geometric-optical techniques in combustion research. *In* "Progress in Combustion Science and Technology" (J. Ducarme, M. Gerstein, and A. H. Lefebvre, eds.), p. 111. Pergamon, Oxford.

Weinberg, F. J. (1963). "Optics of Flames." Butterworth, London.

Wertheimer, A. L., and Wilcock, W. L. (1976). Light scattering measurements of particle distributions. *Appl. Opt.* **15**, 1616.

Westbrook, C. K. (1982). Chemical kinetics of hydrocarbon oxidation in gaseous detonations. *Combust. Flame* **46**, 191.

Westbrook, C. K., and Dryer, F. L. (1979). Comprehensive mechanism for methanol oxidation. *Combust. Sci. Technol.* **20**, 125.

Williams, F. A. (1965). "Combustion Theory." Addison-Wesley, Reading, Massachusetts.

Wilson, E. B., Jr., Decius, J. C., and Cross, P. C. (1955). "Molecular Vibrations, The Theory of Infrared and Raman Vibrational Spectra." McGraw-Hill, New York.

Withrow, L., and Rassweiler, G. M. (1938). Studying engine combustion by physical methods. *J. Appl. Phys.* **9**, 362.

Witze, P. O. (1979). Application of laser velocimetry to a motored internal combustion engine. *In* "Laser Velocimetry and Particle Sizing" (H. D. Thompson and W. H. Stevenson, eds.), p. 239. Hemisphere Publ. Corp., Washington, D.C.

Witze, P. O. (1980a). A critical comparison of hot wire anemometry and laser Doppler velocimetry for I. C. engine applications. *SAE Tech. Pap. Ser.* **800132**.

Witze, P. O. (1980b). Influence of air motion variations on the performance of a direct injection stratified-charge engine. *Proc. Automob. Div., Inst. Mech. Eng. Conf. Stratified Charge Autom. Eng., Pap.* **C394/80**.

Witze, P. O. (1982). The effect of spark location on combustion in a variable swirl engine. *SAE Tech. Pap. Ser.* **820044**.

Witze, P. O., and Vilchis, F. R. (1981). Stroboscopic laser shadowgraph study of the effect of swirl on homogeneous combustion in a spark-ignition engine. *SAE Tech. Pap. Ser.* **810226**.

Yanagi, T., and Mimura, Y. (1981). Velocity-temperature correlation in premixed flame. *Symp. (Int.) Combust. [Proc.]* **18**, 1031.

York, J. L., and Stubbs, H. E. (1952). Photographic analysis of sprays. *Trans. ASME* **74**, 1157.

Yoshida, A. (1981). An experimental study of a wrinkled laminar flame. *Symp. (Int.) Combust. [Proc.]* **18**, 931.

Yoshida, A., and Tsuji, H. (1979). Measurement of fluctuating temperature and velocity in a turbulent premixed flame. *Symp. (Int.) Combust. [Proc.]* **17**, 945.

Yule, A. J., Chigier, N. A., Atakan, S., and Ungut, A. (1977). Particle size and velocity measurement by laser anemometry. *J. Energy* **1**, 220.

Yule, A. J., Seng, C. A., Felton, P. G., Ungut, A., Chigier, N. A. (1981). A laser tomographic investigation of liquid fuel sprays. *Symp. (Int.) Combust. [Proc.]* **18**, 1501.

Yule, A. J., Seng, C., Felton, P. G., Ungut, A., and Chigier, N. A. (1982). A study of vaporizing fuel sprays by laser techniques. *Combust. Flame* **44**, 71.

Zapka, W., Pokrowsky, P., and Tam, A. C. (1982). Noncontact optoacoustic monitoring of flame temperature profiles. *Opt. Lett.* **7**, 477.

Zinn, B. T., ed. (1977). "Experimental Diagnostics in Gas Phase Combustion Systems." Am. Inst. Aeron. Astron., New York.

Zizak, G., and Winefordner, J. D. (1982). Application of the thermally assisted atomic fluorescence technique to the temperature measurement in a gasoline-air flame. *Combust. Flame* **44**, 35.

Zizak, G., Bradshaw, J. D., and Winefordner, J. D. (1981a). Thermally assisted fluorescence: A new technique for local flame temperature measurement. *Appl. Spectrosc.* **35**, 59.

Zizak, G., Horvath, J. J., and Winefordner, J. W. (1981b). Flame temperature measurement by redistribution of rotational population in laser-excited fluorescence: An application to the OH radical in a methane-air flame. *Appl. Spectrosc.* **35**, 488.

COHERENT ANTI-STOKES RAMAN SPECTROSCOPY (CARS): APPLICATION TO COMBUSTION DIAGNOSTICS

Robert J. Hall and Alan C. Eckbreth

United Technologies Research Center
East Hartford, Connecticut

I. Introduction

The application of light-scattering and wave-mixing techniques for analytical purposes has been made possible largely by the development of high-power laser sources. The availability of high-peak-power, pulsed lasers has made spontaneous Raman scattering and nonlinear variants such as CARS potentially attractive for nonintrusive, *in situ* measurements of species concentrations and temperature. Of particular interest has been the application of these techniques to the hostile, though easily perturbed, environments characterizing practical combustion devices such as furnaces, gas turbine combustors, and internal combustion engines. The goals of improved combustion efficiency and cleanliness, important in this age of energy shortage and environmental concern, will be greatly aided by the understanding gained from spatially and temporally precise nonperturbing diagnostic probing.

The linear, spontaneous Raman technique has been intensively investigated for these applications (Lapp and Penney, 1974; Lederman, 1977; Zinn, 1977; Lapp and Hartley, 1976; Roquemore and Yaney, 1979; Williams *et al.*, 1977; Smith, 1979; Setchell, 1978). However, it has become apparent that the smallness of the Raman cross sections and the incoherent nature of the scattering process restrict its applicability to relatively clean, interference-free environments. Interferences from background luminosity (Eckbreth *et al.*, 1979a; Roquemore and Yaney, 1979), fuel-fragment fluorescences (Bailly *et al.*, 1976; Aeschliman and Setchell, 1975; Leonard, 1974; Yaney, 1979), and laser-induced soot incandescence (Eckbreth, 1977; Pealat *et al.*, 1977) can obscure the Raman signals, often by orders of magnitude. With visible-wavelength laser sources, laser-induced soot incandescence is generally the most severe interference in hydrocarbon-fueled diffusion flames (Eckbreth *et al.*, 1979a). Spontaneous Raman scattering is thus often limited in practical situations to restricted fuels, cycles, spatial locations, and operating stoichiometries. Because of this, it has generally been conceded that a stronger diagnostic technique with an inherently higher signal-to-interference ratio is needed for optical diagnostic probing to achieve its full potential.

With very-high-power, pulsed laser sources, optical effects originating in the nonlinear response of molecules to incident electric fields can become important, and one such phenomenon, designated coherent anti-Stokes Raman spectroscopy (CARS), has come to prominence in recent years because of its great promise for, among other things, combustion diagnostics. Although CARS was discovered and understood by Maker and Terhune in 1965, it remained for several years in the domain of nonlinear optics. It attracted attention for combustion diagnostics based upon the pioneering investigations of J. P. Taran and co-workers at ONERA during the 1970s (Regnier and Taran, 1973; Regnier *et al.*, 1973, 1974; Moya *et al.*, 1975, 1977; Pealat *et al.*, 1980). The superiority of CARS over the incoherent spontaneous Raman process derives from its high signal conversion efficiency and the fact that the CARS signal emerges as a coherent, laser-like beam. CARS signals are often orders of magnitude stronger than those produced by spontaneous Raman scattering. Its coherent behavior ensures collection of the entire signal rather than a small fraction as in an incoherent process. Furthermore, the CARS radiation is collected over such a small solid angle that collection of interferences is greatly minimized. CARS thus enjoys advantages in signal-to-interference ratio of several orders of magnitude over spontaneous Raman processes and is capable of probing practical combustion environments over broad operational ranges. Indeed, CARS has been demonstrated in numerous practical devices, including internal combustion engines, furnaces, coal gasifiers, and afterburning jet engine exhausts. CARS is similar to spontaneous Raman scattering in that it is most suited to thermometry and major species concentration measurements, i.e., typically $>0.1\%$. It is thus complementary to laser-induced fluorescence which is appropriate to molecular radical measurements at trace levels, i.e., parts per million.

In the CARS process depicted in Fig. 1, three photons, two at the "pump" frequency ω_1, and one at the "Stokes" frequency ω_2, are mixed to produce a coherent beam at the anti-Stokes frequency $\omega_3 = 2\omega_1 - \omega_2$.

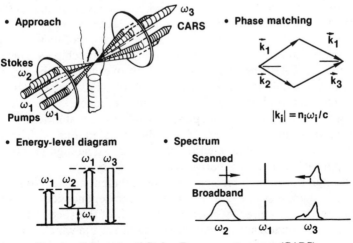

FIG. 1. Coherent anti-Stokes Raman spectroscopy (CARS).

CARS is a three-wave, optical mixing process in which the incident laser fields interact through the third-order nonlinear susceptibility of the medium to induce a polarization which generates the CARS radiation. When the frequency difference ($\omega_1 - \omega_2$) is made to coincide with the frequency of a Raman-active mode in the medium, normally vibrational–rotational transitions, the CARS signal will be resonantly enhanced and the spectral signature will be uniquely characteristic of that species. Measurements of medium properties derive from the density and temperature dependencies contained in the susceptibility. Temperature measurements are performed from the shape of the CARS spectrum, and concentration measurements generally from the intensity or strength of the signal. In certain concentration ranges, however, the spectral shapes are concentration sensitive as well, permitting species measurements from shapes, a unique aspect of CARS spectroscopy. From a diagnostic standpoint, parameter measurements from spectral shapes are always preferable because they are considerably easier to perform. Absolute intensity measurements, even when appropriately referenced, are not trivial and are not as accurate as a measurement from a spectral shape. Stimulated Raman gain (SRG) spectroscopy is another third-order nonlinear process whose physical origin is basically the same as CARS. It has been employed for very-high-resolution molecular spectroscopy (Owyoung, 1978; Owyoung and Esherick, 1980; Rosasco et al., 1983) and for laminar flame probing (Rahn et al., 1979). This technique, however, is more difficult to implement than CARS for "instantaneous" measurements of medium properties. SRG has not yet been developed to the point of practical diagnostic measurements and will not be pursued further here.

Considerable interest centers on CARS because of its potential for very high spectral resolution in molecular studies (Henesian et al., 1977; Harvey, 1981). Resolution is typically limited only by the linewidths of the laser sources employed. Furthermore, the phase-matching alignment requirements of CARS are easily satisfied by overlapping the interacting laser beams in a collinear fashion. For combustion diagnostics, however, where high temporal and spatial precision are generally required, it is common to forgo both the high-spectral-resolution capabilities and simple alignment requirements of CARS as alluded to in Fig. 1. The CARS spectrum can be generated in either of two ways. The high-spectral-resolution approach employs a narrowband Stokes source scanned sequentially through the Raman resonances. For turbulent combustion diagnostics, this is inappropriate due to the nonlinear dependence of CARS on density and temperature. Piecewise generation of the spectrum in a fluctuating environment leads to distorted signatures from which true average properties are not obtainable. This is circumvented by employing a broadband Stokes source (Roh *et al.*, 1976) which generates the entire CARS spectrum with each pulse, allowing instantaneous measurements of medium properties. In laminar flows, the broadband approach can also lead to more expeditious time averaging. For efficient signal generation, the incident laser beams must be aligned such that the wave mixing is properly phased. This can be performed collinearly in gases, but in flames often leads to poor spatial resolution. This is circumvented with crossed-beam phase matching, one variant of which is shown in Fig. 1. CARS is generated only at the mutual intersection of the wave-mixing components leading to high resolution.

In the following sections, the theory and application of CARS to gaseous combustion diagnostics will be discussed. CARS is also finding application to biological and molecular spectroscopy studies, and the interested reader is referred to reviews of these subjects in the books by Durig (1977), Moore (1979), and Harvey (1981). It is the aim of this article to show that CARS is capable of probing with high measurement accuracy the high-interference environments of practical combustion devices. For this and other reasons which will be apparent, it is reasonable to assert that "CARS is now the best optical technique for nonintrusive flame diagnostics" (Pealat *et al.*, 1980), and it is anticipated that CARS will supplant spontaneous Raman scattering for remote, spatially and temporally precise diagnostic probing of combustion processes in many instances.

II. Theoretical Principles and Spectral Synthesis

The physical origin of CARS lies in the nonlinear response of molecules to incident electric fields, giving rise to a macroscopic polarization that is proportional to the third power of electric field amplitude. This polarization in turn acts as a source term in Maxwell's equations for the coherent CARS signal. The third-order nonlinear electric susceptibility relates the macroscopic polarization to the cubic power of electric field and, because it is dependent on molecular properties and medium conditions, it plays a central role in permitting the extraction of diagnostic information from the CARS signals. The ability to calculate CARS spectra therefore plays an important role in deriving temperature and species concentrations from CARS signatures. In this section, the theoretical basis for CARS will be summarized. Treatment of the theory in depth may be found in the original reviews by Armstrong *et al.* (1962), Bloembergen (1965), Maker and Terhune (1965), and Butcher (1965). More recent reviews by Nibler and

Knighten (1979), DeWitt *et al.* (1976), Druet and Taran (1979), Eesley (1982), and Levenson (1982) are also recommended.

A. NONLINEAR POLARIZATION

The nonlinear response of matter subjected to intense optical fields can be expressed as a power series of the polarization in which each term responds to the incident electric fields according to the relation

$$\mathbf{P} = \chi^{(1)}\mathbf{E} + \chi^{(2)}\mathbf{E}^2 + \chi^{(3)}\mathbf{E}^3 + \cdots \tag{1}$$

$\chi^{(1)}$ is the normal linear electric susceptibility describing ordinary refraction and absorption of light waves including such phenomena as Raman and Rayleigh scattering; the second-order term, together with all other even-order terms, is zero in any medium with inversion symmetry (gases, liquids, centrosymmetric crystals). The lowest order nonlinearity in isotropic media is therefore the third-order term $\mathbf{P}^{(3)}$; this term is responsible for CARS, as well as for such phenomena as third-harmonic generation, field-induced second-harmonic generation, optical Kerr effects, and stimulated Raman scattering. Each of these effects derives from a particular way in which different field frequency components are mixed by the third-order electric susceptibility, $\chi^{(3)}$. $\chi^{(3)}$ is a fourth-rank tensor characterized by four distinct frequencies and polarizations and is responsible for the mixing of three independent input fields.

If the input optical electric field is therefore assumed to consist of three distinct frequency components, the total input field may be expressed in plane waves as

$$\mathbf{E} = \sum_1^3 \left(\mathbf{E}(\omega_i)e^{i(\mathbf{k}_i \mathbf{r} - \omega_i t)} + \text{c.c.} \right) \tag{2}$$

where the individual frequency components and wave vectors are denoted by ω_i and $\mathbf{k}_i$ ($|\mathbf{k}_i| = n_i\omega_i/c$, where n_i is the index of refraction at frequency ω_i). Due to the mixing of these fields by $\chi^{(3)}$, frequency components arise in the nonlinear polarization given by

$$\mathbf{P}^{(3)}(\omega_4) = \chi^{(3)}\mathbf{E}_{p_1}(\omega_1)\mathbf{E}_{p_2}(\omega_2)\mathbf{E}_{p_3}(\omega_3)e^{i(\mathbf{k}_4 \mathbf{r} - \omega_3 t)} + \text{c.c.} \tag{3}$$

where

$$p_i = \pm 1, \qquad \omega_4 = \sum_1^3 p_i\omega_i, \qquad \mathbf{k}_4 = \sum_1^3 p_i\mathbf{k}_i$$

and

$$\begin{aligned}\mathbf{E}_{p_i} &= \mathbf{E}(\omega_i) \qquad \text{if} \quad p_i = +1 \\ &= \mathbf{E}^*(\omega_i) \qquad \text{if} \quad p_i = -1\end{aligned} \tag{4}$$

The CARS polarization component is given by

$$\mathbf{P}^{(3)}(\omega_3) = \chi^{(3)}(-\omega_3, \omega_1, \omega_1', -\omega_2)\mathbf{E}(\omega_1)\mathbf{E}(\omega_1')\mathbf{E}^*(\omega_2) + \text{c.c.} \quad (5)$$

where the frequency

$$\omega_3 = \omega_1 + \omega_1' - \omega_2 \quad (6)$$

lies on the anti-Stokes side of ω_1 and ω_1'. The source of the CARS signal thus will be determined by $\chi^{(3)}$. As has been mentioned, $\chi^{(3)}$ is a fourth-rank tensor characterized by four distinct frequencies and polarizations; thus

$$\chi^{(3)} \rightarrow \chi_{ijkl}^{(3)}(-\omega_3, \omega_1, \omega_1', -\omega_2) \quad (7)$$

where the subscripts $ijkl$ denote the polarization orientations of the respective fields in the order given by the frequency arguments. This tensor must be invariant to particular permutations of its indices and fields as well as to all spatial symmetry transformations of the medium. In an isotropic medium, the number of independent tensor elements is thus reduced from 81 to 3; in frequency-degenerate CARS ($\omega_1' = \omega_1$) there are only 2 independent elements.

The symmetry property that $\chi^{(3)}$ be invariant to all permutations of the pairs $j\omega_1$, $k\omega_1'$, and $l\omega_2$ requires that Eq. (5) be multiplied by a factor of 6 when $\omega_1' \neq \omega_1$, and a factor of 3 when $\omega_1' = \omega_1$. That is

$$\mathbf{P}_i^{(3)} = D\chi_{ijkl}^{(3)}(-\omega_3, \omega_1, \omega_1', -\omega_2)\mathbf{E}_j(\omega_1)\mathbf{E}_k(\omega_1')\mathbf{E}_\ell^*(\omega_2) + \text{c.c.} \quad (8)$$

where D is the permutation symmetry factor

$$\begin{aligned} D &= 6 \quad \text{when} \quad \omega_1' \neq \omega_1 \\ &= 3 \quad \text{when} \quad \omega_1' = \omega_1 \end{aligned} \quad (9)$$

Owyoung (1971) has shown that the susceptibility components may be expressed in the form

$$\begin{aligned} \chi_{xxxx} &= \tfrac{1}{24}(3\sigma + 4a + 4b) \\ \chi_{xyyx} &= \tfrac{1}{24}(\sigma + 2b) \\ \chi_{xyxy} &= \chi_{xxyy} = \tfrac{1}{24}(\sigma + 2a + b) \end{aligned} \quad (10)$$

where σ represents a fast-responding, nonresonant electronic contribution, and a and b describe the nuclear response of a molecule. For Raman-type nonlinearities, a and b are resonant terms related to the isotropic and anisotropic parts, respectively, of the polarizability tensor. More detailed expressions for the resonant and nonresonant contributions to $\chi^{(3)}$ will be presented later.

In conventional CARS only one ω_1 input pump beam is employed, $j = k$, and hence the Stokes polarization must be parallel to the pump polarization, $k = 1$. The CARS radiation will then exhibit the same polarization, $i = j$, as the pump and Stokes beams and is generated via χ_{1111}. With two independent pump beams, however, there is independent control of the j and k polarizations. If j and k are parallel, then the Stokes beam must have the same polarization for CARS to be generated, as in conventional CARS. However, if j and k are orthogonally polarized, and the Stokes beam is aligned along one of the pump components, then the CARS radiation will emerge orthogonally polarized to the Stokes beam. The tensor subscripts on $\chi^{(3)}$ will now be dropped from the discussion until Section V, and it can be assumed unless otherwise stated that the pump and Stokes beams have parallel polarizations.

B. GENERATION OF ANTI-STOKES WAVE

If the two input fields (frequency-degenerate CARS with aligned polarizations assumed for simplicity) are represented by plane waves propagating in the z direction

$$\mathbf{E}(\omega_i) = \tfrac{1}{2}A_i^{(0)}e^{i(k_i z - \omega_i t)} + \text{c.c.} \qquad (i = 1, 2) \tag{11}$$

then from Eq. (8) the polarization at $\omega_3 = 2\omega_1 - \omega_2$ is given by

$$\mathbf{P}(\omega_3) = \tfrac{3}{8}\chi^{(3)}A_1^{(0)2}A_2^{(0)*}e^{i(2k_1 z - k_2 z - 2\omega_1 t + \omega_2 t)} + \text{c.c.} \tag{12}$$

which in turn acts as a source for an electric field satisfying the wave equation

$$\nabla \cdot \nabla \cdot \mathbf{E}(\omega_3) - (\omega_3^2/c^2)\varepsilon(\omega_3)\mathbf{E}(\omega_3) = -(4\pi\omega_3^2/c^2)\mathbf{P}(\omega_3) \tag{13}$$

where ε is the optical dielectric constant. For propagation in a charge-free region ($\nabla \cdot \mathbf{E} = 0$) along the z coordinate this becomes

$$\partial^2\mathbf{E}(\omega_3)/\partial z^2 + (\omega_3^2/c^2)\varepsilon(\omega_3)\mathbf{E}(\omega_3) = -(4\pi\omega_3^2/c^2)\mathbf{P}(\omega_3) \tag{14}$$

For an anti-Stokes wave of the form

$$\mathbf{E}(\omega_3) = \tfrac{1}{2}A_3^{(0)}e^{i(k_3 z - \omega_3 t)} + \text{c.c.} \tag{15}$$

Eq. (14), in the slowly varying envelope approximation $\partial^2 A_3/\partial z^2 \ll k_3\partial A_3^{(0)}/\partial z$, becomes

$$\partial A_3^{(0)}/\partial z = (4\pi i\omega_3/cn_3)\mathbf{P}(\omega_3)e^{-i(k_3 z - \omega_3 t)} \tag{16}$$

which can be combined with Eq. (12) and integrated, assuming negligible pump depletion, to give the anti-Stokes or CARS intensity

$$I_3 = \left(\frac{4\pi^2\omega_3}{c^2n_3}\right)^2 I_1^2 I_2 |3\chi^{(3)}|^2 L^2 \left[\frac{\sin(\Delta kL/2)}{\Delta kL/2}\right]^2 \tag{17}$$

where I_i is the intensity of wave i, L is the interaction length, and Δk is the phase mismatch $k_3 - 2k_1 + k_2$. Thus, the efficiency and spectral properties of CARS generation are governed by the squared modulus of the third-order, nonlinear electric susceptibility. If perfect phase matching is achieved, $\Delta k = 0$, and the CARS intensity will vary quadratically with the interaction length. The energy conservation relation $\omega_3 = 2\omega_1 - \omega_2$ does not automatically ensure phase matching, however, because medium dispersion causes the index of refraction to be different at the three frequencies. Because $|k_i| = n_i\omega_i/c$, $\Delta k \neq 0$ in general, and the CARS intensity reaches a succession of progressively weaker maxima. The distance to the first maximum is termed the coherence length and is determined by the relation $\Delta kL_c/2 = \pi/2$. Thus, $L_c = \pi/\Delta k$, and the CARS intensity varies as

$$(2L_c/\pi)^2\sin^2[(L/L_c)(\pi/2)] \tag{18}$$

There are in reality two third-order polarization terms at ω_3. In addition to the source CARS component, there is a depletion term due to stimulated or inverse Raman scattering between ω_3 and ω_1 that is proportional to

$$\chi^{(3)}(+\omega_3, \omega_1, -\omega_1, \omega_3)|E_1(\omega_1)|^2 E_3(\omega_3) \tag{19}$$

This term is usually small compared to the CARS term because $|E_2| \gg |E_3|$ and thus normally results in negligible depletion of the CARS wave.

C. Third-Order Electric Susceptibility

The third-order electric susceptibility can be derived both classically and quantum mechanically. The quantum mechanical approach will be described first because it is the most rigorous and results in a more complete expression for the susceptibility. The problem involves the computation of the expectation value of the electric dipole moment induced by the third power of the electric field, and the general approach is to solve the equation of motion for the density matrix using time-dependent perturbation theory in the impact and isolated line approximations. Because the $\chi^{(3)}$ nonlinearity is a third-order effect, a third-order perturbation expansion is required. In this way, detailed expressions for $\chi^{(3)}$ have been derived algebraically by Butcher (1965), Armstrong et al. (1962), Flytzanis (1975), Lynch (1977), and Bloembergen et al. (1978). Time-ordered Feynman diagrams have also been found to provide a useful bookkeeping technique for keeping track of the large number of terms arising in a perturbation solution, and have been used by Yee et al. (1977) and Druet et al. (1978) to derive a general expression for $\chi^{(3)}$ which is in agreement with the results of Lynch.

The quantum mechanical solution is based on the solution of the equation of motion for the density operator ρ

$$\frac{\partial \rho}{\partial t} = \frac{-i}{\hbar} [H_0 + V(t), \rho] + \frac{\partial \rho}{\partial t}\bigg|_{\text{damping}} \tag{20}$$

with the electric dipole interaction energy $V(t) = -\mathbf{\mu} \cdot \mathbf{E}$, where H_0 and $\mathbf{\mu}$ are the unperturbed molecular Hamiltonian and electric dipole moment operator, respectively. In a perturbation expansion solution

$$\rho(t) = \rho^{(0)} + \rho^{(1)}(t) + \rho^{(2)}(t) + \cdots \tag{21}$$

The CARS polarization will derive from the third-order term

$$\mathbf{P}^{(3)}(\omega_3) = N\text{Tr}[\rho^{(3)}(\omega_3)\mu] \tag{22}$$

where N is the number density of Raman-active molecules, and $\rho^{(3)}(\omega_3)$ is the Fourier component of $\rho^{(3)}$ at $\omega_3 = 2\omega_1 - \omega_2$. If a single Raman-active transition $i \to f$ with frequency ω_{if} is considered, then the full solution to Eq. (20) contains 24 terms proportional to $\rho_{ii}^{(0)}$, an equal number proportional to $\rho_{ff}^{(0)}$, and twice as many proportional to $\rho_{nn}^{(0)}$, where n is an excited electronic energy level (if $\omega_1' \neq \omega_1$, then there are 48 terms proportional to $\rho_{ii}^{(0)}$, etc.). For $\omega_1' = \omega_1$ a total of 16 terms are vibrationally resonant, with the Raman resonance denominator $\omega_{if} - \omega_1 + \omega_2 \simeq 0$. The remaining 80 terms correspond to nonresonant contributions from virtual electronic transitions involving one- and two-photon absorptions, and can be lumped into a nondispersive background contribution, χ^{nr}. Any contributions from remote Raman resonances can be included in χ^{nr} as well. If the probed molecule is a majority species, then χ^{nr} will be relatively small and real if one- and two-photon absorptions are avoided in the diluent molecules. The 16 Raman resonant terms constitute the resonant contribution to $\chi^{(3)}$ (Druet and Taran, 1979) which can be expressed as

$$\chi_{if}^{(3)} = (N/\hbar^3)(\omega_{if} - \omega_1 + \omega_2 - i\Gamma_{if})^{-1}$$

$$\times \left\{ \sum_{n'} \left(\frac{\mu_{in'}\mu_{n'f}}{\omega_{n'i} - \omega_3 - i\Gamma_{n'i}} + \frac{\mu_{in'}\mu_{n'f}}{\omega_{n'f} + \omega_3 + i\Gamma_{n'f}} \right) \right.$$

$$\times \sum_n (\rho_{ii}^{(0)} - \rho_{nn}^{(0)}) \times \left(\frac{\mu_{fn}\mu_{ni}}{\omega_{n'i} + \omega_2 - i\Gamma_{ni}} + \frac{\mu_{fn}\mu_{ni}}{\omega_{ni} - \omega_1 - i\Gamma_{ni}} \right)$$

$$- \sum_{n'} \left(\frac{\mu_{in'}\mu_{n'f}}{\omega_{n'i} - \omega_3 - i\Gamma_{n'i}} + \frac{\mu_{in'}\mu_{n'f}}{\omega_{n'f} + \omega_3 + i\Gamma_{n'f}} \right) \times \sum_n (\rho_{ff}^{(0)} - \rho_{nn}^{(0)})$$

$$\left. \times \left(\frac{\mu_{fn}\mu_{ni}}{\omega_{nf} - \omega_2 + i\Gamma_{nf}} + \frac{\mu_{fn}\mu_{ni}}{\omega_{nf} + \omega_1 + i\Gamma_{nf}} \right) \right\} \tag{23}$$

where Γ represents collisional or Lorentzian damping factors (pressure-broadened linewidths), n and n' denote excited electronic states, and the permutation symmetry factor D has been subsumed into $\chi^{(3)}$.

Proper consideration of the time ordering of the field–molecule interactions in four-wave mixing actually gives rise to vibrationally resonant corrective terms that are not shown in Eq. (23) (Prior et al., 1981). These terms have resonant denominators that correspond, for example, to Raman transitions between initially unpopulated levels in the upper electronic manifold. These resonances, which vanish in the low-pressure limit of radiative damping, have been termed "pressure-induced extra resonances," or "PIER4" resonances, and are attributed to inducement of coherence by dephasing collisions which similarly perturb the initial and final states of the transition (Prior et al., 1981; Grynberg, 1981; Fujimoto and Yee, 1983). For most conditions, however, these corrective terms are expected to be small.

From Eq. (23) it is apparent that the resonant contribution to the CARS susceptibility can be further enhanced if ω_1, ω_2, or ω_3 coincides with electronic absorptions in the probed molecule. This forms the basis for electronic resonance enhancement of CARS, offering the promise of improved minority species detectivity. The technique has been investigated in I_2 vapor by Attal et al. (1978); in C_2 by Gross et al. (1979), Attal et al. (1983), and Greenhalgh (1983a); in NO_2 by Guthals et al. (1979); and in OH by Verdieck et al. (1983); these results will be discussed in Section V,F. Many molecules of practical interest in combustion (N_2, CO, H_2, H_2O, etc.) have electronic absorptions that are not readily accessible by presently available laser sources, however, and thus, with the exception of OH, this technique will not be pursued at great length in this article. For more detailed discussions of electronic resonance enhancement, the reader is particularly referred to Druet *et al.* (1978). An extension of the theory to the nonimpact limit has been given by Mizrahi *et al.* (1983).

If electronic absorptions are avoided, then the population factor $\rho_{nn}^{(0)}$ and electronic damping factors in Eq. (23) can be neglected. With the cross section for spontaneous Raman scattering given by

$$\left(\frac{d\sigma}{d\Omega}\right)_{if} = \frac{\omega_2^4}{c^4} \left|\sum_n \left[\frac{\mu_{fn}\mu_{ni}}{\hbar(\omega_{ni} - \omega_1)} + \frac{\mu_{fn}\mu_{ni}}{\hbar(\omega_{nf} + \omega_1)}\right]\right|^2 \tag{24}$$

Eq. (23) reduces to

$$\chi^{(3)} = \frac{2Nc^4}{\hbar\omega_2^4} \sum_{if} \left(\frac{d\sigma}{d\Omega}\right)_{if} \Delta\rho_{if}^{(0)}/[2(\omega_{if} - \omega_1 + \omega_2) - i\Gamma_{if}] \tag{25}$$

where $\Delta\rho_{if}^{(0)} = \rho_{ii}^{(0)} - \rho_{ff}^{(0)}$, and the contributions of all Raman-active transitions have been summed, a procedure that is valid when adjacent lines are not strongly overlapped. Because $(d\sigma/d\Omega)_{if} = (\omega_2/c)^4\alpha_{if}^2$, where α_{if} is the matrix element of the polarizability, Eq. (25), adding in the background susceptibility, can also be expressed as

$$\chi^{(3)} = \frac{2N}{\hbar} \sum_{if} \frac{\alpha_{if}^2\Delta\rho_{if}^{(0)}}{2(\omega_{if} - \omega_1 + \omega_2) - i\Gamma_{if}} + \chi^{nr} = \sum_j (\chi_j' + i\chi_j'') + \chi^{nr} \tag{26}$$

The expression for the resonant Raman contribution to the susceptibility agrees with the results of a classical calculation based on the polarizability approximation (Moya, 1976; DeWitt *et al.*, 1976). In this model, the vibration amplitude q is represented as a damped harmonic oscillator subjected to a driving force equal to $\frac{1}{2}(\partial\alpha/\partial q)E^2$, where α is the polarizability, and the susceptibility is defined as before from the induced polarization

$$\mathbf{P}^{(3)} = N(\partial\alpha/\partial q)q(\omega_1 - \omega_2)\mathbf{E}_1 \tag{27}$$

where $q(\omega_1 - \omega_2)$ is the component of q modulated at the difference frequency $\omega_1 - \omega_2$.

The resonant contribution to $\chi^{(3)}$ can also be derived from a quantum mechanical analysis based on the polarizability approximation and first-order perturbation expansion of the density matrix (Hall *et al.*, 1980). The equation of motion for the density matrix is solved in this approximation with the interaction energy $-\frac{1}{2}\alpha E^2$. If the density matrix is expanded to first order, then $\chi^{(3)}$ can be defined as before from the appropriate definition of the third-order polarization

$$\mathbf{P}^{(3)} = N\mathrm{Tr}[\rho(\omega_1 - \omega_2)\alpha]\mathbf{E}_1 \tag{28}$$

where $\rho(\omega_1 - \omega_2)$ is the Fourier component of ρ at the difference frequency $\omega_1 - \omega_2$. The result is equivalent to that obtained by third-order perturbation theory in the limit that no field frequencies encroach upon electronic absorptions and the classical polarizability model, Eq. (25). Although the polarizability approach is not useful for analysis of electronic resonance enhancement, it does have the advantage that line overlap or collisional narrowing effects can be readily included. The representation of the resonant part of $\chi^{(3)}$ as a sum of complex Lorentzian lineshapes is strictly valid only for isolated lines because the collisional coupling between off-diagonal density matrix elements has been neglected; the required corrections to the theory will be discussed in Section II,F.

The mixing of two photons at ω_1 and ω_2 can be thought to drive a coherent oscillation in the probed species, producing a modulation of the dielectric constant at the difference frequency $\omega_1 - \omega_2$. The coherent oscillation is dephased by molecular collisions through the linewidths Γ_j. The CARS signal is then the upper sideband resulting from the scattering of a second photon at ω_1 by the dielectric constant modulated at $\omega_1 - \omega_2$. Thus the CARS process is usually thought to involve the destruction of two photons at ω_1 and the creation of photons at ω_2 and ω_3, leaving the molecular state unchanged, and is commonly represented by energy-level diagrams. Such diagrams do have the advantage that the molecular states in resonance with the fields are visualized, but, as Druet and Taran have pointed out, give a misleading picture of energy exchange in CARS. They show that the real and imaginary contributions to $\chi^{(3)}$ correspond respectively to parametric and Raman-type processes. In the parametric process, two ω_1 photons are destroyed and photons at ω_2 and ω_3 are created with no change in material state, but the Raman processes do result in creation of vibrational quanta. If the CARS electric field is small, the Raman process will be the stimulated Raman coupling between ω_1 and ω_2. The rates of the parametric and Raman processes are proportional to $(\chi')^2$ and $(\chi'')^2$, respectively (Druet et al., 1978). The imaginary contribution to $\chi^{(3)}$ displays lineshape or resonance behavior and the real part displays dispersive behavior, as depicted qualitatively in Fig. 2.

The resonant part of $\chi^{(3)}$ contains information about the state of material excitation of temperature through the population factors $\Delta\rho_{if}^{(0)}$ and is directly proportional to the number density N of the probed molecule; thus, examination of the resonant contribution to the CARS signal can yield information about temperature and species concentration. The background susceptibility, on the other hand, is spectrally uninteresting, and will interfere with the desired resonant signal for a weak Raman mode or a minority-probed species. However, the presence of the background susceptibility is not necessarily detrimental. In certain concentration

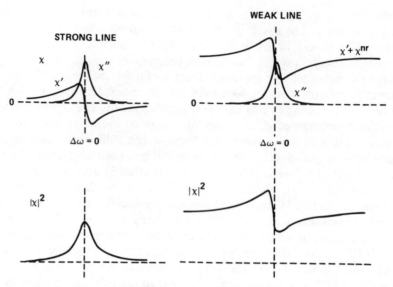

FIG. 2.   Illustration of strong- and weak-line CARS intensity profiles.

ranges, the interference between the resonant and nonresonant signals gives rise to CARS spectral shapes that are concentration sensitive, thus offering the unique capability of performing concentration measurements from spectral shapes. If the probed molecule concentration is sufficiently low, the desired resonant signal will disappear into a dispersionless background (Fig. 2), and no diagnostic information can be retrieved from the CARS signal unless steps are taken to suppress the nonresonant background contributions. There are, however, several techniques for reducing or removing the background signals which offer the promise of improving the minority species detectivity of CARS. Also, because the background susceptibility is proportional to N, the signal associated with it can serve as an *in situ* reference standard for measurements of concentration using integrated intensities. These concepts will be discussed in Section V. Thus, although the background susceptibility was at one time perceived to be a major limitation of CARS, it is now becoming clear that its presence can be exploited to extend the diagnostic range of CARS.

The interpretation of CARS spectra is complicated by constructive and destructive interference effects which occur between adjacent Raman lines and between the resonant and nonresonant contributions to the susceptibility. This complication can be understood by examining the expression for $|\chi^{(3)}|^2$

$$|\chi^{(3)}|^2 = \left| \sum_j (\chi_j' + i\chi_j'') + \chi^{nr} \right|^2$$

$$= \left(\sum_j \chi_j' \right)^2 + \left(\sum_j \chi_j'' \right)^2 + 2\chi^{nr} \sum_j \chi_j' + \chi^{nr2} \tag{29}$$

For a minority species, the pure resonant contributions represented by the first two terms will be small and the CARS spectral distribution will take on the dispersive profile of the real resonant part, $\sum_j \chi_j'$. For active molecule concentrations approaching the lower detectivity limit, the resonant contribution appears as a faint modulation of the dispersionless background signal. Because the real part χ_j' goes from positive to negative in the vicinity of exact tuning ($\omega_1 - \omega_2 = \omega_j$), the interference between the resonant and background contributions gives rise to a characteristic interference minimum, as seen in Fig. 2. For a majority species, the influence of the background becomes progressively weaker and the signature becomes dominated by the two resonant terms, $(\sum_j \chi_j')^2$ and $(\sum_j \chi_j'')^2$. If $\chi_j' + i\chi_j'' = K_j[\Gamma_j/(2\Delta\omega_j - i\Gamma_j)]$, then $|\chi|^2$ will contain isolated line contributions of the form $K_j^2[\Gamma_j^2/(4\Delta\omega_j^2 + \Gamma_j^2)]$, as well as cross terms of the form

$$\frac{K_j K_k \Gamma_j \Gamma_k}{(2\Delta\omega_j - i\Gamma_j)(2\Delta\omega_k + i\Gamma_k)} + \text{c.c.} \tag{30}$$

The cross or interactive components give rise to constructive interference effects between lines if the lines are overlapped. Because they depend on line positions and homogeneous linewidths, they represent an additional complication in the interpretation of CARS spectra. The existence of these terms, which is unique to CARS and which arises from the nonlinear character of the CARS process, generally precludes simplified data reduction techniques. Data reduction in CARS is thus somewhat more complicated than it is for spontaneous Raman scattering, which is simply an incoherent addition of contributions from each transition. The spontaneous Raman signal is roughly proportional to $\sum_j \chi_j''$, and a spectrally integrated measurement will not display sensitivity to homogeneous linewidth, Γ_j. The CARS interference effects are readily handled numerically by computer (Hall, 1979, 1983), however, and therefore do not pose significant problems for data reduction. The sensitivity of CARS to Raman linewidth will be discussed later.

Stimulated Raman coupling between ω_1 and ω_2 can lead to significant population perturbation and measurement inaccuracies if the fields at ω_1 and ω_2 are too intense. The characteristic time for population transfer from $i \rightarrow f$ due to the stimulated Raman process for monochromatic sources (Regnier et al., 1974) is given by

$$\tau^{-1} = -\frac{1}{\Delta\rho_{if}^{(0)}} \frac{\partial\Delta\rho_{if}^{(0)}}{\partial t} = 2 \left(\frac{4\pi}{\hbar c}\right)^2 \alpha_{if}^2 \frac{\Gamma_j}{4\Delta\omega_j^2 + \Gamma_j^2} I_1 I_2 \qquad (31)$$

Population perturbations might be expected if this characteristic time is shorter than the exciting pulse duration. In practice, the intensity product threshold for population perturbation or saturation will be higher than that calculated from Eq. (31) because of collisional relaxation processes which will tend to maintain a thermal equilibrium distribution of states.

D. CARS SELECTION RULES

The vibrational and rotational selection rules for CARS have been found by Nibler and Knighten (1979) and Yuratich and Hanna (1976) to be essentially identical to those for spontaneous Raman scattering. For a detailed discussion of these rules the reader is referred to these reviews and to the review by Weber (1973). The special case of diatomic molecules provides

$$
\begin{array}{lllll}
\text{vibrational} & \Delta v = +1 & \Delta J = \begin{array}{l} = +2 \\ 0 \\ = -2 \end{array} & \begin{array}{l} \text{S-branch} \\ \text{Q-branch} \\ \text{O-branch} \end{array} & (32) \\
\text{CARS} & & & & \\
& & & & \\
\text{rotational} & \Delta v = 0 & \Delta J = \pm 2 & \text{O, S-branch} & \\
\text{CARS} & & & &
\end{array}
$$

The polarizability matrix elements [Eq. (26)] for vibrational CARS are of the form

$$\alpha_{if} = \langle v, J|\alpha|v + 1, J + \Delta J\rangle \qquad \Delta J = 0, \pm 2 \qquad (33)$$

and may be expressed in the harmonic oscillator–rigid rotor approximation as

Q-branch: $|\langle v, J|\alpha|v + 1, J\rangle|^2 = \dfrac{\hbar}{2M\omega_v} [\alpha'^2 + \tfrac{4}{45}b_J^J\gamma'^2](v + 1)$

$$(34)$$

O, S-branch: $|\langle v, J|\alpha|v + 1, J \pm 2\rangle|^2 = \dfrac{\hbar}{2M\omega_v} [\tfrac{1}{15}b_J^{J\pm2}\gamma'^2](v + 1)$

where M is the reduced mass of the molecular oscillator, ω_v is the associated vibrational frequency, the b_j^i are Placzek–Teller coefficients, and α' and γ' are the derivatives of the mean polarizability and anisotropy, respectively, with respect to the internuclear coordinate. A linear variation of α and γ with internuclear coordinate has been assumed. In vibrational CARS a considerable simplification results from the fact that the depolarization of the Raman modes of most molecules of interest is quite small. This means that the O- and S-branch transitions usually make a negligible contribution to the vibrational CARS spectrum, and that the Q-branch is dominated by the isotropic part of the polarizability tensor, α. For example, O- and S-branch transitions make a negligible contribution to the calculated vibrational CARS spectra of N_2, CO, H_2, O_2, H_2O, and CO_2. Corrections to Eq. (34) for vibrational anharmonicity and centrifugal distortion can be made, but normally will be quite small. The analysis of Bovanich and Brodbeck (1976) shows that the Q-branch matrix element is subject to a J-dependent correction factor

$$\left[1 + \frac{3}{4}\left(\frac{2B_e}{\omega_e}\right)^2 (a_1 - 1)J(J + 1) \right] \tag{35}$$

where B_e and ω_e are the usual rotational and vibrational constants and a_1 is the first of the Dunham anharmonic coefficients (Dunham, 1932). For N_2, CO, and O_2 the correction is no more than a few percent even for the large J values important in flames. For H_2, however, the correction will be significant for much smaller values of J and should be included in data reduction. If the effect of vibrational anharmonicity on the vibrational portion of the matrix element is also analyzed by the methods of Bovanich and Brodbeck, it is also found to be small in molecules such as N_2, CO, and O_2 whose CARS spectra contain significant contributions from higher vibrational states (hot bands).

For diatomic vibrational Q-branches, the transition frequency or term value is given by the expression $\omega(vJ \rightarrow v + 1, J) \simeq \omega_{v,v+1} - \alpha_e J(J + 1)$ where $\omega_{v,v+1}$ is the difference in vibrational energies, and α_e gives the linear variation of rotational constant with vibrational state. In a given vibrational band, therefore, there will be a splitting of individual Q-transition components determined by α_e. The origins of neighboring vibrational bands will be separated by approximately $2\chi_e\omega_e$, where $\chi_e\omega_e$ is the familiar vibrational anharmonicity coefficient.

For pure rotational CARS, the derivation of $\chi^{(3)}$ must take into account the selection rules obeyed by the magnetic sublevels m belonging to each rotational state $J(-J \leq m \leq J)$ (Shirley et al., 1980a). If the pump and

Stokes sources have parallel polarizations, $\Delta m = 0$, and the appropriate modulus for the susceptibility is

$$K_{JJ'} = \frac{2N}{\hbar} \left(\frac{4}{45}\right)(b_J^{J'}\gamma_{00}^2)\Delta\rho_{JJ'}^{(0)}\Gamma_{JJ'}^{-1} \qquad (36)$$

where γ_{00} is the anisotropic polarizability, and the normalized population difference factor $\Delta\rho_{JJ'}^{(0)} = \rho_{JJ}^{(0)} - (2J + 1)/(2J' + 1)\rho_{J'J'}^{(0)}$. The spontaneous Raman cross sections for pure rotational scattering are usually larger than those for Q-branch scattering, but this potential signal strength advantage for pure rotational CARS will tend to be offset by the population difference factors, which are much smaller than those for vibrational transitions. This question will be addressed in Section IV,F. The term value for O- and S-branch transitions is

$$\omega_{JJ'} = E_{J\pm2} - E_J \simeq (4J + 6)B_v \qquad \text{(S-branch)}$$
$$\simeq (-4J + 2)B_v \qquad \text{(O-branch)} \qquad (37)$$

Thus, the O- and S-branches will consist of essentially equally spaced lines with a separation equal to $4B_v$, where B_v is the rotational constant.

The selection rules governing the vibrational Q-branch spectra of the triatomics H_2O and CO_2 will be discussed later in discussions describing thermometry using these molecules.

E. CARS Linewidths

The nonlinear nature of the CARS process has been shown to lead to a dependence of theoretical spectra on assumed Raman linewidth. In the pressure-broadened regime, these widths represent the rate at which the coherences driven by ω_1 and ω_2 are dephased by molecular collisions, and are identical to the spontaneous Raman linewidths. Because CARS data reduction depends in most cases on fitting theoretical to experimental signatures, information about these widths is required for high measurement accuracy. The quantum mechanical derivation leading to Eqs. (25) and (26) for $\chi^{(3)}$ is based on the assumption that each Raman line is homogeneously (pressure) broadened by collisions in the binary collision regime. For most molecules of practical interest in combustion, with the exception of H_2, the assumption of pressure broadening will be valid for pressures in excess of 1 atm.

1. Doppler Broadening

The low-pressure or Doppler-broadened limit has been examined by Henesian and Byer (1978) and by Druet et al. (1979). The physical origin of the Doppler effect in CARS is that molecular thermal motions result in a slight, effective detuning of fields nominally tuned to the Raman resonance. If a normalized detuning is defined as $\delta_j = [\omega_j - (\omega_1 - \omega_2)]/\Delta\omega_D$, and the ratio of pressure-to-Doppler linewidths as $E_j = \Gamma_j/2\Delta\omega_D$, then the resonant part of $\chi^{(3)}$ can be expressed (Henesian and Byer, 1978) in terms of the complex error function $w(z)$ as

$$\chi^{(3)} = \frac{N}{\hbar} \frac{\sqrt{\pi}}{\Delta\omega_D} \sum_j \alpha_j^2 \Delta\rho_j^{(0)} iw^*(\delta_j + iE_j) \tag{38}$$

where the Doppler width for forward spontaneous Raman scattering is $\Delta\omega_D = \omega_j \sqrt{kT/M}/c$, M being the mass of the scattering molecule. The forward-scattering Doppler width occurs in CARS because the CARS signal is generated in the forward direction. In the Doppler limit, $\Gamma_j \to 0$, and the imaginary part of $\chi^{(3)}$ is Gaussian with full width at $1/e$ intensity = $2\Delta\omega_D$. The CARS lineshape is not Gaussian because of the dispersive real contribution to $\chi^{(3)}$; as a consequence, the CARS width in the Doppler regime at $1/e$ intensity is about $2.4\Delta\omega_D$ (Henesian and Byer, 1978).

2. Dicke Narrowing

In the transition from the Doppler to the pressure-broadened regime, an initial narrowing of the Doppler profile width occurs with increasing pressure. This effect, termed Dicke narrowing (Dicke, 1953; Wittke and Dicke, 1956), gives rise to a minimum value of linewidth as a function of pressure in H_2 and D_2 (Henesian et al., 1976). The physical explanation for the effect is that elastic collisions cause diffusive molecular motion, absorbing recoil momentum imparted to the active molecule in the photon scattering process. Dicke narrowing can also be interpreted in terms of collisional coupling of different velocity classes within the Doppler profile; when the switching frequency due to momentum-changing collisions exceeds the spacing between velocity classes, the overall lineshape will narrow with increasing pressure (see Section II,F). Dicke narrowing can be included in the lineshape analysis in the manner presented by Galatry (1961); sensibly, this requires that Γ_j be replaced by $\Gamma_j + 4\Pi^2 D/\lambda^2$, where D is the diffusion coefficient for the scattering molecule and λ is the Raman shift wavelength. For relatively heavy molecules such as N_2, CO, and H_2O, Dicke narrowing will not be observable for experimental conditions of practical interest; for light molecules such as H_2 and D_2, it is important for pressures below approximately 10 atm.

Thus, for most practical applications, even at flame temperatures, the individual Raman transitions will be pressure broadened, with widths that increase linearly with pressure. In the limit $\Gamma_j/\Delta\omega_D \gg 1$, $w(z) \to i/\sqrt{\pi} z$, and Eq. (38) reduces to a sum of complex Lorentzian lineshapes.

3. Pressure Broadening

As has been discussed, the ordinary impact pressure-broadening regime is the one of most importance for practical CARS diagnostics. The pressure-broadened Raman linewidth for an isolated transition represents the collisional decay rate of the off-diagonal density matrix element or coherence associated with the transition. In general, contributions to the Raman linewidth will be made by inelastic lifetime-limiting collisions and by elastic collisions which dephase the vibrational or rotational motions (Gordon, 1966b). For a transition $i \rightarrow f$, the coherence dephasing rate in the pressure-broadened regime (Wilson-Gordon et al., 1982) may be represented as

$$\frac{1}{2} \Gamma_{if} = \left(\frac{1}{T_2}\right)_{if} = \frac{1}{2} \left[\left(\frac{1}{T_1}\right)_i + \left(\frac{1}{T_1}\right)_f \right] + \left(\frac{1}{T_2^*}\right)_{if} \tag{39}$$

where T_1 represents lifetimes due to inelastic collisional processes, and T_2^* is a lifetime associated with phase-interrupting collisions. The T_1 terms will arise mainly from rotational redistribution processes if vibrational relaxation rates are slow; although both vibrational and rotational dephasing in general contribute to the pure dephasing T_2^* term, only the vibrational process will make a contribution to the broadening of isotropic Q-branch transitions if vibration–rotation interaction is small. The vibrational dephasing is manifested through a vibrational phase shift η_{if} which can be represented in terms of an integral of the isotropic intermolecular potential V_{ISO} over the collision trajectory (Robert, 1982)

$$\eta_{if} = \hbar^{-1} \int_{-\infty}^{+\infty} dt' \left[\langle v_f | V_{ISO}(t') | v_f \rangle - \langle v_i | V_{ISO}(t') | v_i \rangle \right] \tag{40}$$

If V_{ISO} is not strongly dependent on the vibrational quantum number, then the collision-induced phase shifts will be small. For the vibrational Q-branch transitions of most interest, rotationally inelastic collisions are indeed expected to make the dominant contribution to linewidth.

The impact theory of spectral line broadening has been extensively developed since it was first put forward by Anderson (1949). Fiutak and Van Kranendonk (1963) extended the Anderson theory to describe broadening of Raman lines, and Tsao and Curnutte (1962) extended the theory to a large number of molecular interactions. In the foregoing semiclassical theories, the translational motion is treated classically, in a straight-path approximation, with a quantum mechanical second-order perturbation treatment of the vibrational and rotational motions. The interaction potential in these early models is of a long-range multipole type, with an arbitrary distance of closest approach serving as a catchall for short-range forces. The overall agreement of the early models with experimental Raman linewidth data is fair. Srivastava and Zaidi (1977) extended the calculation to include an averaged short-range or core potential and obtained good agreement for the rotational linewidths of N_2, H_2, and CO_2. Further improvements have been contributed by Robert and Bonamy (1979), who represent the short-range interaction potential as a sum of atom–atom Lennard–Jones potentials, and by Bonamy et al. (1977), who introduce a parabolic trajectory for close collisions instead of a straight path. These authors also extend the calculation to a higher order of perturbation theory and obtain very good agreement for the rotational Raman linewidths of pure N_2, CO_2, and CO. A calculation (Hall, 1980) based on the theory of Bonamy et al. (1977) also gives good agreement with stimulated Raman gain measurements (Rahn et al., 1980a) of N_2 Q-branch widths at 300 K and in a flame at ~1700 K. A quasi-classical model based on an atom–atom pairwise additive short-range potential and Monte Carlo averaging over collision parameters has been presented by Rahn et al. (1980a). This calculation also gives good agreement with the measured N_2 Q-branch linewidths. Summaries of the available experimental data are given in linewidth theory reviews and by Weber (1973).

In general, Raman homogeneous linewidths are in the range 0.05–0.50 cm^{-1}/atm at room temperature and can be expected to have an inverse temperature dependence. For diatomics, the widths will generally decrease with increasing rotational quantum number, reflecting the increasing lifetime of the higher levels whose spacing increases with respect to kT. The rotational quantum number dependence can be expected to become less pronounced at high temperatures. It is also expected that there will be some dependence of linewidth on collision partner. Characterization of the linewidths is thus potentially quite complicated, with different temperature and concentration dependencies expected for each transition. However, the very good agreement of the recent linewidth theories with the experimental data that do exist suggests that theoretical calculations can be used to fill in gaps in the data. Of particular value are calculations such as those of Bonamy et al. (1977), which do not involve the use of adjustable parameters. Also, if lifetime-limiting rotationally inelastic collisions are predominant, then the linewidths should not have a significant dependence on the radiative vibrational and rotational selection rules; that is, the widths for pure rotational Raman, microwave, infrared, and vibrational Raman should all be roughly the same. Thus, for example, a good approximation to unknown Q-branch Raman linewidths might be provided by microwave linewidth data, if available. Measured Q-branch Raman linewidths for CO have indeed been found to be remarkably similar to the associated infrared linewidths for the fundamental and vibrational overtones (Rosasco *et al.*, 1983; BelBruno *et al.*, 1983). The high-resolution inverse Raman (IRS) techniques (Rosasco *et al.*, 1983; Rahn, *et al.*, 1980a) can also be expected to generate a vast amount of basic linewidth data. The sensitivity of calculated CARS spectra to homogeneous linewidth variations will vary from molecule to molecule, and this question will be addressed for each molecule in the discussion on thermometry and concentration measurements (Sections IV and V).

There are a number of other possible contributions to the homogeneous linewidth which should be evaluated for each set of experimental conditions. Among these are transit-time broadening, which arises from the thermal motion of the probed molecules across the finite diameter of the pump and Stokes beams; power broadening, which arises from perturbation of the state populations by stimulated Raman scattering between ω_1 and ω_2; pulse lifetime broadening due to the finite lifetime of the laser sources; and Stark broadening due to the intense pump electric field (Rahn *et al.*, 1980b). In addition, St. Peters (1979) has shown that axial mode dephasing in multimode sources can lead to an augmentation of the homogeneous linewidth by the individual modal widths of the pump and Stokes if these widths are not transform limited. For most experimental conditions, however, these phenomena will make a relatively small contribution to the homogeneous linewidth.

The foregoing discussion of Raman linewidths has implicitly assumed that the pressure is not so high that adjacent transitions are strongly overlapped. As long as this is the case, the lineshape associated with a particular resonance is adequately described by a pressure-broadened linewidth whose value will increase linearly with pressure. However, when the linewidths become comparable to the splitting between adjacent transitions, or when the ability to resolve individual transitions is being lost, the character of the broadening changes and knowledge of individual or isolated linewidths no longer suffices to describe the bandshape. Necessary modifications to the theory in this high-pressure regime are described in the following.

F. Pressure Effects in CARS

Practical considerations require that the effect of pressure on the shape and strength of CARS signatures be evaluated because many practical combustion systems operate at pressures in excess of 1 atm. Examples include gas turbine combustors, jet engines, and internal combustion engines. A naive approach to theoretical analysis would be to apply Eq. (26) for $\chi^{(3)}$ at continuously increasing pressures. The isolated linewidths Γ_j

can be expected to increase linearly with pressure such that, as the pressure increased, the spectrum width would be determined not by frequency splitting between lines, but by some average value of Γ_j. All Q-branches would merge into a single, unresolved band whose width would increase linearly with pressure. The observed pressure dependencies of the N_2 CARS and Raman signatures are much different, however. Experimentally, it is observed that the N_2 CARS spectrum at 300 K begins to narrow immediately as the pressure is increased from 1 atm and then remains sensibly constant from 10 to about 100 atm (Hall et al., 1980). The 300 K spontaneous Raman spectrum of N_2 begins narrowing at a pressure of about 30 atm and continues to about 350 atm (May et al., 1970). These results are a manifestation of the phenomenon of collisional, or pressure-induced, narrowing, which has its origins in collisional energy transfer between energy levels and which becomes important when the pressure has increased to the point where neighboring transitions have begun to overlap substantially. Writing the resonant susceptibility as a sum of contributions with complex Lorentzian lineshape factors, as in Eq. (26), assumes that each transition is independently broadened, but this isolated line approximation is strictly valid only for lines that do not overlap. The isolated linewidth Γ_j associated with each transition represents the rate at which the coherent oscillation driven by ω_1 and ω_2 is dephased by collisions; in the absence of vibrational effects, it arises from lifetime-limiting rotationally inelastic collisions. If the driving frequency difference between the pump and Stokes sources is resonant with a set of neighboring transitions that are substantially overlapped, then the dynamic response of the medium will be affected by collisional energy transfer between the energy levels associated with the transitions. There does not seem to be an entirely satisfactory physical picture for the fact that the spectrum narrows, however. In one picture, it is noted that frequency splitting due to vibration–rotation interaction is the determinant of the bandwidth at low pressure, and that there is therefore a characteristic frequency for the radiation process that is determined by vibration–rotation interaction. When the inelastic collision frequency becomes large compared to this characteristic frequency, the molecules will then be passing through many different rotational states during the radiation process, and they will thus tend to radiate at a frequency characteristic of the more populous rotational states. Thus, the spectrum will tend to coalesce or collapse toward a frequency center of gravity determined by the most populous rotational level. If, however, phase-interrupting collisions are important, ordinary pressure broadening will occur to counteract the narrowing effect of the phase-conserving collisions.

An expression for $\chi^{(3)}$ which accounts for the collisional narrowing phenomenon has been derived by Hall *et al.* (1980) in the polarizability and impact approximations:

$$\chi_3 = iN/\hbar \sum_r \alpha_r \sum_s [G]_{rs}^{-1} \Delta\rho_s^{(0)} \alpha_s \tag{41}$$

where the elements of the G matrix are given by

$$G_{rs} = i(\omega_1 - \omega_2 - \omega_r)\delta_{rs} + [(\Gamma_r/2) - i\Delta_r]\delta_{rs} + \gamma_{rs}(1 - \delta_{rs}) \tag{42}$$

This expression for $\chi^{(3)}$ differs from that which is valid in the isolated line approximation by the appearance of off-diagonal linewidth parameters, γ_{rs}. These parameters describe the rate of collisional energy transfer between states and control the spectrum collapse with increasing pressure. If the collisional scattering-matrix S is defined by

$$\psi' = S\psi \tag{43}$$

where ψ and ψ' are the molecular wave functions before and after a collision event, respectively, then Γ_r, Δ_r, and γ_{rs} can be related to the S matrix (Alekseyev *et al.,* 1968) through the relationships

$$\tfrac{1}{2}\Gamma_r - i\Delta_r = \int (1 - S_{r_1 r_1}^+ S_{r_2 r_2}) F(g)\, dg$$

$$\gamma_{rs} = -\int S_{r_1 s_1}^+ S_{s_2 r_2} F(g)\, dg \tag{44}$$

where g denotes the parameters describing a bimolecular collision (impact parameter, relative velocity, internal state of collision partner, etc.) and $F(g)$ is the rate of g-type collisions.

In molecules such as N_2 and CO it should be a good approximation, as stated earlier, to neglect the contributions of vibrational dephasing and vibrationally inelastic collisions, and to assume that the S matrix is sensibly independent of vibrational level. In this case, for Q-branch transitions

$$\Delta_r = 0$$

$$\tfrac{1}{2}\Gamma_r = \int (1 - |S_{rr}|^2) F(g)\, dg \tag{45}$$

$$\gamma_{rs} = -\int |S_{sr}|^2 F(g)\, dg$$

and γ_{rs} is simply the first-order rate constant for rotational energy transfer between rotational states r and s (in the direction s to r). Conservation of probability in bimolecular collisions requires that

$$\sum_{s \neq r} \gamma_{sr} = -\Gamma_r/2 \tag{46}$$

and the principle of detailed balance can be applied to relate forward and reverse rate constants, viz.

$$\rho_s^{(0)}\gamma_{rs} = \rho_r^{(0)}\gamma_{sr} \tag{47}$$

Thus, to describe the CARS spectrum of overlapping lines, knowledge of inelastic rate processes is required. Even if the Γ_r are known, however, Eqs. (46) and (47) do not suffice to fill out the matrix γ_{rs} parameters. Either these quantities will have to be calculated on an *ab initio* basis, or a phenomenological approach can be taken (Hall *et al.*, 1980; Hall, 1983; Stufflebeam *et al.*, 1983; Rosasco *et al.*, 1983).

Collisional narrowing will be most important in those conditions where lines are strongly overlapped, the overlap being governed by the isolated linewidths and the spacing between adjacent lines. Isolated linewidths have an inverse temperature dependence, and, as the temperature is increased, Q-branch transitions corresponding to higher rotational quantum numbers with a greater splitting between lines become important. Thus, line overlap is expected to be less important at elevated temperatures, but is nevertheless significant for conditions of great practical interest, as will be seen. For moderate pressures where the overlap effects are of first order, a useful approximation for $\chi^{(3)}$ can be derived by expanding G^{-1} to first order in pressure, giving

$$\chi_3 = \frac{2N}{\hbar} \sum_r \frac{\alpha_r^2 \Delta\rho_r^{(0)}}{2\Delta\omega_r - i\Gamma_r} - \frac{iN}{\hbar} \sum_r \frac{\alpha_r}{i\Delta\omega_r + \frac{1}{2}\Gamma_r} \sum_s \frac{\gamma_{rs}\alpha_s \Delta\rho_s^{(0)}}{i\Delta\omega_s + \frac{1}{2}\Gamma_s} \tag{48}$$

In this expression the overlap effects appear as a correction term to the normal isolated line approximation.

One phenomenological model for rotational cross-relaxation that has been found to have widespread utility in correlating measured rates is based on an inverse power relationship between relaxation rate and the rotational energy defect (Brunner and Pritchard, 1982). A simplified statement of the relationship (Stufflebeam *et al.*, 1983) is

$$\gamma_{rs} = -K_0\rho_{rr}^{(0)} \mid \Delta E_{rs}\mid^{-\gamma} \tag{49}$$

where K_0 and γ are constants. The off-diagonal relaxation or linewidth parameters may thus be calculated by least-squares fitting to the measured isolated linewidths. This procedure gives a very good fit to the N_2 and CO Q-branch linewidths measured by Rahn *et al.* (1980a) and Rosasco *et al.* (1983) (Stufflebeam *et al.*, 1983).

It should be stressed that the off-diagonal linewidth elements, which are generally tensor quantities, control the coupling of off-diagonal density matrix elements or coherences by collisions. It is only in the limit that the scattering matrix does not depend on vibrational quantum number that one has the fortuitous result that these coupling or mixing coefficients are equivalent to rotational relaxation rates. The fact that inelastic rotational transfer rates are then being dealt with does not imply that rotational nonequilibrium is important; to the contrary, the first-order perturbation treatment is restricted to conditions in which the initial distribution of molecular states (usually Boltzmann) is undisturbed.

Even though it is necessary to resort in practice to phenomenological estimates of the off-diagonal relaxation linewidth parameters, the foregoing represents what might be termed an "exact" treatment of collisional narrowing effects. The requirement of inverting the G matrix, however, can make this approach very cumbersome and costly for conditions requiring the inclusion of large numbers of transitions. A simpler, approximate approach based on Gordon's extended J-diffusion theory of rotational motion (Gordon, 1966a) results in a simpler algorithm for the CARS susceptibility. Brueck (1977) has applied Gordon's model to the problem of tripling in liquids, and Hall and Greenhalgh (1982) have in turn adapted these results to CARS. By representing the damping term [Eq. (20)] in simple time-constant-relaxation form, it is possible to derive the following expression for the third-order susceptibility for the $v \rightarrow v + 1$ band:

$$\chi_{v,v+1}^{(3)} = \frac{2N}{\hbar} \alpha_{v,v+1}^2 \frac{\sum_j \Delta\rho_{jj}^{(0)}/(2\Delta\omega_j - i\Gamma_j)}{1 + i \sum_j (1 - \phi_j)\Gamma_j\rho_{jj}^{(0)}/(2\Delta\omega_j - i\Gamma_j)} \tag{50}$$

where ϕ_j represents the fractional contribution of pure dephasing processes to the linewidth of transition $Q(j)$. Evaluation of Eq. (50) requires essentially no more run time than an isolated line calculation and generally gives very good results, as will be seen. The Gordon model approach to narrowing problems will undoubtedly play a role in rapid data reduction schemes.

G. Finite Bandwidth Sources

If the pump and Stokes sources are monochromatic, the generated CARS spectrum will be proportional to $|\chi^{(3)}(\omega_1 - \omega_2)|^2$, with ω_2 varied to scan across a vibrational transition. However, the assumption of source monochromaticity is almost never valid in practice; even single-mode pulsed sources typically have finite frequency bandwidths comparable to pressure-broadened Raman linewidths, and the desire to achieve single-pulse generation of CARS spectra requires a broadband Stokes source.

As has been discussed, practical combustion environments are often turbulent or nonstationary, and the nonlinear dependence of the CARS intensity on density and temperature renders the scanning approach impractical in these situations. To achieve the single-shot capability required for unsteady media, a broadband Stokes source must be employed, with a bandwidth sufficiently large to encompass all of the vibrational or rotational resonances. Use of a broadband Stokes source simulates a large number of temporally superimposed, scanned experiments in which each Q-branch transition selects the spectral segment of the Stokes source for which the resonance condition $\omega_1 - \omega_2 \simeq \omega_{vJ}$ is satisfied. In either the scanned or multiplex approaches to spectral synthesis, the finite bandwidths of the laser sources must be accounted for if the data are to be reduced properly.

The definition of $\chi^{(3)}$ is that it relates the Fourier components of the third-order electric polarization to the Fourier components of the driving electric fields, e.g.,

$$\mathbf{P}_3(\omega_3 = \omega_1 + \omega_1' - \omega_2) = \chi^{(3)}(-\omega_3, \omega_1, \omega_1', -\omega_2)\mathbf{E}_1(\omega_1)\mathbf{E}_1(\omega_1')\mathbf{E}_2^*(\omega_2)$$

$$= \chi^{(3)}(\omega_1 - \omega_2)\mathbf{E}_1(\omega_1)\mathbf{E}_1(\omega_3 - \omega_1 + \omega_2)\mathbf{E}_2^*(\omega_2)$$

$$(51)$$

where one pump frequency component (ω_1) is assumed to drive the molecular coherence with ω_2, and the other (ω_1') is scattered by the material excitation. For finite laser bandwidths, the contributions to $\mathbf{P}_3(\omega_3)$ can simply be superimposed, which can be formally expressed as a convolution superposition:

$$\mathbf{P}_3(\omega_3) = \int d\omega_1 \, \mathbf{E}_1(\omega_1) \int d\omega_2 \, \mathbf{E}_1(\omega_3 - \omega_1 + \omega_2)\mathbf{E}_2^*(\omega_2)\chi^{(3)}(\omega_1 - \omega_2) \quad (52)$$

This integral cannot be performed for nonmonochromatic sources, however, because the electric fields will in this case not be deterministic quantities. Even single-mode sources have bandwidths that are far from transform limited due to phase diffusion (Glauber, 1965), and a multimode laser with a large number of independent modes is accurately described as a chaotic or thermal field (Debethune, 1972). Thus, the electric fields must be described in a stochastic sense, and an experimental CARS spectrum is properly represented as an ensemble-averaged quantity

$$I_3(\omega_3) \sim |\overline{\mathbf{P}_3(\omega_3)}|^2 \quad (53)$$

where the bar denotes ensemble average. If the pump and Stokes sources are assumed to be statistically independent, the result of performing the ensemble average is that the CARS intensity depends on a fourth-order electric field correlation (Yuratich, 1979) of the form

$$\overline{E_1(t_1)E_1(t_2)E_1^*(t_3)E_1^*(t_4)} \tag{54}$$

Thus, the theoretical description of CARS generation for finite bandwidth sources requires that the fourth-order electric field correlation of the pump be known. If the pump field fluctuations are assumed to be a stationary, Gaussian (normal) process, then the fourth-order correlation can be represented in terms of second-order correlations

$$\overline{E_1(t_1)E_1(t_2)E_1^*(t_3)E_1^*(t_4)} = \Phi_1(t_1 - t_2)\Phi_1(t_3 - t_4) + \Phi_1(t_1 - t_3)\Phi_1(t_2 - t_4)$$

$$+ \Phi_1(t_1 - t_4)\Phi_1(t_2 - t_3) \tag{55}$$

where

$$\Phi_1(t_a - t_b) \sim \int I_1(\omega)e^{i\omega(t_a - t_b)}\, d\omega$$

and the CARS intensity can be expressed simply in terms of the spectral energy densities I_i of the laser sources

$$I_3(\omega_3) \sim \int d\omega_1\, I_1(\omega_1) \int d\omega_2\, I_1(\omega_3 - \omega_1 + \omega_2)I_2(\omega_2)|\chi^{(3)}(\omega_1 - \omega_2)|^2 \tag{56}$$

Thus, the effect of incoherence in the laser sources is to average out coherence between different contributions to the same polarization frequency component $P_3(\omega_3)$, leading to a representation of the CARS intensity $I_3(\omega_3)$ as an incoherent superposition of monochromatic source solutions. Of course, a pulsed laser source cannot exactly represent a stationary process, but stationarity will be a good approximation if the field is far from transform limited. The representation of the electric field fluctuations as a Gaussian (normal) process also seems reasonable.

For broadband Stokes operation, the convolution integral gives the spectral density of the CARS signal prior to entering the detector apparatus, and the resulting spectrum must be convolved with an effective detector response (slit) function to give the measured CARS spectral distribution. If $\Delta\omega_2 \gg \Delta\omega_1$, as is usually the case, the final result for the multiplex CARS intensity (Hall and Eckbreth, 1978) is

$$I_3(\omega_3) \sim \int d\omega_3'\, T(\omega_3 - \omega_3') \int d\Delta\, I_1(\omega_3' - \Delta)I_2(\omega_1^{(0)} - \Delta)|\chi^{(3)}(\Delta)|^2 \tag{57}$$

where T is the spectrometer slit function, $\omega_1^{(0)}$ is the center frequency of the pump, and the dummy variable Δ runs over all values of the detuning.

In the scanning case, a monochromator is not generally employed, and the spectrally integrated CARS intensity is recorded for each setting of the pump and Stokes. Thus, the intensity measured in the scanning case (Yuratich, 1979) is given by

$$I_3(\omega_1^{(0)} - \omega_2^{(0)}) = \int d\omega_3 \, I_3(\omega_3)$$

$$= \int d\Delta \, I_1 * I_2(\Delta)|\chi^{(3)}(\Delta)|^2 \qquad (58)$$

where $I_1 * I_2$ is the convolution of the pump and the Stokes profiles

$$I_1 * I_2(\Delta) = \int d\omega_1 \, I_1(\omega_1)I_2(\omega_1 - \Delta) \qquad (59)$$

If the pump and Stokes profiles are both Gaussian or both Lorentzian, the convolution will similarly be a Gaussian or Lorentzian with a width given by the root-mean-square (Gaussian) or sum (Lorentzian) of the pump and Stokes widths.

Theoretical investigations of the effect of source coherence and statistical properties on CARS generation will probably be an interesting field of endeavor in the future in much the same way that they have been in other types of spectroscopy. However, because most CARS measurements derive from spectral shapes and not absolute intensities, it is most important to know how the bandwidths of the sources affect the observed anti-Stokes profile. As will be seen, the simple convolutions [Eqs. (57) and (58)] over the intensity profiles of the laser sources give such good agreement with experiment that there can be little doubt that the assumptions underlying the expressions are correct.

III. Experimental Considerations

A. LASER SELECTION

Although CARS has no threshold per se and can be generated with CW laser sources (Barrett and Begley, 1975), high-intensity pulsed laser sources are required for the probing of high-interference environments in order to generate CARS signals well in excess of the various sources of interference. For applications in fluctuating environments where it is desired to make "instantaneous" measurements of medium properties, the individual laser pulses must be energetic enough to provide a statistically significant number of CARS signal photons during each pulse. For gas-phase diagnostics, pulsed, Q-switched, solid-state lasers are generally required to satisfy the above criteria. Calculations (Eckbreth et al., 1979a) of CARS signal magnitudes for representative laser equipment indicate that interferences, either occurring naturally from combustor background luminosity or laser induced, should not be significant compared to CARS signal levels. CARS thus appears capable of probing high-interference environments with little difficulty and this has indeed been borne out in practice. In regard to signal level calculations, Barrett and Begley (1975), in a CW CARS experiment, were able to obtain experimental signals in good agreement (within 25%) with the equations given earlier. In pulsed CARS experiments, it is common to obtain experimental signal levels about an order of magnitude lower than those predicted by these equations (Nibler and Knighten, 1979; Druet and Taran 1979; Eckbreth, 1980), and this should be kept in mind when assessing feasibility for a given application.

There are a number of factors which potentially limit the laser intensities employable in a CARS experiment, thus restricting the signal level which can be generated. One factor is stimulated Raman gain in which the Stokes wave at ω_2 is amplified via a Raman resonance by an intense ω_1 pump laser. This can perturb the Stokes wave and lead to distorted spectra. This effect is generally unimportant at atmospheric pressure but can be important at several tens of atmospheres and must be avoided. Another limitation involves a perturbation of the population difference between the two quantum states involved in the wave mixing (Moya et al., 1977). As described by Eq. (31), the population difference will not be perturbed if τ_Δ is sufficiently larger than the laser pulsewidth. For broadband Stokes waves, the spectral intensity within the Raman linewidth should be used in Eq. (31). Optical breakdown is also a limitation, i.e., ionization of the medium by the intense electric fields of the incident light waves (Smith and Meyerand, 1974). In particle-laden combustion media, this generally limits the focal fluxes to a few tens of gigawatts/cm^2 at best. Another factor is optical Stark effect broadening and splitting (Rahn et al., 1981; Farrow and Rahn, 1982). Splitting occurs only in transitions involving a change in rotational quantum number, i.e., Q-branch vibrational transitions do not split, and can set in at intensities a few times 10^{10} W/cm^2. The Stark shifts become significant at higher intensities and are generally on the order of 0.1 cm^{-1} per 10^{12} W/cm^2. In general, these effects are probably more important in high-resolution spectroscopy than in diagnostic applications and become important at intensities generally above the "dirty" breakdown or saturation thresholds.

Although many of the early CARS investigations used ruby lasers (Regnier and Taran, 1973; Roh et al., 1976), most CARS systems today employ frequency-doubled neodymium: YAG (2 × Nd) lasers for a number of reasons. 2 × Nd lasers can be operated at a repetition rate generally an order of magnitude higher than ruby, i.e., at least 10 pulses per second (pps) versus 1 pps. Nd : YAG systems with rates up to 50 pps are commercially available. The higher repetition rate obviously expedites data collection. It also permits the use of boxcar averagers. At 10 pps or better, the experiment behaves in a nearly CW manner, permitting "tweaking" of the optical adjustments while the apparatus is running, an important feature for the critical alignment requirements of CARS. If a portion of the pump laser is split off to pump the Stokes dye laser, which is generally the case, 2 × Nd lasers at 5320 Å can pump very efficient dyes in the 5500- to 7000-Å region of the spectrum. Ruby lasers, on the other hand, must pump lower efficiency and generally less stable near-IR dyes. The CARS radiation for equivalent laser powers is slightly stronger from 2 × Nd than from ruby due to the quadratic dependence of the CARS generation efficiency on anti-Stokes frequency [Eq. (17)]. Furthermore, the CARS radiation from 2 × Nd resides in a region of higher detection quantum efficiency than does CARS generated from 6943 Å.

The Stokes beam is typically provided by a dye laser pumped by splitting a fraction of the 2 × Nd laser. The Stokes dye laser is made to operate in either a narrow- or broadband mode depending on the spectral resolution or temporal response desired. For high spectral resolution, a narrowband, tunable dye source is employed which is scanned to generate the CARS spectrum piecewise. The spectral resolution is then nominally limited by the linewidths of the laser sources and no monochromator is strictly required, although one is often used for filtering. For laminar flame studies, maximum species sensitivity is achieved because all of the Stokes power is available to drive the individual Raman resonances, in contrast to broadband CARS where the Stokes spectral intensities are considerably lower. For nonstationary and turbulent combustion diagnostics, however, spectral scanning is inappropriate due to the nonlinear dependence of CARS on temperature and density. Generating the spectrum piecewise in the presence of large density and temperature fluctuations leads to distorted signatures weighted toward the high-density, low-temperature excursions from which true medium averages cannot be obtained. The alternate approach mentioned earlier (Roh *et al.*, 1976) is to employ a broadband Stokes source. This leads to weaker signals, but generates the entire CARS spectrum with each pulse, permitting, in principle, instantaneous measurements of medium properties. From a time series of instantaneous measurements, the probability distribution function of a given parameter can be assembled from which the true time average can be obtained as well as the magnitude of the turbulent fluctuations. In laminar, i.e., time-steady flows, the broadband approach can lead to more expeditious averaging and higher quality spectra, as will be explained in a discussion on experimental layout (Section III,C).

B. Phase Matching

For efficient signal generation, the incident beams must be aligned such that the three-wave mixing process is properly phased. This ensures that the CARS generated at a certain point will be in phase with the CARS generated at a subsequent point, leading to a constructive buildup of the signal. The general phase-matching diagram for degenerate three-wave mixing is shown in Fig. 3a and requires that $2\mathbf{k}_1 = \mathbf{k}_2 + \mathbf{k}_3$. $\mathbf{k}_i$ is the wave vector at frequency ω_i with absolute magnitude equal to $\omega_i n_i/c$, where n_i is the refractive index at frequency ω_i. Because gases are virtually dispersionless, i.e., the refractive index is nearly invariant with frequency, the photon energy conservation condition $2\omega_1 = \omega_2 + \omega_3$ indicates that phase matching occurs when the input laser beams are aligned parallel to each other (Fig. 3b). Collinearity, although easy to implement, possesses a problem in regard to spatial resolution. Because the CARS signal is coherent and undergoes an integrative growth process, the spatial resolution cannot be well defined by imaging techniques. CARS signal generation scales as the intensity product $I_1^2 I_2$ [Eq. (17)], and, consequently, the incident laser beams are generally tightly focused for diagnostic purposes when collinear phase matching is employed. In this manner, CARS generation occurs primarily in the focal region, presumably with good spatial resolution. However, depending on the specific diagnostic circumstance (the focusing-lens-to-measurement-point separation, laser beam quality etc.), the spatial resolution may be worse than desired. For diffraction-limited beams, the interaction is assumed to occur primarily within a

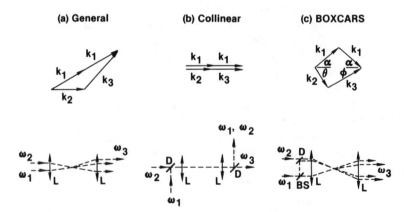

FIG. 3. Illustration of CARS phase-matching approaches. Subscripts denote beams: 1, pump; 2, Stokes; 3, anti-Stokes.

cylindrical focal volume of diameter φ and length 6ℓ (Regnier *et al.*, 1974) given by

$$\varphi = 4\lambda f/\pi D, \qquad \ell = \pi\varphi^2/2\lambda \qquad (60)$$

where f is the focusing lens focal length; D, the beam diameter incident on the lens; and λ, the laser wavelength. In Table I, the probe volume focal diameter and measurement length are tabulated for various focal length lenses for a 1-cm-diameter beam at 5320 Å. Many laser beams are not diffraction limited and lead to poorer spatial resolution than shown in Table I. For a "three-times diffraction-limited" beam divergence angle, the linear resolution would be an order of magnitude poorer. In the presence of density gradients, the resolution degrades further because the CARS radiation scales nominally as the square of the gas density. For probing hot flames surrounded by regions of lower temperature and higher density, significant contributions to the total CARS signal may originate from the outer regions. In certain instances the desired measurement location, i.e., the focal region, is not "seen" at all because the contributions from the outer regions completely dominate.

Clearly, it is desirable to avoid beam overlap and potential three-wave mixing in all regions except the desired measurement location. This can be accomplished using crossed-beam phase matching (Eckbreth, 1978) as depicted in Fig. 3c. Based upon the general shape of the phase-matching diagram, the technique has been termed BOXCARS. In this approach, the ω_1 pump beam is split into two components which are crossed at a half angle of α which depends on the spatial resolution desired. The ω_2 Stokes beam is introduced at specified angle θ, dependent upon α, producing phase-matched CARS at angle φ. CARS is generated where all three beams intersect, leading to fine and unambiguous spatial resolution. Reso-

TABLE I

COLLINEAR PHASE-MATCHED CARS
PROBING VOLUME

Focal length (cm)	Focal diameter[a] (cm)	Sample length[a] (cm)
10	6.11(−4)	7.32(−2)
20	1.22(−3)	2.93(−1)
50	3.06(−3)	1.83
100	6.11(−3)	7.32

[a] Powers of 10 shown parenthetically.

lution of a millimeter or less is readily achieved. There are many variations to the basic BOXCARS scheme. If $\alpha + \theta$ is set to 180°, counterpropagating CARS (Laufer and Miles, 1979; Compaan and Chandra, 1979)
is obtained in which one ω_1 component and ω_2 are directed in opposition to
each other. Phase matching need not be confined to a plane and can be
performed three dimensionally (Prior, 1980; Shirley et al., 1980b). Three
dimensional phase matching is illustrated in Fig. 4 and can be envisioned
as a folding of planar crossed-beam phase matching, hence the term
"folded BOXCARS." There are also two-beam BOXCARS approaches
(Marko and Rimai, 1979; Klick et al., 1981) in which multiple regions of a
single ω_1 beam wave-mix with ω_2. The latter variations possess experimental advantages, some of which will be described later. If the ω_1 beams
are focused into sheets using cylindrical lenses, spatially resolved CARS
transverse to a line can be generated (Murphy et al., 1979) which can be of
utility in diagnosing turbulent jets. Experimentally, phase matching is
readily achieved. With the many possible phase-matching geometries, an
optical approach to a given application is generally not a difficult problem.

The main drawback that phase matching imposes is the general requirement for line-of-sight optical access. This is due to the fact that CARS is
generally generated in the forward direction, i.e., the direction in which

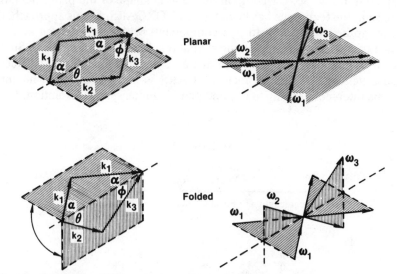

FIG. 4.   CARS crossed-beam phase-matching approaches showing the phase-matching
diagram and actual geometry of the optical beams for planar and folded BOXCARS. Subscripts denote beams: 1, pump; 2, Stokes; 3, anti-Stokes. [Reprinted from Shirley et al.
(1980b) with the permission of The Optical Society of America.]

ω_1 and ω_2 propagate. A question often asked in regard to CARS is whether it can be generated in the backward direction; the answer is "not simply." There is no closely packed scheme, from an angular sense, in which the three wave-mixing beams enter about an axis with the CARS beam emerging along or near the same axis. In counterpropagating CARS, Compaan and Chandra (1979) used a retroreflector to obtain ω_1 and ω_2 in opposition, and the CARS beam emerged in the backward direction at a small angle. Nevertheless, line-of-sight access is required for the retroreflector. If the BOXCARS diagram (Fig. 3) is contorted such that $\theta + \varphi = 180°$, then the CARS beam propagates counter to the Stokes beam, but the ω_1 pump components must be introduced at a large angle to the $\omega_2-\omega_3$ axis. If large-angle folded BOXCARS is viewed from the ω_1 component plane, it is seen that the CARS beam emerges into the same hemisphere from which the wave-mixing beams emanate. Such a scheme may prove attractive for measurements near a single, larger optical port, e.g., through a transparent cylinder head in an internal combustion engine (Dale, 1980; Greenhalgh, 1983b).

C. EXPERIMENTAL LAYOUT

A typical broadband CARS arrangement for combustion diagnostics (Eckbreth and Hall, 1981) is schematically illustrated in Fig. 5. The neodymium : YAG laser provides two second-harmonic beams at 5320 Å by frequency doubling the fundamental 1.06-μm output from the laser and subsequently frequency doubling the residual 1.06 μm emerging from the first doubler. The primary second harmonic, i.e., after the first doubler, is typically 200 mJ, while the secondary, i.e., from the second doubler, is about an order of magnitude less. The secondary passes through a slab of KG3 Schott glass placed at the Brewster angle to absorb any remaining 1.06 μm. The secondary is then directed by mirrors and focused to pump, slightly off axis, the Stokes dye cell oscillator. The output from the dye laser is amplified in a second dye cell, in flow series with the first, pumped by splitting a portion of the primary (typically 33%) at a beamsplitter, as shown in Fig. 5. The Stokes laser and 5320-Å pump component each pass through Glan laser polarizers to ensure polarization purity, which is important when performing polarization-sensitive CARS experiments. Each beam passes through expansion or contraction telescopes whose function will be detailed later. The Stokes beam passes through a rotatable optical flat and then, in folded BOXCARS, directly to the focusing lens. In planar BOXCARS (Eckbreth, 1978) the Stokes beam would pass through a dichroic mirror used to reflect one of the primary pump components as depicted in Fig. 3. One advantage of folded BOXCARS is the elimination

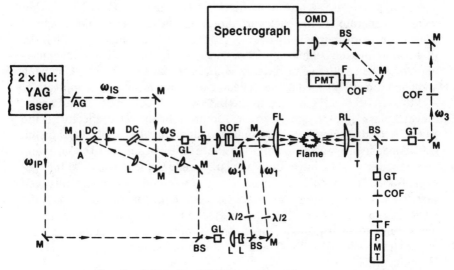

FIG. 5.   Schematic of CARS experimental arrangement.

of this dichroic element. Another advantage of folded over planar BOX-CARS is the complete angular and spatial separation of the CARS beam which eliminates the need to disperse or filter the CARS beam from one of the pump components. The primary pump beam is split at a second beamsplitter (50%) to produce the two components for BOXCARS. These beams pass through low-order half-wave plates which control the polarization orientation of the pump components. The pump components are directed by mirrors to the focusing/crossing lens. If the three beams are aligned parallel to each other, they will by definition cross at the focal point of the lens. However, due to wavefront sphericity and chromatic aberration, they may not necessarily waist at the crossing point. The adjustable telescopes are used to position the beam waists at the crossing point. At high laser intensities, the beams are readily visualized near the focal region via the room-air Rayleigh scattering. The primary component waists will coincide only if the distances traveled by each pump component from the second beamsplitter to the focusing lens are equal. The telescopes, depending on their magnification, also permit the focal diameters to be varied. The diameters can be readily estimated employing a scanning knife-edge test and measuring the distance traversed between full illumination of, and total extinction on, a target plate. As a rough rule of thumb, the Stokes and pump beam diameters are generally set equal to one another. Recent experiments with collinearly phase-matched CARS (Byer, 1980) indicate better signal generation efficiencies if the Stokes focal diameter is less than the pump diameter to exploit the quadratic dependence on pump intensity. This has not been closely examined for BOX-CARS and requires further examination. Rotation of the optical flat in the Stokes beam adjusts its displacement on the focusing/crossing lens, permitting the phase-matching angle θ to be varied. After passing through the crossing point, the four beams (i.e., CARS at ω_3, ω_1, ω_1', and ω_2) are recollimated by a second lens of generally the same focal length as the focusing lens. The wave-mixing components are trapped and the spatially and angularly separated CARS beam is filtered off and split into two components, not necessarily equivalent, by a beamsplitter. The component split off at right angles passes through a Glan Thomson polarization analyzer set normal to the polarization of the resonant-mode CARS signal. Thus, only the nonresonant CARS signal (actually some fraction thereof) is transmitted and is monitored after spectral filtering by a photomultiplier. This signal is used for *in situ* referencing, which will be treated in detail later. The portion of the CARS radiation transmitted by the beamsplitter passes through a polarization analyzer and is then directed to a spectrograph fitted with an optical multichannel detector (OMD) or a monochromator equipped with a photomultiplier for spectral scanning. The second polarization analyzer may allow all contributions to the CARS radiation (i.e., isotropic and anisotropic resonant modes, nonresonant signals) to pass, or it may be oriented in concert with the half-wave plates to suppress detection of one or two of the modes or signals. The manner in which this occurs will be explained subsequently. A small fraction of the CARS beam is split off at a glass slide before the spectrograph and sent to a photomultiplier tube fitted with an appropriate narrowband interference filter. This signal is averaged on a boxcar averager and used to monitor and "tweak" the alignment of the CARS system. It is relatively immune to angular and translational motions of the CARS beam which occur during peaking up of the alignment. The alignment of a CARS system should never be monitored through a narrow angular and spatial acceptance, such as a slit, because signal loss due to slight steering can be misinterpreted for alignment detuning. The spatial resolution of a given configuration is readily checked by generating CARS from within a translatable, thin microscope slide cover, with the laser beams suitably attenuated to prevent optical damage to the glass. CARS generation results from the nonresonant susceptibility of the glass. Microscope slide covers are generally less than 200 μm thick and allow the pointwise CARS signal contribution to be ascertained. Upon integration of the pointwise contributions, the spatial resolution can be found. In general, it is far better practice to measure the spatial resolution than to calculate it.

One advantage of broadband CARS resides in the simplicity of the

Stokes dye laser, which is devoid of elements for tuning and spectral condensation. A flowing dye cell oriented at Brewster's angle resides within a planar Fabry–Perot oscillator cavity. Because the dyes amenable to 5320-Å pumping typically exhibit high gain, slightly off-axis pumping works quite well and leads to very good beam quality, i.e., low angular beam divergence, in contrast to that often obtained with transverse pumping. The dye spectrum is centered at the desired wavelength by selecting the dye appropriately and by adjusting its concentration. Solvent tuning can also be employed.

With the planar Fabry–Perot dye oscillator arrangement, bandwidths vary from 100 to 200 cm^{-1} depending on the pump energy and dye employed. Binary dye mixtures are often used to improve dye conversion efficiency. For single-pulse CARS diagnostics or for special scanning of laminar situations, it is important that the dye spectrum be smooth and reproducible from pulse to pulse. The dye cavity is purposely designed to have a high Fresnel number and to accommodate as many modes as possible to "fill in" the spectrum. The pulse-to-pulse spectral stability of the dye laser has been examined on the OMD and found to be fairly good. Single-pulse dye spectra display an irregular, fine structure with an amplitude variation of $\pm 5\%$, presumably due to spatial "hole burning" and incomplete mode filling in the dye laser. This fine structure averages out in time to produce spectrally smooth profiles. High-quality, averaged CARS spectra are readily achieved, but single-pulse CARS spectra may exhibit some spurious structure, particularly at moderate spectral resolution (<1 cm^{-1}). At lower resolutions (≥ 3 cm^{-1}), this fine structure is spectrally smeared and is not a problem. In broadband CARS, either the linewidth of the pump laser or the resolution of the monochromator/spectrograph determines the ultimate resolution of the spectrum. Monochromators of 1 m typically have a limiting resolution of about 0.5 cm^{-1} in the visible. $2 \times$ Nd lasers with intracavity etalons have linewidths in the range 0.1–0.4 cm^{-1}, and 0.8 cm^{-1} without intracavity etalons. CARS spectra thus obtained therefore have a limiting resolution between 0.5 and 1 cm^{-1}. This moderate resolution is generally more than sufficient for diagnostics. Furthermore, it is important to note that the large pump laser linewidth is not detrimental in regard to the strength of the CARS radiation. For broadband CARS generation, it is easy to show that the spectrally integrated intensity of a CARS transition is independent of the pump laser linewidth.

Narrowband CARS systems for combustion diagnostics are conceptually similar to broadband CARS systems (Farrow et al., 1981) with the obvious exception of the tunable Stokes source and the normalization approach. This latter aspect of CARS will be discussed later. For narrowband Stokes work, a variety of dye laser configurations exist and are commercially available or readily assembled. The oscillator sections are generally transversely pumped to obtain gains high enough to overcome the insertion losses engendered by the spectral condensation scheme employed. The Hansch design (1972) of a circular telescope, large, two-dimensional grating has generally given way to one-dimensional expansion schemes employing a grazing incidence grating (Littman, 1978) or multiple-prism beam expanders (Duarte and Piper, 1980). The one-dimensional expanders require a grating large in one dimension only and greatly reduce cost.

D. NORMALIZATION

When recording a CARS spectrum over a period of time, it must be considered how the spectrum is to be normalized. Normalization is the procedure which accounts for pulse-to-pulse laser power fluctuations, longer term power drifts, spectral variations in dye power, temporal changes in optical alignment, etc., in the CARS spectrum. In narrowband, scanned CARS work, reference cells are generally employed. The reference cell can be optically positioned either in series with or parallel to the sample region. In the latter case, a small fraction of the wave-mixing beams are split off and used to generate the normalization CARS signal in a reference leg. In the former case, all of the wave-mixing laser energy is used to generate the reference CARS, which is separated prior to the sample region with a dichroic mirror. If the reference is placed after the sample, then the sample CARS is separated out with a dichroic prior to reference cell CARS generation. In addition to accounting for the effects just described, reference cells also serve for calibration purposes because they are filled with a known gas, typically inert and at a fixed pressure and temperature, whose CARS signal magnitude is readily calculable. From this and from the relative strength of the sample and reference signals, the absolute magnitude of the sample signal can be determined. Care must be taken to ensure that the signals from the sample region and reference cell respond to optical misalignments in exactly the same way. The CARS signal from the reference cell photomultiplier is used to normalize the CARS signal from the sample, typically in a two-channel boxcar averager or in a minicomputer after digitization. For broadband CARS with broadband detection, no normalization is required on a shot-to-shot basis as long as the dye spectrum is reproducible from pulse to pulse, which is generally the case; if it is not, its shape must be monitored, typically by generating a CARS signal nonresonantly in an inert gas reference sample. If a broadband CARS spectrum is being scanned with a monochromator, the spectrally selected monochromator signal is readily normalized by sampling the spectrally integrated CARS signal. The latter is obtained by splitting off a small fraction of the CARS beam entering the monochromator and monitoring it through an interference filter of adequate bandwidth. Absolute signal intensity determinations, of course, require employment of a calibrating reference cell.

There are several problems associated with reference cells. First, the normalization is not particularly good. Because there is no easy way to maintain the same longitudinal phasing between the pump and Stokes waves in the reference and sample, temporal mode beating leads to fluctuations in the *normalized* CARS signal on the order of ±25% (Taran, 1976; Goss *et al.*, 1980b) if multimode laser sources are employed. If a single-mode pump laser is used, these normalized fluctuations can be reduced to the level of 5–7%. Clearly, this imposes an accuracy limitation on the use of CARS for instantaneous density measurements when reference cells are used. Second, normalization is not possible when the sample is a turbulent combustion region in which the wave-mixing laser beams can possibly be steered and defocused. Clearly, a presample reference cell experiences no steering or defocusing and thus cannot account for these effects. A postsample reference would experience those effects, but not in exactly the same way as the sample/measurement volume. The postsample reference would indicate a perturbed measurement, but there is no rigorous way to "correct" the raw data from the sample region. Beam attenuation from soot and/or fuel droplets is also problematical in this regard and is only "correctable" if assumptions are made for a spatial and size distribution for the particulates, certainly not very rigorous for quantitative measurements. To overcome these difficulties with separate reference cells, considerable attention is being devoted to *in situ* referencing concepts. These concepts are being developed to permit density measurements in turbulent, particle-laden media and will be discussed in greater detail in Section V. These concepts should have general applicability for spectral studies in the laboratory as well, as recently demonstrated (Farrow *et al.*, 1981; Oudar and Shen, 1980).

IV. Thermometry

It was noted in Section II,C that the resonant contribution to the CARS intensity contains implicit information about the state of internal excitation of the probed molecule through the normalized population difference factors $\Delta\rho_{if}^{(0)} = \rho_{ii}^{(0)} - \rho_{ff}^{(0)}$. If the rotational and vibrational modes can be assumed to be in mutual equilibrium with the translational degree of freedom, and if saturation of the medium by the exciting fields is avoided, then the population factors can be calculated from simple Boltzmann equilibrium relationships. For example, if i represents the vibrational and rotational quantum numbers v and J, respectively, then $\rho_{ii}^{(0)}$ is given by

$$\rho_{ii}^{(0)} = \frac{g_J(2J + 1)}{Q_V Q_R^{(v)}} e^{-E_v/kT} e^{-B_v J(J+1)/kT} \tag{61}$$

where g_J is the nuclear spin statistical weight, Q_V and $Q_R^{(v)}$, respectively, are the vibrational partition function and rotational partition function for level v, E_v is the vibrational energy, B_v is the rotational constant $= B_0 - \alpha_e(v + 1/2)$, k is Boltzmann's constant, and T is temperature. In vibrational CARS, the Raman-active modes for molecules of practical interest in combustion are isotropic to a good approximation, such that the CARS signature will be dominated by Q-branches whose frequency is given by (for diatomics)

$$\omega_{if} = E_{v+1} - E_v - \alpha_e J(J + 1)$$
$$\simeq \omega_e - 2\omega_e \chi_e(v + 1) - \alpha_e J(J + 1) \tag{62}$$

Thus, the shape of the resonant CARS spectrum will be determined by the temperature and frequency splitting due to vibration–rotation interaction α_e. If the vibration–rotation interaction constant is large, as in H_2, neighboring Q-branch transitions will be well separated, and temperature can be readily extracted from relative peak heights in the spectrum. Usually, however, there is significant line overlap, with attendant interference effects, that renders simple data reduction impractical. In this case, the data must be reduced by fitting of computer-generated spectra to experimental signatures (Hall, 1979, 1983). If the probed molecule is a majority species, the CARS spectral signatures will be qualitatively similar to those for spontaneous Raman scattering, and CARS thermometry can be performed from spectral shapes, as in the spontaneous case. A probed molecule will be most useful for thermometry if it is a majority species or has a large Raman cross section because interference from the nonresonant background will then be kept to a minimum. Although it is possible in principle to extract both temperature and concentration from one CARS spectrum, the CARS signatures do tend to lose temperature sensitivity for low molecular concentrations.

In the discussion to follow it will be shown that CARS thermometry using diatomic probed molecules is now capable of high accuracy, particularly in the case of N_2. Good progress is also being made in understanding the important pressure dependence of diatomic CARS spectra. For triatomics such as H_2O and CO_2, the agreement between experiment and theory is good, but further work is necessary before these molecules can be used for reliable diagnostics. Whereas good progress is thus being made on all of the major products of hydrocarbon–air and hydrogen–air combustion, very little quantitative modeling has been devoted to the hydrocarbons alone, and further work in this area is needed.

A. NITROGEN

Nitrogen is the dominant constituent of air-fed combustion processes and will be present everywhere in large concentrations. Making temperature measurements from N_2 CARS spectra provides information on the location of the combustion heat release and, to some degree, on the extent of chemical reaction. Nitrogen vibrational CARS has therefore received the most attention for thermometry thus far. The molecule has the added advantage that the spectroscopic and linewidth parameters needed for very accurate measurements are by now relatively well known.

The potential of N_2 vibrational CARS for thermometry can be appreciated by examining the calculated temperature dependence of the N_2 CARS spectrum shown in Fig. 6. The calculations were carried out using the computer model described by Hall (1979), and correspond to multiplex (broadband Stokes) CARS with a 0.8-cm^{-1}-bandwidth pump, a 150-cm^{-1}-bandwidth Stokes, and a 1-cm^{-1} instrumental slit function. It is apparent that the spectra display pronounced temperature sensitivity, particularly at the higher temperatures, and that a great deal of spectral detail is apparent even though the instrumental resolution is not particularly high. At low temperatures, an unresolved fundamental band corresponding to $v = 0 \rightarrow v = 1$ Q-branch transitions is observed whose width increases dramatically with temperature. At about 1000 K, the $v = 1 \rightarrow v = 2$ "hot band" appears and individual Q-branch transitions can be observed in the fundamental band. At high temperatures, the distinct peaks correspond to even-J Q-branch transitions in the approximate range Q(20)–Q(40). The odd-J Q-branches, which have a nuclear spin statistical weighting equal to half that of the even-J transitions, are reduced in intensity by a factor of 4 and do not stand out. For fundamental Q-branch transitions beyond Q(40), overlap with the hot-band transitions occurs, giving rise to two prominent peaks in the hot band due to close spectral coincidences. A second hot band corresponding to $v = 2 \rightarrow v = 3$ transitions is important for $T > 2000$ K (not shown in Fig. 6). For the moderate resolution parameters of Fig. 6, the accuracy of N_2 CARS thermometry is highest at elevated temperatures. If the instrumental resolution was higher, however, spectral detail would be observable at lower temperatures, and the accuracy would improve at these lower temperatures. A

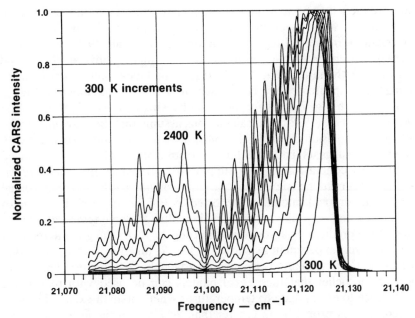

FIG. 6. Temperature variation of the CARS spectrum of N_2 for a pump linewidth of 0.8 cm^{-1}, a Stokes bandwidth of 150 cm^{-1}, and a 1-cm^{-1} slit width. [Reprinted from Hall and Eckbreth (1981) with the permission of the Society of Photo-Optical Instrumentation Engineers.]

pump bandwidth of 0.4 cm^{-1}, for example, makes possible the resolution of individual Q-transitions even at 300 K. At lower spectral resolution, ~3-cm^{-1} slit width for the laser parameters of Fig. 6, the fine structure is lost, but the spectra still exhibit good temperature sensitivity through the width of the fundamental and the relative strength of the hot band. It is also to be noted that for the assumed 70% N_2 concentration of Fig. 6, there is little evidence of interference from the real background contribution. The background susceptibility actually does play a small role that will be discussed later.

1. N_2 Raman Linewidths

Because temperature must be inferred from N_2 CARS signatures by computer fitting theoretical spectra to experimental data, it is important that the calculations be based on parameters that are relatively well known. At the time that the earliest hot N_2 CARS spectra were taken (Regnier et al., 1974; Hall, 1979), little was known about the pressure-broadened Q-branch linewidths, and sensitivity studies indicated a moderate sensitivity of calculated spectra to assumed linewidth if the uncertainties in the widths were a factor of 2 to 3 (Hall, 1979). Rahn *et al.* (1980a) subsequently measured the N_2 Q-branch widths at 300 K and in an air–methane flame at $T \simeq 1700$ K using high-resolution stimulated Raman gain spectroscopy. These measured widths are shown in Fig. 7. Also shown in Fig. 7 are theoretical widths which are based on the model of Bonamy *et al.* (1977). The calculations assume that the linewidth arises solely from rotationally inelastic collisions, and the good agreement with experiment suggests that contributions from vibrational effects are indeed small. This conclusion is also supported by the fact that Rahn *et al.* (1980a) observed no significant variation of linewidth with vibrational band. A theoretical calculation reported by Rahn *et al.* (1980a) also gives good agreement with the experimental results. At flame temperatures the theoretical predictions agree to within roughly 10%. The good agreement obtained with experiment as shown in Fig. 7, together with the fact that the theoretical calculations involve no adjustable parameters, lends support to the belief that theoretical models can be used with reasonable confidence to extrapolate to temperatures where no data exist. The associated error is probably no greater than 20% and at flame temperatures

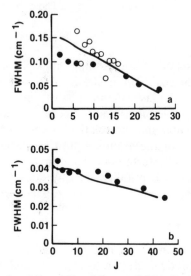

FIG. 7.   Pressure-broadened linewidths of Raman Q-branch transitions in N_2. (a) Room air, $T = 295$ K; (b) air–methane flame, $T \simeq 1730$ K. Experimental data: ●, from A. Owyoung and L. Rahn; ○, from W. Fletcher; line denotes theoretical calculations. [Reprinted from Hall (1980) with the permission of the Society for Applied Spectroscopy.]

may be as good as 10%. As will be discussed later, linewidth uncertainties of this magnitude give rise to very minor errors in inferred temperature. Least-squares fits to the theoretical predictions shown in Fig. 7 give the following simple functional form for the N_2 Raman linewidths; this is recommended for use in any medium where N_2 is the dominant constituent:

$$\Gamma(J) = 8T^{-0.71} - (18.6T^{-1.45})J \quad \text{(FWHM: cm}^{-1}\text{/atm)} \quad (63)$$

As flame temperatures are approached, the temperature- and J-dependencies of the widths are predicted to become relatively weak; the weak J-dependence is a consequence of the flattening rotational distribution of collision partners.

Although the recommended N_2 Q-branch linewidth expression is for N_2 self-broadening, it should provide a good approximation in air-fed flame mixtures where products such as H_2O and CO_2 have appreciable concentrations. The theoretical calculation shown in Fig. 7 implies that linewidth concentration dependencies are not large for N_2, whereas the theoretical calculation carried out by Rahn et al. (1980a) suggests that they can give rise to roughly a 10% linewidth error. The more recent N_2 linewidth measurements of Rosasco et al. (1983) are in good agreement with those of Rahn et al. (1980a).

2. Clean Flame Thermometry

The accuracy of broadband N_2 CARS thermometry has been examined in premixed flat flames by comparison with radiation-corrected fine-wire thermocouples (Eckbreth and Hall, 1981). The radiation corrections were experimentally calibrated at different flame temperatures by sodium line reversal. In Fig. 8 is shown a scanned N_2 CARS spectrum in a clean, premixed, air–methane flame at a point where the temperature was determined to be 2110 K. The instrumental parameters correspond to those of Fig. 6. Shown as the dotted curve is a theoretical least-mean-squares fit to the experimental signature; the best fit temperature is 2104 K with a standard deviation of 9 K.

The need to reproduce theoretically the prominent peaks on the vibrational hot band which result from close overlaps of fundamental and hot-band Q transitions requires that the spectral constants used to calculate the N_2 vibrational–rotational energy levels be very accurate. Slightly different sets of spectral constants which give good results have been reported by Gilson et al. (1980), Rahn (1979), and Hall (1979). Interference with the nonresonant background susceptibility will not significantly affect N_2 thermometry for N_2 concentrations above approximately 50% (Fig. 9); for values lower than this, the background should either be

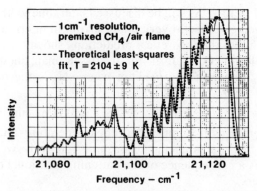

FIG. 8.   Comparison of experimental (solid line) and theoretical least-squares-fit (dashed line) N_2 CARS spectra at a measured temperature of 2110 K. [Reprinted from Hall and Eckbreth (1981) with the permission of the Society of Photo-Optical Instrumentation Engineers.]

cancelled (see Section V,B) or the N_2 mole fraction should be included as a fit parameter. Caution should be exercised particularly in fuel-rich combustion zones.

Accurate N_2 thermometry in flames has also been reported by Farrow *et al.* (1981) using tuned, narrowband pump and Stokes sources. An example of the high-resolution spectra obtained by these investigators is shown in Fig. 10, where the Q-branch spectrum of the fundamental band is displayed at a point where a thermocouple gave a reading of 1830 ± 60 K. The theoretical least-squares fit, given by the solid line, corre-

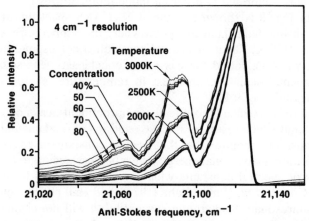

FIG. 9.   Calculated concentration dependence of hot N_2 CARS spectra. Overall resolution, 4 cm^{-1}.

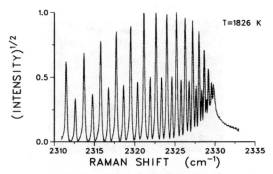

Fɪɢ. 10. Experimental (solid line) and theoretical (dashed line) spectra of the nitrogen Q-branch bandhead measured 5.0 cm above the center of the burner. The best least-squares fit was obtained for a temperature of 1826 K, compared to a thermocouple reading of 1830 ± 60 K. [From Farrow *et al.*, 1981; reprinted by permission of the authors.]

sponds to a temperature of 1826 K. These investigators were also able to obtain high-resolution N_2 spectra at 620 K and to obtain temperatures from N_2 O-branch transitions that were in good agreement with those deduced from the Q-branch.

3. Sooting Flame Thermometry

CARS has also been used for temperature measurement in highly sooting flames (Eckbreth and Hall, 1979; Beattie *et al.*, 1978). Although incoherent and coherent interferences from C_2 Swan bands can be encountered for a 5320-Å pump laser, they are largely suppressible by reducing the Stokes laser bandwidth and through the use of a polarization filter (Eckbreth and Hall, 1979). C_2 will be produced by laser vaporization of soot even on the very short time scale (10^{-8} sec) of the pump pulse duration. Figure 11 shows N_2 CARS spectra obtained at two different locations in a highly sooting, laminar, propane diffusion flame (Eckbreth and Hall, 1979). A comparison of these spectra with the clean flame spectrum shown in Fig. 8 shows that the sooty flame spectra are of equivalent quality. The spatial resolution in these experiments was about 0.3 × 1 mm. Also shown in Fig. 11 are theoretical fits and the temperatures inferred from them. Data such as these have been employed to determine the axial and radial temperature profiles in the sooty propane diffusion flame, as shown in Fig. 12. Laser-modulated particulate incandescence (Eckbreth, 1977) would give rise in these experiments to a signal-to-interference ratio for spontaneous Raman scattering of $O(10^{-3})$, thus precluding any successful measurements using that technique. High-quality single-pulse spectra were also obtained in these experiments at laser energies

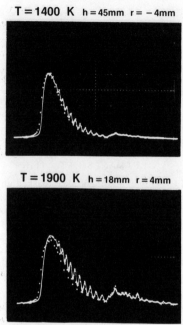

FIG. 11. Comparison of theoretical and experimental N_2 CARS signatures in sooty laminar flame. Dots denote experimental points; solid lines denote the theoretical predictions. The signatures are labeled according to radial (r) coordinate and height (h) in millimeters above burner surface and inferred temperature. Resolution, 1 cm^{-1}. [Reprinted from Hall (1980) with the permission of the Society for Applied Spectroscopy.]

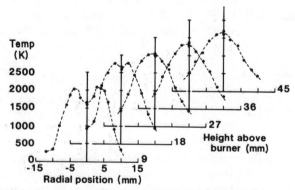

FIG. 12. Radial temperature profiles determined by CARS in a laminar propane diffusion flame. [Reprinted from Eckbreth and Hall (1979) with the permission of the publishers, Elsevier North Holland.]

an order of magnitude lower than those typically employed for single-pulse spontaneous Raman measurements in clean flames. In turbulent media, such single-shot temperature measurements could lead to determination of the temperature probability distribution function (pdf). Examples of single-shot CARS spectra will be presented later in a discussion dealing with thermometry in combustion tunnels and jet engine exhausts.

4. Sensitivity to Linewidth

The foregoing examples have shown that N_2 CARS thermometry is capable of high accuracy, particularly because the spectroscopic and linewidth parameters which go into the theoretical calculations are by now relatively well known. However, the sensitivity of the inferred temperatures to reasonable variations in these parameters, particularly the linewidths, should be addressed. Using the $T = 2104$ K calculation of Fig. 8 as a basis, the sensitivity to linewidth variations was examined and is shown in Fig. 13. The uncertainty in linewidth was simulated by ±50% variations in the values calculated from Eq. (63). As discussed, an uncertainty of this size is probably too pessimistic; the real uncertainty is probably in the range 10–20%. It can be seen that the calculated sensitivity is not large; it has been estimated that the temperature uncertainty for ±20% linewidth variation is no more than 25 K, and even less for the ±10% case.

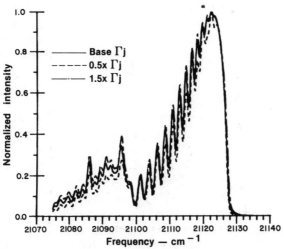

FIG. 13. Sensitivity of predicted N_2 CARS spectrum ($T = 2104$ K) to ±50% variation in isolated linewidths.

5. *Pressure Effects in N_2 CARS Thermometry*

To achieve its maximum usefulness as a diagnostic tool, CARS must be capable of measurements in environments well in excess of 1 atm pressure. High-pressure combustion is important in such practical contexts as gas turbine and rocket propulsion and in ballistic compression. The changes that are expected to occur in the character of the line broadening when adjacent Raman lines are strongly overlapped were discussed in Section II,F, where the modifications required in the theoretical formulation to account for collisional narrowing were also presented. Modeling of high-pressure CARS spectra requires that a matrix of so-called off-diagonal linewidth parameters γ_{rs} be included. γ_{rs} is equal to minus the first-order rate constant for collisional energy transfer from rotational state s to rotational state r if the scattering matrix is independent of vibrational state. Because collisional narrowing arises from line overlap, a good criterion for judging whether it is of importance is to determine by calculation of $|\chi^{(3)}|^2$ whether the ability to resolve individual Q-transitions is being lost due to the large, isolated linewidth.

The effect of pressure on N_2 CARS spectra has been investigated experimentally and theoretically over the pressure range 1–100 atm at 300 K (Hall *et al.*, 1980). A similar experimental study has been performed by Roland and Steele (1980b) for pressures up to 30 atm. Rosasco *et al.* (1983) have seen significant nonadditivity effects in the 300 K N_2 IRS spectrum for pressures as low as 1 atm. The experimental results of Hall *et al.* show that after an initial narrowing of the profile within the first 5–10 atm, the bandwidth remains sensibly constant with increasing pressure out to 100 atm. The observed bandwidth at 100 atm is slightly less than the 1 atm value. Figure 14 shows scanned N_2 spectra at 1 and 100 atm. At 1 atm, the instrumental resolution is sufficiently good such that individual Q-branch transitions can be resolved, but at 100 atm the individual transitions have coalesced into a more symmetrical band shape. Also shown in Fig. 14 are theoretical calculations based on the generalized G-matrix expression for $\chi^{(3)}$ that is valid for overlapping spectral lines. As can be seen, the agreement with the experiment is very good over the entire pressure range. The Gordon model calculation [Eq. (50)] also gives good agreement (Hall and Greenhalgh, 1982). Even though the high-pressure spectrum is only slightly narrower than that at 1 atm, collisional narrowing is playing a very significant role, as can be seen in Fig. 15. There the theoretical spectrum which includes narrowing effects is compared to one based on the isolated line approximation. At 100 atm, the isolated Raman linewidths will be approximately 10 cm^{-1}, and the calculated spectrum would have an overall width of about this magnitude if collisional narrow-

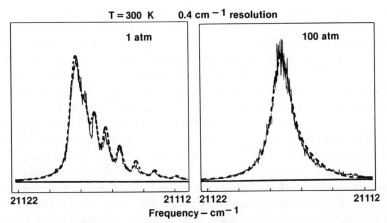

FIG. 14.   CARS signatures of 300 K N_2 at 1 and 100 atm pressure. The solid line is the experimentally scanned spectrum in each case, the dashed line the theoretical prediction. [Reprinted from Hall *et al.* (1980) with permission of North-Holland Publishing Company.]

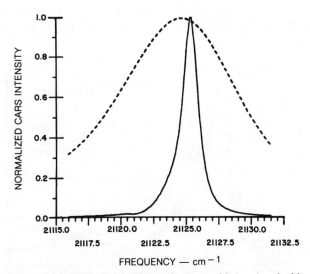

FIG. 15.   Comparison of calculated 100-atm signature with (——) and without (- - -) collisional narrowing. [Reprinted from Hall *et al.* (1980) with permission of North-Holland Publishing Company.]

ing played no role. Examination of Figs. 14 and 15 shows that the spectrum collapse associated with collisional narrowing has resulted in a contraction of the CARS bandwidth by about a factor of 10. The experimental investigations of Roland and Steele (1980b) are consistent with the foregoing results over the pressure range they considered. At the highest pressures, the agreement with experiment is improved by including an elastic contribution to linewidth of a few percent (Hall, 1983).

The effect of pressure on high-temperature N_2 CARS spectra has also been investigated (Stufflebeam et al., 1983). The expectation is that narrowing effects will become less important at high temperature because the dominant Q-branch transitions will be more widely spaced and the isolated linewidths will have an inverse temperature dependence at constant pressure. Figure 16 shows a 103-atm signature of N_2 obtained in an internally heated high-pressure vessel at a measured temperature of approximately 1700 K. At this pressure the isolated linewidths are sufficiently large such that individual Q-lines on the hot side of the fundamental have all but disappeared. Note, however, that the hot band is still resolvable, thus the ability to perform thermometry has not been lost. A theoretical calculation which includes narrowing and employs the inverse power law model for the off-diagonal parameters is also shown. The agreement is quite good, as can be seen. The role played by collisional narrowing in obtaining this agreement can be appreciated by considering the great differences between the isolated line and collisional narrowing calculations. It is apparent that line overlap effects play a significant role and that an

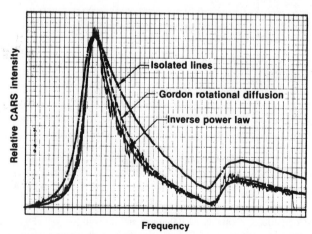

FIG. 16.   CARS spectrum of N_2 at 1700 K and 103 atm. Predictions of various theoretical models are shown. Overall resolution, 0.4 cm^{-1}.

isolated line calculation could not be used for accurate thermometry at these conditions. Note also that a calculation based on the Gordon model gives good agreement; significantly, it is more than one order of magnitude faster than the G-matrix inversion algorithm at these conditions. The effect of pressure-induced narrowing of N_2 CARS spectra will undoubtedly need to be accounted for in, for example, diesel engines and high-pressure gas turbines.

B. CARBON MONOXIDE

The pressure dependence of the 300 K spectrum of pure CO has also been recorded over the pressure range from 1 to approximately 100 atm (Stufflebeam *et al.*, 1983). As shown in Fig. 17, the observed dependence is not much different from that of N_2, a result that is probably a consequence of the similar vibration–rotation spectral constants of the two molecules. Although it might be expected that CO would narrow substantially faster than N_2 because there are considerations of nuclear spin

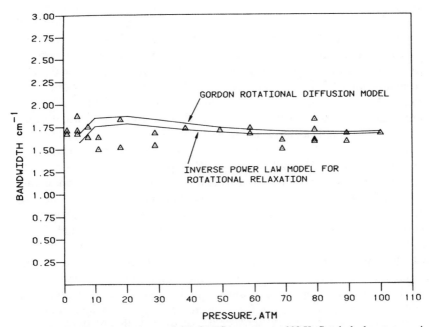

FIG. 17. Pressure dependence of CO CARS spectrum at 298 K. Symbols denote experimental data; solid lines are theoretical predictions. Overall resolution, 0.3 cm^{-1}; 5% vibrational dephasing contribution to linewidth assumed.

statistics in N_2 that preclude collisional mixing of odd and even rotational J levels, it is evident that this is not the case. Theoretical calculations bear this out, predicting faster narrowing for CO only at very low pressures where the collisional narrowing process is just beginning to be important. When the pressure has increased to the point where there is very strong line overlap, the details of the line-to-line spacing and energy transfer apparently become less important. This point is also borne out by the two theoretical predictions shown in Fig. 17, which show that the Gordon model calculation and the inverse power law approach give similar results that asymptotically approach one another with increasing pressure. The isolated linewidths employed in these calculations were those measured by Rosasco *et al.* (1983). As in N_2 (Hall, 1983), there is a need for a small elastic or pure dephasing contribution to linewidth which becomes important at the higher pressures.

C. HYDROGEN

H_2, when present, is ideal for combustion thermometry because of the simplicity of its vibrational Q-branch spectrum. Because the rotational constant B_e and the vibration–rotation interaction constant α_e are both very large for H_2, theoretical CARS spectra show only a few, well-separated lines even at flame temperatures (Fig. 18). The spacing between vibrational states in H_2 is also so large that the vibrational hot band makes no significant contribution even at very high flame temperatures. The observed intensity alternation between odd and even rotational components again reflects the ortho and para modifications of H_2 in which states with odd rotational quantum numbers have three times the nuclear spin statistical weight of states with even quantum numbers. A number of other factors make H_2 well suited for thermometry; because the H_2 Raman cross section is so large and so few rotational levels are populated, the CARS spectrum is relatively free from background interference even for H_2 concentrations down to a few percent, and H_2 can be used for thermometry even as a minority species. Another advantage of H_2 CARS is that the wide spacing between Q-branch transitions means that the calculational problems associated with collisional narrowing will be minimized until very high pressures are reached. Also, the H_2 Q-branch lines will be primarily Doppler broadened, at least at 1 atm, making them readily calculable [see Eq. (38)], thus removing uncertainty about the widths. For example, the Doppler width of the Q(1) line at 300 K is 0.037 cm^{-1}, and the pressure-broadened linewidth (self-broadening) is approxi-

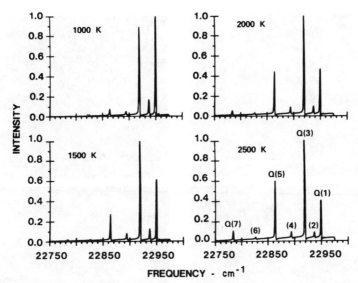

FIG. 18.   Theoretical CARS spectra of H_2 over the temperature range 1000 K $\leq T \leq$ 2500 K.

mately 0.0015 cm^{-1}/atm (Henesian and Byer, 1978). Corrections to the widths for Dicke narrowing can be applied, depending on the pressure (Shirley *et al.*, 1979b). Thus the spectrum peak heights in Fig. 18 are sensibly proportional to the squares of Boltzmann populations, and temperature can be obtained from ratios of line intensities.

Experimental H_2 BOXCARS signatures at various locations in a flat H_2–air diffusion flame verify the expected CARS signatures, as shown in Fig. 19 (Shirley *et al.*, 1979a,b). Temperatures were deduced from the ratios of the Q(1), Q(3), and Q(5) peak intensities, with a resulting standard deviation of 2–8% in the three intensity ratios at each location over the range 900–2100 K. Better accuracy would probably result from using ratios of integrated line strengths. The results of a temperature survey of a flat diffusion flame are shown in Fig. 20, where temperatures deduced from H_2 and O_2 CARS spectra are compared to measurements with a radiation-corrected thermocouple. Extraction of temperatures from O_2 spectra follows along the lines reported for N_2; the details will not be described here. As seen, the temperatures agree well in the cooler portions of the flame. The larger discrepancies occurring at the higher temperatures are probably due to the lower concentrations and poorer signal-to-noise ratios.

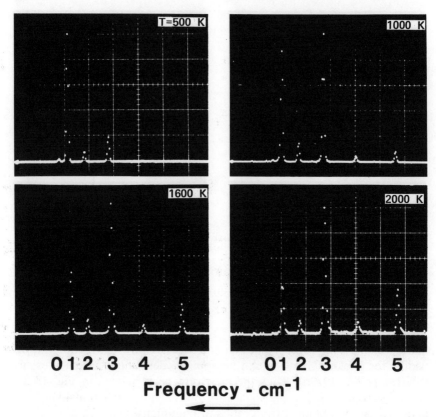

FIG. 19.   CARS spectra of H_2 in H_2–air diffusion flame at temperatures determined from relative strengths of indicated Q-branch transitions. Frequency scale corresponds to 0.60 cm^{-1} per dot.

D. WATER VAPOR

Water vapor is the major product of air-fed hydrogen-fueled combustion and often of hydrocarbon-fueled combustion as well. Measurement of its concentration serves as a gauge of the extent of chemical reaction and of overall combustion efficiency. At locations where it is sufficiently plentiful, it can also be used for thermometry.

H_2O is an asymmetric top possessing three vibrational modes, one of which, the ν_1 symmetric stretch with a Raman shift of 3657 cm^{-1}, is strongly Raman active. In addition to the quantum number J giving the total rotational angular momentum, each rotational state is characterized by a pseudoquantum number τ whose value lies in the range $-J < \tau <$

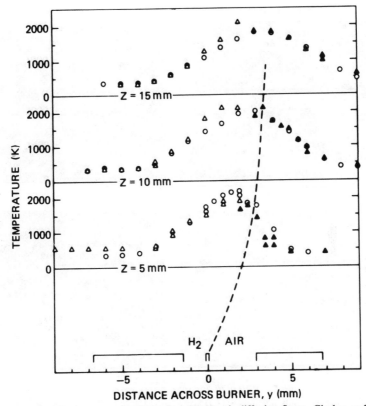

FIG. 20.   Temperature measurements in a flat H_2–air diffusion flame. Circles, radiation-corrected thermocouple; open triangles, H_2 CARS; solid triangles, O_2 CARS. Dashed-line curve is locus of maximum temperature.

$+J$. Each J_τ substate has the usual $2J + 1$ magnetic substates or orientation degeneracy. The Raman selection rules for H_2O are potentially quite complex, but the fact that the depolarization of the ν_1 mode is relatively small (≤ 0.06) means that isotropic Q-branch transitions with the simplified selection rules $\Delta J = \Delta \tau = 0$ will dominate the CARS spectrum. The calculation of H_2O CARS spectra has been described by Hall *et al.* (1979) and Hall and Shirley (1983). Such a calculation places very stringent requirements on vibration–rotation energy-level accuracy because of strong spectral interference effects arising from line overlap. In general, use of the energy-level data of Flaud *et al.* (1976) has been found to give good agreement with experimental spontaneous Raman (Bribes *et al.*, 1976) and CARS spectra (Hall *et al.*, 1979; Hall and Shirley, 1983).

Figure 21 shows a comparison of measured and calculated CARS spec-

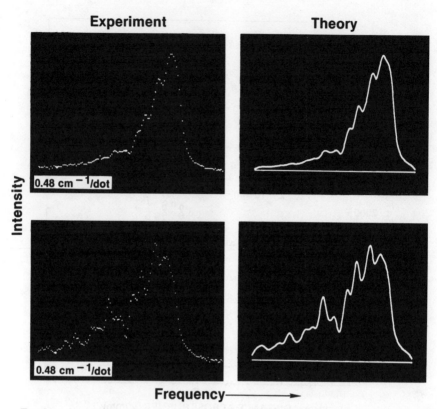

FIG. 21. Comparison of calculated CARS spectra of pure H_2O with experimental spectra measured in a heated cell at 473 (upper graphs) and 773 K (lower graphs).

tra of pure water vapor in a heated cell at 473 and 773 K (Shirley *et al.*, 1980a). The measured spectra were obtained on an optical multichannel detector using broadband CARS. The good agreement shown resulted in the inference that the self-broadened H_2O Q-branch linewidths must be extremely large, ~0.5 cm^{-1}/atm at low to moderate temperatures. The calculated spectra are somewhat sensitive to assumed linewidth, and use of smaller widths resulted in poor agreement with experiment. There is evidence to suggest that linewidths of this magnitude are reasonable; self-broadened H_2O linewidths for microwave and infrared transitions are as large as 1 cm^{-1} (Benedict and Kaplan, 1964; Mandin *et al.*, 1982). Further, the size of the width is consistent with the intuitive assumption that H_2O–H_2O inelastic collisions would be unusually efficient. The particular energy-level compilation used in these calculations is due to Professor R. Gaufres of Montpelier and is based on earlier, unpublished data of J. M.

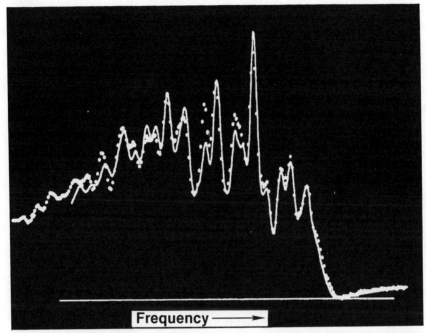

FIG. 22. Comparison of experimental (dots) and theoretical (line) CARS spectra of water vapor in a flame at 1700 K. A best-fit pressure-broadened linewidth $\Gamma = 0.19$ cm^{-1} was inferred for all transitions in the calculation.

Flaud, C. Camy-Peyret, and J. P. Maillard. A total of 977 Raman lines are considered, with vibrational hot bands such as (010) → (110) and (100) → (200) accounted for. The energy-level data are sufficiently complete to permit calculations over a range extending to about 70 cm^{-1} less than the band origin at 3657 cm^{-1}.

Measurements have also been made of the H_2O spectrum in a premixed methane/air flame at a temperature determined from thermocouple measurements to be 1700 K. Figure 22 shows the experimental spectrum and a least-squares theoretical fit. In the least-squares fit, the Raman linewidth was taken as the fitting parameter, and the best fit was obtained for a value of 0.193 cm^{-1}. This value is consistent with the pure H_2O value deduced from the Fig. 21 fits and with the expectations that the linewidth should have an inverse temperature dependence and could be smaller for broadening by diluents like N_2. As can be seen, the agreement is generally quite good. A few discrepancies are probably due to the assumption of constant Raman linewidth; the linewidth will in reality have a dependence on Q-branch transition, that is, on J and τ. CARS calculations would probably benefit a great deal from theoretical Raman linewidth calculations similar to those performed by Benedict and Kaplan for microwave transitions or from high-resolution inverse Raman measurements. In the theoretical calculation of Fig. 22, an H_2O concentration of 20% was assumed, and the degree of modulation near the bandhead arising from background susceptibility interference appears to be calculated accurately. Comparison of Figs. 21 and 22 reveals the dramatic changes that occur in the H_2O CARS spectrum as a function of temperature. H_2O should thus prove useful for thermometry; at the present time, with limited available information about Raman linewidths, H_2O CARS thermometry is probably slightly less accurate than N_2 CARS, but the accuracy can be expected to improve as more linewidth information becomes available. The CARS spectrum of high-pressure steam has also been measured experimentally over the ranges 1–16 atm and 200–500°C and has been interpreted in terms of the rotational diffusion model (Greenhalgh et al., 1983b).

E. Carbon Dioxide

The other major product of hydrocarbon combustion is carbon dioxide. BOXCARS spectra of CO_2 have been obtained by Eckbreth (Hall and Eckbreth, 1981) in several propane–air and CO–air flames, as shown in Fig. 23, where a measured thermocouple temperature is given for each spectrum. The experimental signature displays a fundamental band at a Raman shift of about 1388 cm^{-1}, corresponding nominally to the transition between the ground vibrational state and the first symmetric stretch state. A number of hot bands originating in vibrationally excited initial states appear at larger shifts. The relative strength of the hot bands is seen to be moderately sensitive to temperature, making the molecule potentially attractive for thermometry.

CO_2 is a linear triatomic characterized by the four vibrational quantum numbers v_1, v_2, v_3, and ℓ; these denote the number of quanta associated with symmetric stretching, bending, asymmetric stretching, and vibrational angular momentum, respectively. A particular vibrational eigenstate is thus commonly denoted by $(v_1 v_2^{\ell} v_3)$. Nominally only the symmetric stretch (ν_1) mode is Raman active; however, due to the accidental coincidence of ν_1 and $2\nu_2$, the two modes are mixed by Fermi resonance. Because of this near degeneracy, anharmonic terms in the intramolecular potential can cause a large perturbation of the associated eigenstates, and a mixing of the wave functions belonging to the two modes occurs. The Fermi resonance will occur between all states with the same v_3 and ℓ and

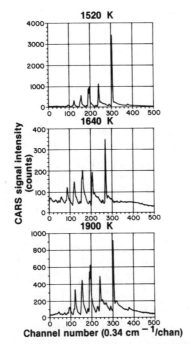

FIG. 23. CARS spectra of CO_2 at various temperatures in laboratory flames.

for values of v_1 and v_2 given by the relationship $(0, v_2{}^\ell, v_3)$, $[1, (v_2 - 2)^\ell, v_3]$, $[2, (v_2 - 4)^\ell, v_3]$.... The fact that the CO_2 spontaneous Raman spectrum has a relatively small depolarization means that attention can again be restricted to vibrational Q-branch transitions for which the selection rules are $\Delta v_1 = 1$, $\Delta v_3 = \Delta \ell = \Delta J = 0$. The $\Delta v_1 = 1$ selection rule connects the harmonic oscillator eigenfunction components $|v_1\rangle$ and $|v_1 + 1\rangle$ in the initial- and final-state wave functions.

The Fermi resonance mixing gives rise to a similar but slightly weaker sequence of vibrational bands whose fundamental occurs at a Raman shift of 1285 cm^{-1} and whose hot bands occur at smaller shifts. This band sequence was not probed in the experiments of Fig. 23 but it has been observed by Eckbreth *et al.* (1979b). The spectra in Fig. 23 show little rotational splitting due to the smallness of the vibration–rotation interaction constant in CO_2. Each vibrational band represents the contributions of O(100) transitions at flame temperatures.

Figure 24 compares the experimental CO_2 CARS spectrum at 1520 K with a theoretical spectrum based on the isolated line approximation (Hall and Eckbreth, 1981). As can be seen, the agreement is fairly good except

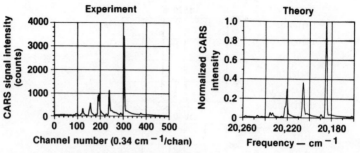

FIG. 24. Comparison of experimental and theoretical CARS signatures of CO_2 in a premixed CO–air flame at 1520 K. [Reprinted from Hall and Eckbreth (1981) with the permission of the Society of Photo-Optical Instrumentation Engineers.]

for the highest hot bands. In the theoretical calculation, vibrational energy levels, rotational B constants, and polarizability matrix elements were computed following the treatment of Courtoy (1957). This basically involves diagonalizing the system Hamiltonian for each set of resonating states. It was found, however, that better agreement with experiment resulted from the use of tabulated values for the rotational constants (Rothman and Benedict, 1978). Similar agreement was obtained with the other experimental spectra of Fig. 23. Further investigations will be needed before CO_2 CARS can be regarded as a reliable diagnostic tool. The calculations display a particular sensitivity to the rotational B constants, and the present best B values for the excited vibrational states may not be sufficiently precise for CARS. The significant degree of line overlap within each vibrational band indicates the isolated line approximation may not be valid even at 1 atm, and the pressure dependence of the 1388-cm^{-1} band at 300 K has been investigated (Hall, 1983).

F. Pure Rotational CARS

Investigations of pure rotational ($\Delta v = 0$, $\Delta J = \pm 2$) CARS spectra were originally motivated in part by the fact that the Raman cross sections for pure rotational processes are somewhat larger than those for vibrational processes (Fenner et al., 1973; Penney et al., 1974; Kondilenko et al., 1980), thus offering the promise of higher signal levels. Also, because adjacent pure rotational lines in a diatomic will be well separated (~8 cm^{-1} in N_2), it was thought that data reduction would thereby be simplified. Computer calculations (to be described later) show for N_2, however, that the advantage of a larger Raman cross section is completely offset by a smaller population difference factor [Eq. (36)]. There is therefore no signal strength advantage to pure rotational CARS, particularly because the recent measurements of pure rotational Raman cross sections for N_2 and O_2 by Kondilenko *et al.* suggest that the cross-section advantage for rotational CARS is not as large as previously thought. Whereas pure rotational CARS does not present any advantage over vibrational CARS in terms of data reduction at low pressures, because the interference effects arising in vibrational CARS are accurately handled by computer, it does have a data reduction advantage at larger pressures. This is due to the fact that the rotational lines should remain well separated or isolated at pressures where collisional narrowing effects are important in the vibrational spectrum. Also, the pure rotational approach may be preferable for thermometry at lower temperatures, for example, below 800 K, because in broadband vibrational CARS it is more difficult to resolve individual Q-branches at low to moderate temperatures, particularly when single-pulse capability is desired.

The main experimental difficulty associated with rotational CARS is the fact that the CARS and pump components are very nearly frequency degenerate due to the smallness of the Raman shift, thus complicating the separation and collection of the CARS signal. However, this difficulty can be circumvented by using the folded BOXCARS technique described by Shirley *et al.* (1980b), or by using planar BOXCARS and ensuring that the overlapped CARS and pump components have orthogonal polarizations (Goss *et al.*, 1980a).

The S-branch ($\Delta J = +2$) rotational CARS spectra of air and N_2 have been obtained by scanning the Stokes laser (Roland and Steele, 1980a; Goss *et al.*, 1980a) and by scanning the monochromator using a broadband Stokes source (Shirley *et al.*, 1980b). Single-pulse spectra of cold N_2 and N_2 in flames have been obtained by Murphy and Chang (1981). Figure 25 shows a scanned, broadband spectrum of N_2 at 300 K and 1 atm pressure (Shirley *et al.*, 1980b). Adjacent states show the 4 : 1 intensity alternation expected on the basis of nuclear spin statistics. Also shown as the dotted line is a theoretical calculation; as can be seen, the agreement with experiment is fairly good. The smaller discrepancies may be a result of uncertainties in the S-branch linewidths; the sensitivity of inferred temperatures for N_2 rotational CARS to these linewidths is not known at this time and is a subject that should be addressed. Murphy and Chang (1981) made single-pulse temperature measurements over the range 135–296 K and obtained very good agreement with thermocouple readings. These investigators have also obtained single-shot spectra in flames, an example of which is shown in Fig. 26. A reference cell is used in these experiments to correct the CARS signal for random variations of dye laser wavelength; the signal from the reference cell is also shown in Fig. 26. High-temperature pure rotational spectra of O_2 have also been obtained by these investigators.

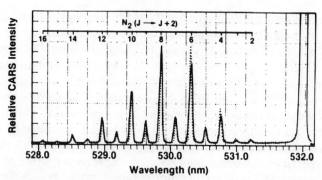

FIG. 25. Rotational CARS spectrum of N_2 at normal temperature and pressure with perpendicular pump and Stokes laser polarizations. The solid curve shows the measured spectrum and the dotted curve shows the theoretical prediction. The calculation is arbitrarily truncated such that transitions beyond $J = 12 \rightarrow 14$ are not shown. [Reprinted from Shirley *et al.* (1980b) with permission of the Optical Society of America.]

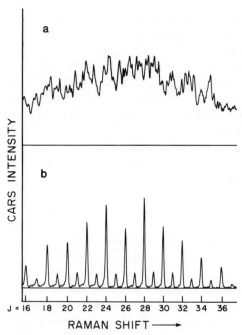

FIG. 26. Pure rotational CARS spectrum of hot N_2 obtained at tip of diffusion flame (b) and simultaneously recorded nonresonant CARS spectrum of Ar (a). (Reprinted with the permission of D. V. Murphy and R. Chang.)

TABLE II

RELATIVE PEAK INTENSITIES OF VIBRATIONAL AND
ROTATIONAL N_2 CARS

	$\Delta\omega_1 = 0.8$ cm^{-1} (300 K, 80% N_2)	$\Delta\omega_2 = 150$ cm^{-1} (1800 K, 70% N_2)
Rotational	0.8	1.0×10^{-3}
Vibrational	1.0	2.5×10^{-3}

The question of possible signal strength advantages of pure rotational CARS compared to vibrational CARS has been addressed by computer calculations. The relative signal strengths of pure rotational and vibrational CARS for N_2 have been calculated at 300 and 1800 K; the results are shown in Table II. In the calculations, the rotational-to-vibrational cross-section ratio determined by Penney *et al.* (1974) was employed, and, because the measured 300 K rotational S-branch and vibrational Q-branch linewidths are about the same (Rahn *et al.*, 1980a; Jammu *et al.*, 1966), it was assumed that the S-branch linewidths at 1800 K could be approximated by the known Q-branch linewidths at that temperature. It is apparent that there is no signal strength advantage at 300 K and that the rotational signal strength decreases more rapidly with increasing temperature than does the vibrational signature. As the temperature increases, the rotational populations tend to equalize, with the result that the population difference factor in rotational CARS is driven rapidly to zero. It has also been assumed in the calculations of Table II that the intensity product $I_1^2 I_2$ is the same for the vibrational and rotational cases; experimentally, this might be difficult to achieve because of poorer dye-pumping efficiency in the rotational case, i.e., assuming a 3 × Nd pump laser.

Thus, although rotational CARS will not enjoy signal strength advantages over vibrational CARS, it does have promise for low-temperature thermometry where signal strength is not a primary concern and has advantages for thermometry at elevated pressures because its spectra will be freer from the complications of collisional narrowing. It could therefore prove to be an attractive approach for CARS diagnosis of internal combustion engines and gas turbine combustors.

V. Species Concentration Measurements

CARS measurements of species concentration are typically performed in either of two ways: from the intensity of the spectrally integrated resonant CARS signal, or, in certain concentration ranges, from the shape of the spectrum. The latter approach, as discussed previously, results when the resonant signal interferes with the background nonresonant electronic contribution. At very low concentrations, each approach is hampered by the presence of the background because both the shape and strength of the CARS signal become relatively insensitive to the species concentration (Hall, 1979; Roh and Schreiber, 1978; Begley *et al.*, 1974). There are a number of approaches which have been demonstrated for reducing or suppressing the nonresonant background. The most straightforward and exact of these is polarization-sensitive CARS whereby nearly complete elimination of the background can be achieved. Other schemes will be briefly reviewed later. Although the background susceptibility contribution can be suppressed using polarization-sensitive CARS, this occurs at considerable expense to the resonant signal level, as will be demonstrated. Due to this signal loss, species detectivity may actually be decreased if the background is suppressed in certain instances. In this section, species concentration measurements using CARS spectral shapes will first be described. Background suppression using polarization-sensitive CARS will be explained, a technique which is often necessary for concentration determinations based on spectrally integrated intensities. Because these measurements are based upon the strength of the signal they must be appropriately referenced. For turbulent, three-phase combustion diagnostics, external reference cells lead to ambiguous results, however, and *in situ* referencing approaches which circumvent these problems and which can lead to experimental simplifications will be described.

A. Concentrations From Spectral Shapes

Unlike most spatially precise diagnostic approaches, CARS offers the unique potential for species concentration measurements over limited concentration ranges from the *shape* of the CARS spectrum. This occurs, as described earlier [see Eq. (29) and Fig. 2], when the species concentration decreases to the point where the resonant and background susceptibilities become comparable. In Fig. 27, computer calculations are shown at various CO concentrations in an 1800 K flame and illustrate this "modulated" spectral behavior quite vividly. The rounded nature of the spectral calculations arises from the Gaussian shape assumed for the broadband Stokes dye laser. As is evident, the spectral shape is a sensitive indicator of concentration in the range from approximately 0.5 to 30%. At very low concentrations ($\sim <0.5\%$), the "signal," i.e., the modulation,

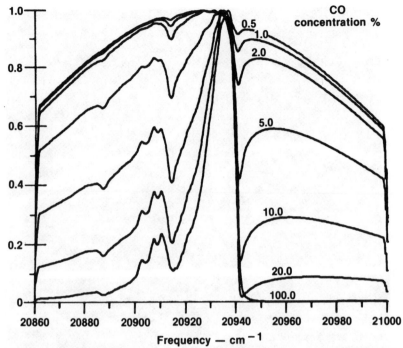

FIG. 27. Computed CARS spectral behavior of CO at various concentrations at 1800 K. [Reprinted from Eckbreth and Hall (1981) with permission of Gordon & Breach Science Publishers, Ltd.]

disappears into essentially a baseline level derived from the nonresonant susceptibility. When this occurs, the species is nominally no longer measurable. This was once believed to be a limitation of CARS for species diagnostics. With background suppression the limitation is circumvented and detectivity then becomes "photon" rather than "background" limited. In situations where the nonresonant susceptibility is small compared to the resonant susceptibility of the species of interest (e.g., high concentration, background suppression), concentration measurements can be performed only from the absolute intensity of some portion or integral of the CARS spectrum.

That spectral curve fitting is indeed viable is illustrated by the calibrations shown in Fig. 28 (Eckbreth and Hall, 1981). (a) Shows the experimental CARS signature (dots) from 2.1% CO in a calibrated CO–Ar gas mixture at room temperature together with the CARS theoretical prediction at the same concentration. (b) Shows a high-temperature calibration performed in the postflame region of a premixed Ar, CO, O_2, and CH_4

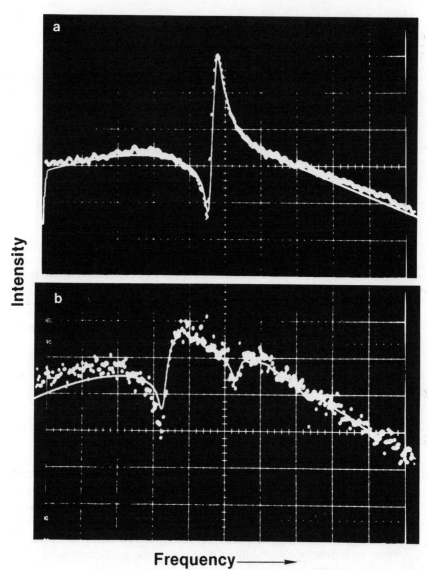

FIG. 28.   Comparison of experimental (dots) and computed (solid line) CARS spectra at (a) 300 K (2.1% CO) and (b) 2100 K (3.6% CO). [Reprinted from Eckbreth and Hall (1981) with permission of Gordon & Breach Science Publishers, Ltd.]

flame. Shown is the experimental CARS signature and theoretical overlay at a CO concentration of 3.6% measured with a quartz microprobe sampling system. The broadband CARS spectrum was averaged for 50 sec or approximately 500 laser pulses. The noise arises from the quantum statistical uncertainty in the subtraction of the intensified vidicon dark current. The agreement between the CARS and probe measurements is quite good in each case.

Spectral-fitting concentration measurements have been used in a number of flame studies. Moya *et al.* (1977) performed CO measurements in this manner through the combustion region of a spherical porous plug burner in which ethylene glycol was combusted. A 15-cm^{-1} spectral region near the destructive interference dip was examined and measurements were made down to the 1% level. Shirley *et al.* (1979a,b) examined O_2 decay through a flat H_2/air diffusion flame. O_2 concentrations were ascertained from O_2 spectral shapes which are qualitatively quite similar to the CO spectra shown in Fig. 27. Eckbreth and Hall (1979) examined N_2 concentrations on the centerline of an axisymmetric, laminar propane diffusion flame. In Fig. 22 it is clear that spectral fitting can also be employed for measuring low H_2O vapor concentrations as well.

The concentration range appropriate to spectral curve fitting for concentration measurements will vary from molecule to molecule and with gas composition and temperature. For diatomic molecules with closely spaced rotational transitions such as CO, O_2, and N_2, the range is approximately 0.1–20% at low temperatures and 0.5–30% at flame temperatures. For molecules with only a few significantly populated levels and narrow linewidths, e.g., H_2, it is considerably lower. At very low concentrations, due to the weakness of the modulation, very stringent requirements exist for the pulse-to-pulse stability of the normalized CARS signal for spectrally scanned studies of temporally steady flames. With broadband detection, e.g., for turbulent flames or for averaging in steady flows, high spectral resolution and smooth, broadband dye envelopes are required if weak modulations are not to be lost due to spectral smearing and noise.

Modulated CARS spectra are not necessarily unique to the three parameters of concentration, temperature, and nonresonant susceptibility. In the limit of small concentration, Eq. (29) can be rewritten as

$$|\chi|^2 \approx \chi^{nr}(\chi^{nr} + 2\chi'_r)$$
$$\approx \chi^{nr2}[1 + (2N/\chi^{nr})\bar{\chi}'_r] \qquad (64)$$

where $\bar{\chi}'_r$ is the per molecule real resonant susceptibility. Thus, there is a family of solutions determined by a constant value of N/χ^{nr} with the same CARS spectral shape. Thus, curve fitting to obtain species concentration levels is only as accurate as the degree to which the background nonresonant susceptibility is known. The commonly used tabulation of nonresonant susceptibilities given by Rado (1967) appears to require correction by about a factor of 2.5, as explained in detail in Eckbreth and Hall (1981).

Revised susceptibility values, verified by experimental calibration in a number of instances, are listed in Table III.

The nonresonant susceptibility of a gas mixture is the mole-fraction-weighted average of the nonresonant susceptibilities of the individual molecular constituents and, thus, will vary with the composition change accompanying combustion. However, for air-fed combustion, where N_2 is the dominant constituent, the nonresonant susceptibility will not vary greatly with combustion. For example, in stoichiometric methane–air combustion, the density-normalized nonresonant susceptibility increases by about 14% between the reactants and products. Based upon the measured temperature and/or location in the combustor or flame, intelligent estimates can be made of the nonresonant susceptibility value best employed to minimize the measurement uncertainty. Because hydrocarbon species generally have nonresonant susceptibilities considerably larger than N_2 and O_2, measurement uncertainties due to composition variations will be greatest in fuel-rich flame regions.

B. POLARIZATION-SENSITIVE CARS

The nonresonant contributions to the susceptibility have primarily an electronic origin through two-photon processes involving virtual states, in contrast to the resonant Raman terms which arise through the nuclear response of the molecule. As a consequence, the nonresonant susceptibil-

TABLE III

REVISED BACKGROUND
SUSCEPTIBILITY VALUES[a]

Gas	χ^{nr} ($\times 10^{18}$ cm^3/erg)
He	2.78
Ar	11.63
D_2	9.75
N_2	10.13
O_2	9.75
NO	31.5
CH_4	22.13
C_2H_6	46.5
CO_2	15.0

[a] To be used in the following expression for $\chi^{(3)}$:

$$\chi^{(3)} = \frac{2c^4}{\hbar\omega_2^4} N \sum_j \frac{\Delta_j(\partial\sigma/\partial\Omega)_j}{2\Delta\omega_j - i\Gamma_j} + \chi^{nr}$$

ity contributions exhibit symmetry properties different than the resonant terms and hence respond differently to the polarizations of the incident laser fields. Bunkin *et al.* (1977) have shown by proper orientation of the laser field and CARS polarizations that it is possible to suppress the nonresonant electronic contributions to the CARS signal. This then permits, assuming an adequate signal level, detection of species to considerably lower concentrations. Unfortunately, for isotropic Raman modes, a factor of 16 resonant-mode signal loss accompanies the suppression of the nonresonant contributions. Demonstrations of background suppression have occurred in flames from CO using collinear phase matching (Rahn *et al.*, 1979) or BOXCARS (Farrow *et al.*, 1980; Eckbreth and Hall, 1981), and from CO_2 using BOXCARS (Attal *et al.*, 1980). Liquid-phase investigations have been reported by Song *et al.* (1976), Akhmanov *et al.* (1978), and Oudar *et al.* (1979).

The theory of background suppression by polarization selection has been set forth in several of the reports just mentioned. The basic idea is that due to the resonant and nonresonant contributions to the third-order electric susceptibility having different tensor properties, the source polarizations for the two contributions will have different orientations in space in the general case where the pump and Stokes polarizations are not aligned. There are certain pump and Stokes orientations for which this angular separation will be maximized such that the use of a polarization analyzer normal to the nonresonant vector will suppress the nonresonant signals with minimized loss of the resonant-mode signal.

This is readily illustrated analytically. Consider pump and Stokes beams propagating with the Stokes polarization aligned with the x axis and the pump component(s) polarizations aligned at an angle θ to the Stokes. For simplicity, it will be assumed in BOXCARS that both pump component polarizations are similar. The induced polarization may then be written in terms of its x and y components as

$$\mathbf{P}_x = 3[\chi_{1111} \cos^2 \theta + \chi_{1221} \sin^2 \theta]\mathbf{E}_1^2\mathbf{E}_2^* \tag{65}$$

and

$$\mathbf{P}_y = 3[\chi_{1212} \cos \theta \sin \theta + \chi_{1122} \sin \theta \cos \theta]\mathbf{E}_1^2\mathbf{E}_2^*$$

The induced nonresonant polarization resides at an angle β inclined relative to the Stokes beam polarization, given by

$$\beta = \tan^{-1}\mathbf{P}_y^{nr}/\mathbf{P}_x^{nr} = \tan^{-1}\frac{\chi_{1212}^{nr} \cos \theta \sin \theta + \chi_{1122}^{nr} \sin \theta \cos \theta}{\chi_{1111}^{nr} \cos^2 \theta + \chi_{1221}^{nr} \sin \theta} \tag{66}$$

From Eq. (10), it can be seen that

$$\chi_{1212}^{nr} = \chi_{1122}^{nr} = \chi_{1221}^{nr} = \tfrac{1}{3}\chi_{1111}^{nr} \tag{67}$$

thus

$$\beta = \tan^{-1}\left\{\frac{2\cos\theta\sin\theta}{2\cos^2\theta + 1}\right\} \tag{68}$$

The resonant polarization components can be expressed as

$$\mathbf{P}_x^r = 3\chi_{1111}^r\cos^2\theta\ \mathbf{E}_1^2\mathbf{E}_2^*$$
$$\mathbf{P}_y^r = 3\chi_{1122}^r 2\cos\theta\sin\theta\ \mathbf{E}_1^2\mathbf{E}_2^* \tag{69}$$

The component of the induced resonant polarization passing through a polarization analyzer aligned at right angles to the nonresonant-induced polarization, thus suppressing viewing of the nonresonant component, can be shown to be, recalling from Eq. (10) that $\chi_{1122}^r = \tfrac{1}{2}\chi_{1111}^r$

$$\mathbf{P}_{pol}^r = 3\chi_{1111}^r\mathbf{E}_1^2\mathbf{E}_2^*[(\sin\beta)/2] \tag{70}$$

Thus, the background-free, induced resonant polarization sampled will be maximized when $(\sin\beta)/2$ is maximized, i.e., when the pump field is rotated to maximize the separation between the nonresonant induced polarization and the Stokes field. β is a function of θ through Eq. (68). From Eq. (68) it can be shown that

$$(\sin\beta)/2 = (\cos\theta\sin\theta)/(8\cos^2\theta + 1)^{1/2} \tag{71}$$

Upon differentiating the left side of Eq. (71) with respect to θ and setting the resulting derivative equal to zero, $(\sin\beta)/2$ will be maximized when $\theta = 60$ or $120°$, at which time $(\sin\beta)/2$ becomes equal to 0.25. When $\theta = 60°$, the induced nonresonant polarization then resides at an angle, $\beta = 30°$, relative to the Stokes field polarization. The polarization analyzer must then be oriented at $120°$ to suppress detection of the nonresonant contribution. The CARS signal received through the polarizer is proportional to the square of the induced resonant polarization which is viewed, i.e., $\mathbf{P}_{pol}^{r2}$. With aligned polarizations, the CARS signal would be proportional to the square of Eq. (69). Comparing the square of $\mathbf{P}_x^r$ with $\mathbf{P}_{pol}^r$, it is seen that they differ by the factor $[(\sin\beta)/2]^2$, or 0.0625. Thus, with background suppression, the resonant-mode signal is decreased by at least a factor of 16. If the individual pump component polarizations are not aligned, there are many more angular combinations which also suppress the background. Unfortunately, these also reduce the resonant CARS signal by a factor of 16 (Eckbreth and Hall, 1981). For a symmetric Raman mode the loss occurs for two reasons. The resonant-mode signal decreases in intensity as the pump and Stokes polarizations are angularly separated, and not all of the resonant signal generated is detected, rather only that fraction passed by the polarization analyzer. The term "suppression" is used deliberately because CARS is generated from the nonresonant background, but its detection is suppressed. This is always the case when using a single pump beam. In BOXCARS, however, there exist angular orientations in which no CARS arises from the nonresonant susceptibility, leading to true cancellation. In BOXCARS there are also more angular combinations for background suppression relative to two-beam CARS. Neither of these differences is significant, however, because the minimum signal reductions are unaltered. There are many angular combinations of the wave-mixing fields and CARS analyzer which lead to background suppression with minimum signal loss. Listed in Table IV are a few of the more convenient combinations. The data of Table IV assume the Stokes beam to be horizontally polarized; θ and φ are the angles of the ω_1 pump components relative to the Stokes field; α is the angle of the CARS analyzer relative to the Stokes wave required to suppress the nonresonant background.

Figure 29 presents an example of nonresonant background suppression from CO in a premixed, flat methane–air flame (Rahn et al., 1979). Collinear phase matching was employed and the CARS spectra were obtained by scanning a narrowband Stokes source. With aligned polarizations, the characteristic modulated spectrum is obtained. With the polarizations aligned as noted, the background-free spectrum in (b) was obtained. In this particular instance, the background was reduced by over three orders of magnitude. Also apparent in the spectrum are O-branch CARS lines from N_2. These can be used for temperature measurements (Farrow et al., 1981), as subsequent experiments using folded BOXCARS in methane diffusion flames have demonstrated.

TABLE IV

ORIENTATION ANGLES FOR MAXIMUM
CARS GENERATION WITH
NONRESONANT SUSCEPTIBILITY
SUPPRESSION

Stokes wave horizontally polarized, 0°		
$\theta(\omega_1)$	$\varphi(\omega_1')$	$\alpha(\omega_3)$
90	45	135
60	60	120
60	120	No analyzer required

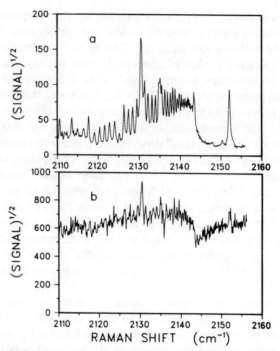

FIG. 29. Background nonresonant susceptibility suppression by polarization rotation. Spectra of flame CO with (a) and without (b) polarization-sensitive background rejection. [Reprinted from Rahn *et al.* (1979) with the permission of the authors.]

C. DETECTIVITY COMPARISONS

With background suppression, the decrease in the CARS signal is generally considerably greater than the resonant-mode loss factor of 16. For a species in low concentration, the CARS signal arises predominantly from the nonresonant susceptibility, with only a minor contribution from the resonant mode of the species. Thus, background suppression in these cases results in signal losses far greater than a factor of 16. Eckbreth and Hall (1981) found signal decreases of over two orders of magnitude when suppressing background from CO. There are experimental situations where concentration sensitivity actually decreases with suppression of background. Whether the background should or should not be suppressed depends upon whether the concentration sensitivity is background or photon limited. In the former, the signal levels are adequate, but sensitivity is lost when the degree of modulation in the CARS spectral shape is no longer accurately determined. This is the classic situation in CARS where the nonresonant background limits the detectivity, thus the term "background" limited. In the photon-limited case, detectivity hinges solely on the signal level being adequate. With high spatial and temporal resolution, e.g., BOXCARS and broadband Stokes sources, Eckbreth and Hall (1981) found CO concentration sensitivity in atmospheric pressure flames to be photon limited and thus obtained better concentration sensitivity *without* suppressing the background. This is illustrated in Fig. 30 for CO concentration measurements in a CO/air/Ar flame at 2000 K. With the background suppressed, the CO concentration sensitivity was limited to about 10% due to count limitations. By not suppressing the background, the nonresonant background could be followed to the point where the modulation was no longer easily measured, i.e., about 1% CO. In Fig. 30 it should be noted that the signal strength of the data with aligned polarizations is understated by about a factor of 4 due to the relative efficiency of the spectrograph grating to s and p polarizations. At lower temperatures, CARS with aligned polarizations was more sensitive by about a factor of 5, i.e., 0.1 versus 0.5%.

At lower spatial resolution, with narrowband CARS and in low-temperature high-density regions, it would be suspected that concentration sensitivity would be background limited. Thus, detectivity would be enhanced by background suppression. This cannot be generalized, and each case must be analyzed separately. For example, Rahn *et al.* (1981), using narrowband Stokes waves, calculated CARS concentration sensitivity for CO to be about 2% at 2000 K and 1 atm pressure with background suppression. Because CARS spectra for CO at flame temperatures are not background limited until about 0.5% (see Fig. 27), there seems to be no advantage to suppressing background in this case. At higher pressures, however, e.g., >1 atm, background suppression would lead to greater sensitivity.

D. *In Situ* REFERENCING

Performing concentration measurements from spectral shapes is the simple approach from an experimental standpoint. It is not necessary to worry explicitly about the strength of the spectrum as long as it is adequate. For concentration measurements from spectra where the nonresonant background is not evident, i.e., high species concentrations or with background suppression, absolute intensity levels must be measured. Experimentally, this is a far more difficult task. As mentioned previously, absolute intensity measurements are generally performed relative to a known standard using a separate, external reference cell. In addition to

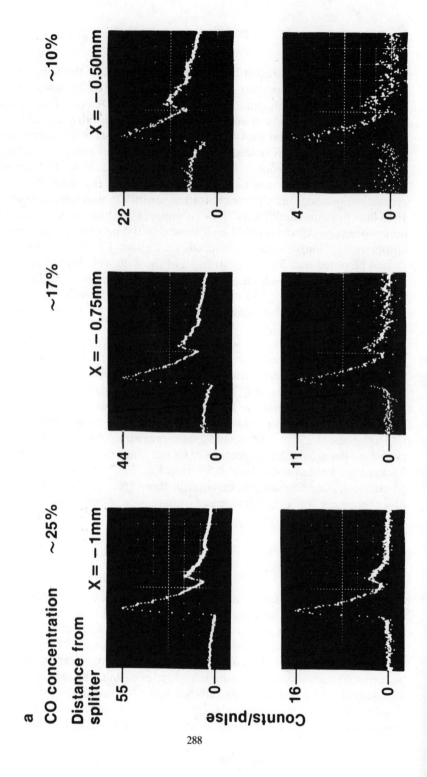

288

b

CO concentration ~5% ~2% ~1%

Distance from splitter X = – 0.25mm X = 0mm X = 0.25mm

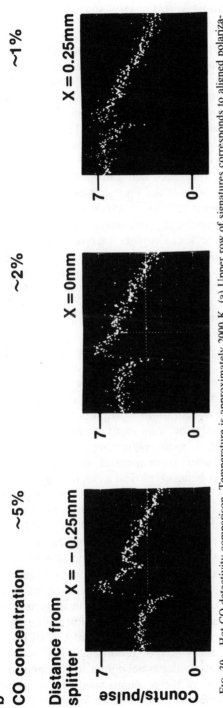

Counts/pulse

Fig. 30. Hot CO detectivity comparison. Temperature is approximately 2000 K. (a) Upper row of signatures corresponds to aligned polarizations, lower row to suppressed background. Inferred CO concentrations are shown. (b) Continuation of CO detectivity evaluation with background present. [Reprinted from Eckbreth and Hall (1981) with permission of Gordon & Breach Science Publishers, Ltd.]

serving as a calibration standard by housing a gas at known temperature and pressure, the reference cell also normalizes for pulse-to-pulse laser power fluctuations and can account for small optical misalignments. The latter results only if great care has been taken to ensure that the reference cell and measurement volume respond to optical misalignment in exactly the same way. Even with a normalizing reference cell and a single-mode pump laser, normalized CARS signals are subject to fluctuations of 5–7% even from a quiescent gas. This occurs due to a slightly different time ordering of the laser pulses in the reference cell and measurement volume, and the temporal irreproducibility of the laser pulses.

In combustion systems, potentially serious problems confront CARS density measurements made from spectrally integrated intensity levels. Beam attenuation from soot and/or fuel droplets together with turbulence-induced steering and defocusing must be accounted for in some manner. One approach, although not exact, would be to employ reference cells both before and after the test region, clearly at the expense of experimental complexity. Even in this case it is not possible to distinguish between a signal decrease due to attenuation, which is potentially correctable, and that due to refractive index nonuniformities, which are not correctable, leading to uncertainty in how to "correct" or normalize the raw data.

Based upon these difficulties, it becomes necessary to generate a reference signal at the *same* location at which the measurement is being made. In this way extinction effects due to soot and/or droplets and refractive effects affect the reference in exactly the same way as the CARS measurement. This, then, is the concept of *in situ* referencing. Actually, when performing concentration measurements from spectral shapes, the background nonresonant susceptibility is being employed as an *in situ* calibrating standard. With *in situ* referencing, the nonresonant background susceptibility is employed as the reference. There are currently two primary ways of doing this. In one approach (Shirley *et al.*, 1980a), in addition to the normal broadband Stokes wave centered on or near the species resonance of interest, a second narrowband Stokes beam is employed. This second Stokes is tuned by a grating considerably off the high-frequency side of the resonant bandhead to generate a reference CARS signal level from the background susceptibility. Both Stokes beams can be generated from the same dye laser by using an intracavity polarization splitter or a grating–mirror combination side by side. The nonresonant CARS reference and resonant CARS signal beams would be recorded simultaneously within the spectral field of the optical multichannel detector. Temperature would be ascertained from the shape of the resonant spectrum as usual; the total gas density would follow from the gas law, knowing the pressure, which would thus calibrate the nonresonant CARS reference. The concentration of the resonant species would in turn be determined by comparison with the strength of the nonresonant reference. If the two Stokes components exhibit relative power fluctuations from pulse to pulse, their respective energies would be recorded and normalized appropriately. This approach has been demonstrated experimentally and worked well (Switzer *et al.*, 1980).

The other *in situ* referencing approach involves polarization orientation and has been demonstrated by Oudar and Shen (1980) and Farrow *et al.* (1981). It is particularly useful when performing minority species measurements with nonresonant susceptibility suppression. In this approach, a beamsplitter is introduced into the CARS beam prior to the usual polarization analyzer, as shown in Fig. 5. The split-off portion of the CARS beam is sampled through a polarization analyzer set to view only the nonresonant CARS signal, i.e., the analyzer is aligned normal to the direction of the resonant CARS signal. The nonresonant signal level thus sampled is proportional to the total gas density and can be employed in the usual manner to determine the resonant species concentration level. The latter is measured through a polarizer normal to the nonresonant background polarization. An analog of this approach applies to major species concentration levels, i.e., when the nonresonant susceptibility has little effect on the resonant CARS spectrum of interest. Polarization rotation of the pump relative to the Stokes is employed to produce angular separation of the nonresonant and resonant contributions to the CARS signal. The nonresonant signal is sampled as above, i.e., through a polarizer set normal to the resonant polarization. The resonant signal is collected directly after the beamsplitter because the nonresonant contribution is slight and need not be filtered out. Optimal angular orientations are those which minimize the resonant-mode signal decrease but, at the same time, permit a substantial nonresonant signal to be sampled. *In situ* referencing schemes are schematically summarized in Fig. 31.

The major inaccuracy in *in situ* referencing resides in composition changes which cause the total nonresonant susceptibility to vary. As mentioned earlier, for air-fed combustion processes the susceptibility undergoes relatively minor variations in the composition change accompanying the transition from reactants to products. Based upon the measured temperature and/or location in the combustor, intelligent estimates can be made of the nonresonant susceptibility value which will minimize the measurement uncertainty. Aside from practical combustor measurements, it may be possible to configure fundamental turbulent combustion experiments to minimize the reactant-to-product variation in the nonresonant background. This can be achieved using excess N_2 as diluent or by

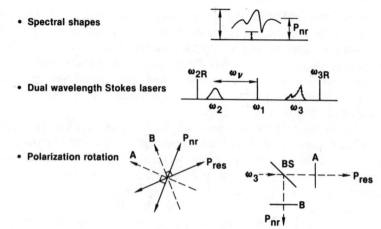

- Spectral shapes

- Dual wavelength Stokes lasers

- Polarization rotation

FIG. 31. Approaches to CARS concentration measurements using *in situ* referencing.

substituting a diluent with a background susceptibility higher than that of N_2, e.g., CO_2. This is analogous to the composition tailoring used in Rayleigh scattering diagnostics to achieve a nearly composition-independent, total Rayleigh cross section (Rambach *et al.,* 1979).

E. OTHER APPROACHES TO BACKGROUND SUPPRESSION

As mentioned earlier, there are other approaches to background suppression which result in a reduction of the nonresonant contribution and/or are limited in their general applicability. These will be discussed here for completeness. There are other techniques to suppress background (Yacoby *et al.,* 1980) which will not be treated here because they are generally inapplicable to combustion diagnostics per se.

1. *Double-Resonance CARS*

In this approach, demonstrated in liquids (Lynch *et al.,* 1976), nonfrequency-degenerate CARS is employed by introducing three beams at frequencies ω_0, ω_1, and ω_2 into the medium under observation. ω_1 and ω_2 are tuned as usual to the Raman resonance of interest. ω_0 is tuned in such a way that $\omega_0 - \omega_2$ approaches a resonance from another constituent. For example, in a combustion application, tuning might be near N_2. Writing the susceptibility as $\chi = \chi_1 + \chi_2 + \chi^{nr}$, where the subscript 1 refers to the resonance being probed and 2 to the resonance being exploited, the square of the modulus of the susceptibility is

$$|\chi|^2 = \chi_1'^2 + 2\chi_1'\chi_2' + 2\chi_1'\chi^{nr} + \chi_2'^2 + 2\chi_2'\chi^{nr} + \chi^{nr^2} + \chi_1''^2 + 2\chi_1''\chi_2'' + \chi_2''^2$$

(72)

If $\omega_0 - \omega_2$ is tuned in such a way that $\chi_2' < 0$, then cross terms such as $2\chi_2'\chi^{nr}$ and, potentially, $2\chi_1'\chi_2'$ can be used to cancel the nonresonant contributions and contributions from the exploited resonances. Using such a technique, the resonance curve of cyclohexane in a benzene background was enhanced nearly an order of magnitude above the background (Lotem et al., 1976). As can be seen, the suppression is imperfect in contrast to polarization-sensitive CARS and it has not been demonstrated in the gaseous phase.

2. Time Delay

In this approach three separate beams are employed (as in BOXCARS) and the pump frequencies may be the same as in normal frequency-degenerate CARS, although this is not necessary. One of the ω_1 pump components and the Stokes at ω_2 are temporally in phase, but the second pump component is appropriately delayed. The simultaneously applied ω_1 and ω_2 beams coherently excite the molecules and then terminate. The nonresonant electronic motions dephase rapidly relative to the molecular vibrations. The second pump pulse is delayed until the electrons dephase but prior to the dephasing of the vibrational modes. CARS is thus generated only from the slowly dephasing Raman modes and not the rapidly dephased electronic motions, leading to a significant reduction in the nonresonant background. For gases at atmospheric pressure, the electrons dephase in a time on the order of 10^{-14} whereas the vibrational motions exhibit a dephasing time on the order of 10^{-10} sec. Clearly, the laser pulses must be comparable to the electronic dephasing times, thus requiring mode-locked laser sources. The general area of molecular dephasing using picosecond laser pulses has been extensively investigated (Laubereau 1979; Zinth et al., 1978). This technique of background suppression has been demonstrated in liquids by Kamga and Sceats (1980), with background levels being reduced by a factor of approximately 10^2. Some resonant-mode signal loss occurring due to partial dephasing of the nuclear motions would be anticipated, but it was not clear how large this loss was. Thus there may be signal advantages to be gained relative to polarization-sensitive CARS. Also, due to the intensity dependence of the CARS generation process, it may be possible to achieve enhanced CARS signal levels for the mode-locked pulse trains over a Q-switched laser pulse of the same energy. This potential advantage may be mitigated by saturation and/or optical Stark-broadening considerations. In broadband picosecond CARS investigations in N_2, however, Shirley (1983) encountered beating between neighboring rotational transitions which caused the coherently excited Raman modes to decay nearly as rapidly as the nonresonant background. With narrowband short pulses, similar problems would be anticipated because the transform-limited linewidth would lead to excitation of several transitions simultaneously. Thus, for the molecules of combustion interest with closely spaced transitions, this approach does not appear promising. Furthermore, aside from detector beating, the technique is not applicable at high pressures due to reduced vibrational dephasing times. It thus appears to offer little feasibility for combustion diagnostics. This is an intriguing area requiring further investigation vis-à-vis its practical utility for combustion diagnostics.

3. Background Subtraction

Background subtraction techniques have been investigated using reference *nonresonant* samples (Molectron Corporation, 1976) but only modest reductions (factor of 3) were achieved. Performing *in situ* background subtraction using a three-frequency approach where $\omega_0 - \omega_2$ is not tuned to a second resonance or using polarization techniques to sample the nonresonant background could be considered as approaches. Such subtraction approaches, however, cannot eliminate interference effects between the resonant and nonresonant susceptibilities and are thus of questionable utility.

F. ELECTRONIC RESONANCE ENHANCEMENT

Discussion of CARS to this point has been restricted to the Raman resonantly enhanced situation, i.e., when $\omega_1 - \omega_2 = \omega_v$. However, CARS can be further enhanced electronically when one or several frequencies ω_1, ω_2, or the CARS frequency ω_3 are close to a one-photon electronic transition frequency. If the resonant-mode CARS signal can be electronically resonantly enhanced relative to the nonresonant background, then species sensitivity can be extended to considerably lower levels than normally permitted either by background or photon limitations. Electronic resonance CARS has been extensively analyzed (Druet *et al.*, 1978, 1979; Oudar and Shen, 1980). Electronically enhanced CARS has been demonstrated in liquids by numerous investigators and, as previously mentioned, in the gas phase in I_2 (Attal *et al.*, 1978), in OH (Verdieck *et al.*, 1983), in NO_2 (Guthals *et al.*, 1979), and in C_2 (Gross *et al.*, 1979; Attal *et al.*, 1983; Greenhalgh, 1983a). In the I_2 work, I_2 at 1 mbar pressure was detected in a 1 atm pressure air background. With the pump laser tuned to an I_2 electronic resonance, the I_2 CARS signal was about twice as intense as the O_2 Q-branch peak contained in the cell at 200 times the pressure. The most impressive achievement of electronically resonant CARS thus far has undoubtedly been the detection of C_2 at a concentration of 10^8 cm^{-3} by Attal *et al.* (1983).

The technique may become a powerful tool for gas-phase spectroscopy and analytical chemistry because of its signal strength and capability to produce spectra with reduced inhomogeneous broadening. It may be of limited utility for the diagnostics of major molecular species of combustion interest because they possess electronic transitions in relatively inaccessible spectral regions below 2000 Å. Flame radicals, whose absorptions are generally accessible in the ultraviolet, are indeed detectable in this manner, however, and the electronically resonantly enhanced spectrum of the OH radical has been observed by Verdieck *et al.* (1983) in a methane–oxygen flame. Because of the important role that OH is thought to play in many combustion mechanisms, these experimental and theoretical investigations will now be described.

1. *Hydroxyl Radical (OH)*

In Section II,C it was pointed out that the most general expression for the vibrationally resonant contribution to the third-order electric susceptibility [Eq. (23)] contains a large number of terms, some of which can become large when any of the electric field frequencies become resonant with one-photon electronic transitions in the probed molecule. The interpretation of any molecule's spectrum in terms of Eq. (23) is potentially very complex, but the analysis can normally be drastically simplified in practice. If the pump source is tuned to a particular electronic absorption, and if the populations of all levels can be neglected except those in the ground electronic and vibrational state (most likely a good assumption for OH), then the dominant contribution to the resonant susceptibility for Raman transition $i \to f$ in the vicinity of the one-photon resonance will be

$$\chi_{if}^{(3)} \cong \frac{N}{\hbar^3} (\omega_{if} - \omega_1 + \omega_2 - i\Gamma_{if})^{-1} \times \rho_{ii}^{(0)}$$

$$\times \frac{\mu_{fn}\mu_{ni}}{\omega_{ni} - \omega_1 - i\gamma_{ni}} \times \sum_{n'} \frac{\mu_{in'}\mu_{n'f}}{\omega_{n'i} - \omega_3 - i\Gamma_{n'i}} \tag{73}$$

In practice, often only one anti-Stokes ($i \to n'$) transition is important as well. For the pump tuned to one $i \to n$ transition, the allowed Raman vibrational resonances will then be determined by those final states f which simultaneously satisfy the selection rules governing $n \rightarrow f$ electronic and $i \rightarrow f$ Raman selection rules. This corresponds to a type of "double resonance"; another results from a double electronic resonance of the type $i \rightarrow n$, $i \rightarrow n'$, with no vibrational resonance. There is also the possibility of "triple resonances," when all resonances of the kinds $i \rightarrow n$, $i \rightarrow n'$, and $i \rightarrow f$ are simultaneously satisfied. The condition for the latter is that $\omega_{if} = \omega_{nn'}$; in OH, computer simulations for pump coincidence with various transitions in the 0–0 or 1–0 A–X electronic systems show that these triple resonances should not be expected (Verdieck et al., 1983). The dominant features in the electronically resonantly enhanced CARS of OH are expected to be double resonances of the $i \rightarrow n$, $i \rightarrow f$ type; because of the large difference in vibrational-state spacing between the two electronic states in OH, double electronic resonances are usually considerably displaced in frequency with respect to these. If the pump is tuned to a Q-branch electronic absorption, selection-rule considerations show that a triplet of vibrational resonances (P, Q, and R transitions in the 0–1 X vibrational band) is expected as the Stokes source is tuned. Pump coincidence with P and R electronic absorptions also is expected to result in triplets, with O and S vibrational resonances appearing. The large rotational constant in OH ($B \simeq 19$ cm^{-1}) will give rise to a splitting of the components of the triplet that is always very large; for values of rotational quantum number $J \simeq 10$, it will be on the order of hundreds of cm^{-1} and thus might be difficult to observe experimentally.

Because the A–X electronic absorptions lie in the ultraviolet, Verdieck et al. (1983) have employed two frequency-doubled, narrowband, tunable dye lasers for the pump and Stokes sources. Considerations of dye power output and doubling efficiency dictated that the 1–0 A–X system be probed by the pump, with the Stokes source then tuned to locate the Raman resonances. The results of an experiment in a methane–oxygen flame in which the pump was tuned to the $Q_1(2)$ electronic absorption and the Stokes was tuned over a wide frequency range are shown in Fig. 32. One disadvantage of probing 0–1 vibrational Raman transitions in OH when water vapor is present is the potential interference with the normal CARS spectrum of the symmetric stretch mode of water vapor; this is clearly evident in Fig. 32. Nevertheless, the Raman Q-branch portion of the OH electronically resonantly enhanced CARS spectrum is clearly visible as the prominent feature lying just to the low-frequency side of the water vapor bandhead. There are also neighboring satellite features whose spacing seems to coincide with the spacing of the Q-branch resonances in the normal Raman spectrum of OH. The appearance of these features is not yet understood, although saturation of the pump reso-

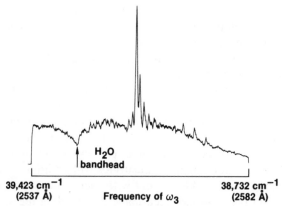

39,423 cm^{-1} (2537 Å)   Frequency of ω_3 38,732 cm^{-1} (2582 Å)

FIG. 32. Electronic resonance enhancement of the CARS spectrum of OH in a CH_4-O_2 flame. Pump is tuned to the $Q_1(2)$ transition of the 1–0 A–X system of OH. $T \approx 2500$ K.

nances is a possibility. Collisional redistribution of radiation by rotationally inelastic transfer is also possible, although it is generally thought that no such redistribution effects occur in CARS. At lower frequencies there is also clear evidence of the expected P-branch component of the triplet, although again with satellite structure. The observed spectra are strongly dependent on the tuning of the pump, ruling out the possibility that the normal off-resonant CARS spectrum of OH is being observed. The spectra also are dependent on the pump intensity, which means that saturation effects are indeed playing some role.

The OH experiments had just begun at the time of writing this article, thus no conclusions about detectivity gains from electronic resonant enhancement in OH could be drawn. It is clear, however, that the first results hold great promise for the development of a valuable diagnostic tool for combustion research. They also raise some interesting theoretical questions about the role of saturation and collisional redistribution of radiation in electronic resonant CARS.

G. APPLICABILITY

In principle, CARS can detect any species possessing Raman resonances. This even includes atomic species such as sulfur, carbon, and oxygen which possess degenerate, ground electronic states with well-separated energy levels (Eckbreth et al., 1979a). This idea was first suggested by Schlossberg (1976) in connection with fluorine atom detection. The Raman cross section of ground-state oxygen atoms has recently been measured (Dasch and Bechtel, 1981) and CARS generation from O atoms has been demonstrated (Teets and Bechtel, 1981). Practically speaking, many species which occur in combustion systems exist at too low a concentration to be measured by CARS with high spatial and temporal resolution because of *photon limitations*. CARS is thus generally limited to measurements of the major species, i.e., those occurring in abundances greater than approximately 0.1%. In some high-temperature flames, certain radical species such as OH may therefore be directly measurable. In other situations electronic resonance enhancement may permit detection of selected radical species. At the higher pressures typical of gas turbines and internal combustion engines, sensitivity will be improved due to the nonlinear scaling of the CARS signal with increasing pressure. As seen previously, good progress has been made in modeling the spectra of the major constituents in combustion, i.e., N_2, O_2, H_2, CO, H_2O, and CO_2. Extension to other diatomic molecules, e.g., NO, which predominate in some combustion systems should be readily achievable. Because hydrocarbon species are Raman active, CARS should be capable of monitoring these as well. Because the C–H stretch spectral region in most hydrocarbons is small, interferences between the various constituents will be common and it might be difficult to quantitatively measure the individual hydrocarbon species. However, because the Raman cross section of many hydrocarbons (Stephenson, 1974) divided by the total number of hydrogen atoms is nearly constant, CARS may be able to measure the total number of C–H bonds with little sensitivity to the specific composition of the hydrocarbons. With broadband detection and moderate spectral resolution, the dominant hydrocarbons may even be identifiable. In the presence of soot, measurements of certain species will most likely be perturbed. Laser vaporization of soot (Eckbreth, 1977), which occurs even on a nanosecond time scale at high laser intensities, will produce copious quantities of C_n species, where $n = 1, 2, 3$, and so on. Thus measurements of these species, and potentially some hydrocarbons, depending on the soot hydrogen content, are likely to be perturbed.

VI. Practical Applications

In late 1978 and during 1979, the potential of CARS for practical, "real-world" applications was realized with measurement demonstrations in a variety of practical devices. Stenhouse *et al.* (1979) used CARS to perform measurements in both motored (with propane) and gasoline-fired internal combustion engines. A narrowband Stokes laser was scanned synchronously with the engine cycle to generate the CARS spectrum piecewise, but this was nevertheless temporally resolved vis-à-vis the engine cycle. Determination of cycle-to-cycle fluctuations would of course require a broadband Stokes source. Collinear phase matching was employed and spectra of N_2 and propane were obtained at various points in the engine cycle. Measurements were interrupted during certain portions of the engine cycle presumably due to refractive index disturbances occasioned by flame-front passage. CARS measurements have also been performed in an actual diesel engine (Kajiyama *et al.*, 1982) at Komatsu Corporation in Japan and in gasoline engines at Ford Scientific Research Laboratories (Klick *et al.*, 1981) and at Sandia National Laboratories (Rahn *et al.*, 1982). Measurements in internal combustion engines using CARS are scheduled for the mid-1980s at many laboratories throughout the world.

CARS measurements in combustion tunnels were performed in the late 1970s at ONERA, Wright–Patterson AFB, and the United Technologies Research Center. At ONERA, late in 1978, Taran and co-workers (Attal *et al.*, 1980; Pealat *et al.*, 1980) performed time-averaged, collinear CARS measurements in the exhaust of a simulated turbomachine combustor burning kerosene. Noise levels near the rig were 110 dB. Numerous spectra of N_2 and CO_2 were recorded by scanning a narrowband Stokes source through the Raman resonances of interest and averaging over several laser pulses at each spectral location. Figure 33 shows the time-averaged spectrum of N_2 points (dots) obtained in the exit plane of the combustor. The data reduction fit (solid line) gave a temperature of 1150 ± 50 K and a N_2 concentration of $78 \pm 5\%$, in good agreement with thermocouple and gas chromatograph measurements giving 1050 K and 78%, respectively. CO_2 concentration sensitivities were estimated at 1000 ppm based on the strength of the experimental measurements. Recently, Taran and Pealat (1982) and Greenhalgh *et al.* (1983a) have successfully performed CARS measurements inside gas turbine combustion cans in tunnels.

At WPAFB, a compact and hardened version of a ruby-laser-based CARS system has been employed to perform temperature and species concentration measurements in a bluff-body stabilized diffusion flame (Switzer *et al.*, 1980) contained in a 25-cm-diameter tunnel operated at 1 atm pressure. Fuels have ranged from gaseous propane to shale-derived JP-8. Collinear phase matching has been employed with an axial spatial resolution of 2 cm. Scanned, broadband-generated N_2 CARS spectra have been used to obtain average temperatures. Single-pulse integrated CARS measurements of N_2 and O_2 were made and reduced in conjunction with weighted average temperatures to obtain concentration profiles. The minimum detectable concentration capabilities of the CARS system were estimated at 2% in N_2 and 0.5% for O_2. Shown in Fig. 34 is a comparison of O_2 concentration measurements made by CARS and a gas sampling

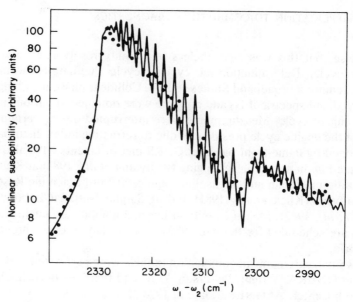

FIG. 33.   Time-averaged spectrum of N_2 in the exit plane of combustor; 10 laser shots are averaged at each point (dots); line, theory; 78 ± 5% N_2; $T = 1150 ± 50$ K. [Reprinted from Pealat *et al.* (1980) by permission of IPC Science and Technology Press Ltd.]

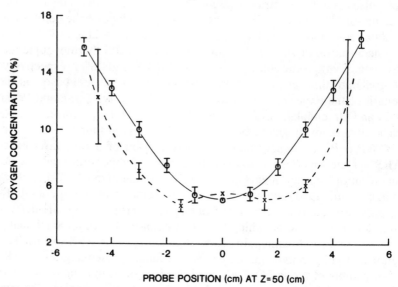

FIG. 34.   Comparison of oxygen concentrations in flame determined by CARS (×) and microprobe sampling (○). [Reprinted from Switzer *et al.* (1980) by permission of the American Institute of Aeronautics and Astronautics.]

probe. The lower CARS concentration measurements in the strong gradient regions may be due to a possible systematic error. Agreement on the centerline is quite good, as seen. Single-pulse temperature measurements using broadband CARS have also been demonstrated in combustion tests using the liquid fuels JP-4 and a shale-oil-derived JP-8. The hardened CARS system has since been upgraded by replacement of the ruby laser with a $2 \times$ Nd:YAG laser system.

In experiments at UTRC, Eckbreth (1980) used BOXCARS in a 50-cm-diameter tunnel fitted with either a swirl burner or a JT-12 combustor. Broadband generation and detection were used. CARS temperature measurements from N_2 were performed in the primary zone of a highly swirled flame fueled with propane or Jet A and in the exhaust of a JT-12 can burning Jet A. A portable CARS apparatus similar in concept to that shown in Fig. 5, but employing planar BOXCARS, was used. Delicate instrumentation such as the spectrograph and optical multichannel analyzer was housed in the control room adjacent to the burner test cell. The CARS signals upon exiting from the tunnel were piped out of the test cell to the control room using a 20-m-long, 60-μm-diameter fiber-optic guide (Eckbreth, 1979). Both average and single-pulse thermometry were performed at various locations in the tunnel at several stoichiometries. Shown in Fig. 35 are averaged CARS spectra at two locations down-

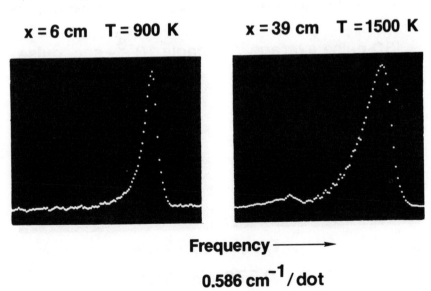

x = 6 cm   T = 900 K   **x = 39 cm   T = 1500 K**

Frequency ───────▶

0.586 cm^{-1}/dot

FIG. 35. Spatial variation of temperature from averaged CARS spectra of N_2 in swirl burner with Jet A fuel. [Reprinted from Eckbreth (1980) by permission of Elsevier North Holland.]

stream of the swirl burner fired with Jet A fuel. The CARS spectra were averaged on the optical multichannel detector for 10–15 sec, corresponding to 100–150 laser pulses. The measurement 6 cm downstream of the burner exit plane was made through the fuel spray just 11 cm downstream of the fuel nozzle. Figure 36 shows a comparison of averaged and single-pulse CARS spectra in the swirl burner fitted with a refractory back wall to simulate a furnace. Although of slightly lower quality due to photon statistics, the single-pulse spectrum is of high enough quality to allow instantaneous temperature measurements. Assembly of a statistically significant sample of instantaneous measurements would permit determination of the probability distribution function (pdf) for temperature from which the mean temperature and turbulent fluctuation magnitudes could be found. Fuji *et al.* (1983) have performed single-shot N_2 CARS thermometry in combustion tunnels using a remotely located CARS system. The CARS equipment was located in a control room adjacent to the combustion tunnel and the laser beams were "piped" to the combustor. Fiberoptic links were used to transmit the CARS signal back to the equipment area.

Recently, CARS has been applied to measurements of temperature and species in the exhaust of actual, afterburning jet engines (Eckbreth *et al.,* 1983). The CARS instrument consists of a transmitter and receiver mounted near the engine on an XY-traversing system which permits map-

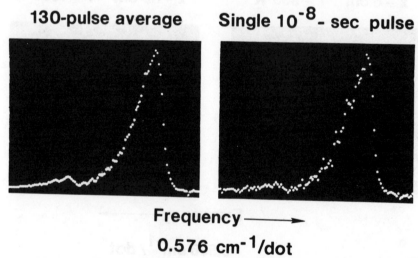

130-pulse average   **Single 10^{-8}- sec pulse**

Frequency ⟶

0.576 cm^{-1}/dot

FIG. 36. Comparison of averaged ($T = 1450$ K) and single-pulse N_2 CARS spectra in swirl burner with refractory back wall fueled with Jet A at an overall equivalence ratio of 0.8. [Reprinted from Eckbreth (1980) by permission of Elsevier North Holland.]

ping over the entire engine exhaust. The transmitter houses the frequency-doubled neodymium:YAG laser and a carriage of five dye cells for measurements of the species of interest, N_2, O_2, CO, H_2O, and unburned hydrocarbons. A single Fabry–Perot broadband dye laser cavity is used with the desired dye cell translated into place. Fiber-optic links are used to transmit the CARS signals to the delicate spectrographic and computer instrumentation located in a nearby control room. Personnel are restricted from the vicinity of the jet engine when it is operating, requiring complete remote control of its operation including critical mirror and other optics adjustments. In the initial demonstrations, time-averaged measurements were made based on detector availability. Figure 37 shows the N_2 CARS spectrum from a fully augmented (i.e., complete afterburning) jet engine. The spectrum was obtained from a 15-sec average corresponding to 150 laser pulses. Temperature is inferred from a least-squares regression with an excellent match between the theoretical and experimental spectra. Following the initial successful demonstrations, the instrument has been upgraded to permit instantaneous measurements of medium properties at a 20-Hz repetition rate. In Fig. 38, a temperature histogram indicative of this capability is shown in an engine exhaust at full augmentation.

There is also considerable interest in applying CARS to investigations of high-pressure propellant burning. McIlwain and Harris (1980) used broadband BOXCARS to obtain single-pulse and averaged temperature measurements from N_2 in nitrate ester propellant flames freely burning at atmospheric pressure. Measured single-shot temperatures gave temperatures consistent with thermochemical calculations. CARS has also been

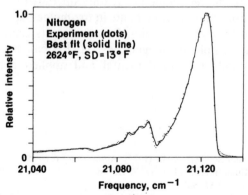

FIG. 37. CARS spectrum of N_2 in exhaust of afterburning jet engine at full augmentation. Overall resolution ≈ 2 cm^{-1}.

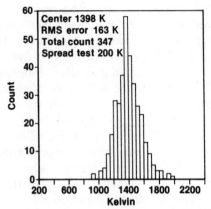

FIG. 38. Temperature histogram of single-pulse measurements obtained at 20 Hz in augmented jet engine exhaust.

successfully applied to energy-related combustion devices. CARS measurements of temperature and concentration of CO_2, O_2, CO, and H_2O have been performed in a semiindustrial furnace measuring 3 m on a side (Ferrario *et al.*, 1983). CARS has also been used to make measurements of CO and H_2S in actual coal gasifiers (Taylor, 1983).

Interest in CARS for measurements in practical situations is increasing rapidly as laboratories throughout the world gear up to apply CARS to internal combustion engines, furnaces, gas turbine combustors, and propellant burning for propulsion and ballistics applications. All of the equipment required for CARS is presently commercially available and the technique can be engineered for application in practical environments. The capabilities of the technique are readily upgraded via minor experimental modifications and improvements in the molecular spectral synthesis routines used to reduce the data. Based upon its capabilities relative to conventionally employed diagnostics, CARS can be anticipated to increase understanding of fundamental and applied combustion processes.

References

Aeschliman, D. P., and Setchell, R. E. (1975). *Appl. Spectrosc.* **29**, 426.

Akhmanov, S. A., Bunkin, A. F., Ivanov, S. G., and Koroteev, N. I. (1978). *Sov. Phys.—JETP (Engl. Transl.)* **47**, 667.

Alekseyev, V., Grasiuk, A., Ragulsky, V., Sobelman, I., and Faizulov, F. (1968). *IEEE J. Quantum Electron.* **QE-4**, 654.

Anderson, P. W. (1949). *Phys. Rev.* **76**, 647.

Armstrong, J. A., Bloembergen, N., Ducuing, J., and Pershan, P. S. (1962). *Phys. Rev.* **127**, 1918.

Attal, B., Schnepp, O. O., and Taran, J. P. E. (1978). *Opt. Commun.* **24**, 77.

Attal, B., Pealat, M., and Taran, J. P. E. (1980). *AIAA Pap.* **80-0282**.

Attal, B., Debarre, D., Muller-Dethlefs, K., and Taran, J. P. E. (1983). *Rev. Phys. Appl.* **18** (in press).

Bailly, R., Pealat, M., and Taran, J. P. E. (1976). *Opt. Commun.* **17**, 68.

Barrett, J. J., and Begley, R. F. (1975). *Appl. Phys. Lett.* **27**, 129.

Beattie, I. R., Black, J. D., and Gilson, T. R. (1978). *Combust. Flame* **33**, 101.

Begley, R. F., Harvey, A. B., and Byer, R. L. (1974). *Appl. Phys. Lett.* **25**, 287.

BelBruno, J. J., Gelfand, J., and Rabitz, H. (1983). *J. Chem. Phys.* **78**, 3990.

Benedict, W. S., and Kaplan, L. D. (1964). *J. Quant. Spectrosc. Radiat. Transfer* **4**, 453.

Bloembergen, N. (1965). "Nonlinear Optics." Benjamin, New York.

Bloembergen, N., Lotem, H., and Lynch, R. T. (1978). *Indian J. Pure Appl. Phys.* **16**, 151.

Bonamy, J., Bonamy, L., and Robert, D. (1977). *J. Chem. Phys.* **67**, 4441.

Bovanich, J. P., and Brodbeck, C. (1976). *J. Quant. Spectrosc. Radiat. Transfer* **16**, 153.

Bribes, J. L., Gaufres, R., Monan, M., Lapp, M., and Penney, C. M. (1976). *Appl. Phys. Lett.* **28**, 336.

Brueck, S. R. J. (1977). *Chem. Phys. Lett.* **50**, 516.

Brunner, T. A., and Pritchard, D. (1982). "Advances in Chemical Physics, Dynamics of the Excited State" (K. P. Lawleg, ed.). Wiley, New York.

Bunkin, A. F., Ivanov, S. G., and Koroteev, N. I. (1977). *Sov. Tech. Phys. Lett.* (*Engl. Transl.*) **3**, 182.

Butcher, P. N. (1965). "Nonlinear Optical Phenomena." Ohio State Univ. Eng. Publ., Columbus.

Byer, R. L. (1980). Stanford University, Stanford, California (private communication).

Compaan, A., and Chandra, S. (1979). *Opt. Lett.* **4**, 170.

Courtoy, C. P. (1957). *Can. J. Phys.* **35**, 608.

Dale, B. (1980). AERE Harwell, England (private communication).

Dasch, C. J., and Bechtel, J. H. (1981). *Opt. Lett.* **6**, 36.

Debethune, J. L. (1972). *Nuovo Cimento B* **12**, 101.

DeWitt, R. W., Harvey, A. B., and Tolles, W. M. (1976). *NRL Memo. Rep.* **NRL-MR-3269**.

Dicke, R. H. (1953). *Phys. Rev.* **89**, 472.

Druet, S., and Taran, J. P. E. (1979). "Chemical and Biological Applications of Lasers" (C. B. Moore, ed.), Vol. 4. Academic Press, New York.

Druet, S. A. J., Attal, B., Gustafson, J. K., and Taran, J. P. E. (1978). *Phys. Rev. A* **18**, 1529.

Druet, S., Taran, J. P. E., and Borde, C. (1979). *J. Phys.* (*Orsay, Fr.*) **40**, 819.

Duarte, F. J., and Piper, J. A. (1980). *Opt. Commun.* **35**,100.

Dunham, J. L. (1932). *Phys. Rev.* **41**, 713.

Durig, J. R., ed. (1977). "Vibrational Spectra and Structure," Vol. 6. Elsevier, Amsterdam.

Eckbreth, A. C. (1977). *J. Appl. Phys.* **48**, 4473.

Eckbreth, A. C. (1978). *Appl. Phys. Lett.* **32**, 421.

Eckbreth, A. C. (1979). *Appl. Opt.* **18**, 3215.

Eckbreth, A. C. (1980). *Combust. Flame* **39**, 133.

Eckbreth, A. C., and Hall, R. J. (1979). *Combust. Flame* **36**, 87.

Eckbreth, A. C., and Hall, R. J. (1981). *Combust. Sci. Technol.* **25**, 175.

Eckbreth, A. C., Bonczyk, P. A., and Verdieck, J. F. (1979a). *Prog. Energy Combust. Sci.* **5**, 253.

Eckbreth, A. C., Hall, R. J., and Shirley, J. A. (1979b). *AIAA Pap.* **79-0083**.
Eckbreth, A. C., Dobbs, G. M., Stufflebeam, J. H., and Tellex, P. A. (1983). *AIAA Pap.* **83-1294**.
Eesley, G. L. (1981). "Coherent Raman Spectroscopy." Pergamon, Oxford.
Farrow, R. L., and Rahn, L. A. (1982). *Phys. Rev. Lett.* **48**, 395.
Farrow, R. L., Mattern, P. L., and Rahn, L. A. (1980). *Sandia Lab.* [*Tech. Rep.*] **SAND80-8640**.
Farrow, R. L., Mitchell, R. E., Rahn, L. A., and Mattern, P. L. (1981). *AIAA Pap.* **81-0182**.
Fenner, W. R., Hyatt, H. A., Kellam, J. M., and Porto, S. P. S. (1973). *J. Opt. Soc. Am.* **63**, 73.
Ferrario, A., Garbi, M., and Malvicini, C. (1983). *Conf. Lasers Electro-Opt.* (CLEO '83). *Pap.* **WD2**.
Fiutak, J., and Van Kranendonk, J. (1963). *Can. J. Phys.* **41**, 21.
Flaud, J. M., Camy-Peyret, C., and Maillard, J. P. (1976). *Mol. Phys.* **32**, 499.
Flytzanis, C. (1975). "Quantum Electronics: A Treatise" (H. Rabin and C. Tang, eds.), Vol. 1A. Academic Press, New York.
Fujii, S., Gomi, M., and Eguchi, K. (1983). *J. Fluids Eng.* **105**, 128.
Fujimoto, J. G., and Yee, T. K. (1983). *IEEE J. Quantum Electron.* **QE-19**, 861.
Galatry, L. (1961). *Phys. Rev.* **122**, 1218.
Gilson, T. R., Beattie, I. R., Black, J. D., Greenhalgh, D. A., and Jenny, S. N. (1980). *J. Raman Spectrosc.* **9**, 361.
Glauber, R. J. (1965). "Quantum Optics and Electronics" (C. De Witt *et al.*, eds.). Gordon & Breach, New York.
Gordon, R. G. (1966a). *J. Chem. Phys.* **44**, 1830.
Gordon, R. G. (1966b). *J. Chem. Phys.* **45**, 1649.
Goss, L. P., Fleming, J. W., and Harvey, A. B. (1980a). *Opt. Lett.* **5**, 380.
Goss, L. P., Switzer, G. L. and Schreiber, P. W. (1980b). *AIAA Pap.* **80-1543**.
Greenhalgh, D. A. (1983a). *Appl. Opt.* **22**, 1128.
Greenhalgh, D. A. (1983b). *J. Raman Spectrosc.* **14**, 150.
Greenhalgh, D. A., Porter, F. M., and England, W. A. (1983a). *Combust. Flame* **49**, 171.
Greenhalgh, D. A., Hall, R. J., Porter, F. M., and England, W. A. (1983b). *J. Raman Spectrosc.* (in press).
Gross, K. P., Guthals, D. M., and Nibler, J. W. (1979). *J. Chem. Phys.* **70**, 4673.
Grynberg, G. (1981). *J. Phys. B* **14**, 2089.
Guthals, D. M., Gross, K. P., and Nibler, J. W. (1979). *J. Chem. Phys.* **70**, 2393.
Hall, R. J. (1979). *Combust. Flame* **35**, 47.
Hall, R. J. (1980). *Appl. Spectrosc.* **34**, 700.
Hall, R. J. (1983). *Opt. Eng.* **22**, 322.
Hall, R. J., and Eckbreth, A. C. (1978). *Proc. Soc. Photo-Opt. Instrum. Eng.* **158**, 59.
Hall, R. J., and Eckbreth, A. C. (1981). *Opt. Eng.* **20**, 494.
Hall, R. J., and Greenhalgh, D. A. (1982). *Opt. Commun.* **40**, 417.
Hall, R. J., and Shirley, J. A. (1983). *Appl. Spectrosc.* **37**, 196.
Hall, R. J., Shirley, J. A., and Eckbreth, A. C. (1979). *Opt. Lett.* **4**, 87.
Hall, R. J., Verdieck, J. F., and Eckbreth, A. C. (1980). *Opt. Commun.* **35**, 69.
Hansch, T. W. (1972). *Appl. Opt.* **11**, 895.
Harvey, A. B., ed. (1981). "Chemical Applications of Nonlinear Raman Spectroscopy." Academic Press, New York.
Henesian, M. A., and Byer, R. L. (1978). *J. Opt. Soc. Am.* **68**, 648.
Henesian, M. A., Kulevskii, L., Byer, R. L., and Herbst, R. L. (1976). *Opt. Commun.* **18**, 255.
Henesian, M. A., Duncan, M. D., Byer, R. L., and May, A. D. (1977). *Opt. Lett.* **1**, 149.
Jammu, K. S., St. John, G. S., and Welsh, H. L. (1966). *Can. J. Phys.* **44**, 797.

Kajiyama, K., Sajiki, K., Kataoka, H., Maeda, S., and Hirose, C. (1982). *SAE Tech. Pap.* **821036.**

Kamga, F. M., and Sceats, M. G. (1980). *Opt. Lett.* **5,** 126.

Klick, D., Marko, K. A., and Rimai, L. (1981). *Appl. Opt.* **20,** 1178.

Kondilenko, I. I., Korotov, P. A., Klimenko, V. A., and Golubeva, N. G. (1980). *Opt. Spectrosc.* **48,** 411.

Lapp, M., and Hartley, D. L. (1976). *Combust. Sci. Technol.* **13,** 199.

Lapp, M., and Penney, C. M. (1974). "Laser Raman Gas Diagnostics." Plenum, New York.

Laubereau, A. (1979). *Philos. Trans. R. Soc. London, Ser. A* **293,** 441.

Laufer, G., and Miles, R. B. (1979). *Opt. Commun.* **28,** 250.

Lederman, S. (1977). *Prog. Energy Combust. Sci.* **3,** 1.

Leonard, D. A. (1974). "Field Tests of a Laser Raman Measurement System for Aircraft Engine Exhaust Emissions," *Air Force Aero Propul. Lab. Rep.* **AFAPL-TR-74-100.**

Levenson, M. D. (1982). "Introduction to Nonlinear Laser Spectroscopy." Academic Press, New York.

Littman, M. G. (1978). *Opt. Lett.* **3,** 138.

Lotem, H., Lynch, R. T., Jr., and Bloembergen, N. (1976). *Phys. Rev. A* **14,** 1748.

Lynch, R. T. (1977). Ph.D. Thesis, Harvard University, Cambridge, Massachusetts (unpublished).

Lynch, R. T., Jr., Kramer, S. D., Lotem, H., and Bloembergen, N. (1976). *Opt. Commun.* **16,** 372.

McIlwain, M. E., and Harris, L. E. (1980). *Proc. JANNAF Conf. Combust., 17th,* (CPIA Publ. 329) **II,** 379.

Maker, P. D., and Terhune, R. W. (1965). *Phys. Rev.* **137,** A801.

Mandin, J. Y., Camy-Peyret, C., Fland, J. M., and Guelachvili, G. (1982). *Can. J. Phys.* **60,** 94.

Marko, K. A., and Rimai, L. (1979). *Opt. Lett.* **4,** 211.

May, A. D., Stryland, J. C., and Varghese, G. (1970). *Can. J. Phys.* **48,** 2331.

Mizrahi, V., Prior, Y., and Mukamel, S. (1983). *Opt. Lett.* **8,** 145.

Molectron Corporation (1976). *Appl. Note* **113.** Sunnyvale, California.

Moore, C. B., ed. (1979). "Chemical and Biological Applications of Lasers," Vol. 4. Academic Press, New York.

Moya, F. S. (1976). "Application de la Diffuson Raman Anti-Stokes Coherente aux Mesures de Concentrations Gaseuses dans les Ecoulments," Ph.D. Thesis, University of Paris (Orsay) (available as ONERA Tech. Rep. No. 1975-13).

Moya, F. S., Druet, S. A. J., and Taran, J. P. E. (1975). *Opt. Commun.* **13,** 169.

Moya, F. S., Druet, S. A. J., Pealat, M., and Taran, J. P. E. (1977). *Prog. Astronaut. Aeronaut.* **53,** 549. New York.

Murphy, D. V., and Chang, R. K. (1981). *Opt. Lett.* **6,** 233.

Murphy, D. V., Long, M. B., Chang, R. K., and Eckbreth, A. C. (1979). *Opt. Lett.* **4,** 167.

Nibler, J. W., and Knighten, G. V. (1979). *Top. Curr. Phys.* **3.**

Oudar, J. L., and Shen, Y. R. (1980). *Phys. Rev. A* **22,** 1141.

Oudar, J. L., Smith, R. W., and Shen, Y. R. (1979). *Appl. Phys. Lett.* **34,** 758.

Owyoung, A. (1971). *Air Force Off. Sci. Res. [Tech. Rep.] AFOSR-TR (U.S.)* **AFOSR-TR-71-3132.**

Owyoung, A. (1978). *IEEE J. Quantum Electron.* **QE-14,** 192.

Owyoung, A., and Esherick, P. (1980). *C.R.—Conf. Int. Spectrosc. Raman, 7th, 1980,* p. 656.

Pealat, M., Bailly, R., and Taran, J. P. E. (1977). *Opt. Commun.* **22,** 91.

Pealat, M., Taran, J. P. E., and Moya, F. (1980). *Opt. Laser Technol.* **12,** 21.

Penney, C. M., St. Peters, R. L., and Lapp, M. (1974). *J. Opt. Soc. Am.* **64,** 712.

Prior, Y. (1980). *Appl. Opt.* **19**, 1741.

Prior, Y., Bogdan, A. R., Dagenais, M., and Bloembergen, N. (1981). *Phys. Rev. Lett.* **46**, 111.

Rado, W. G. (1967). *Appl. Phys. Lett.* **11**, 123.

Rahn, L. A. (1979). Private communication to A. C. Eckbreth.

Rahn, L. A., Zych, L. J., and Mattern, P. L. (1979). *Opt. Commun.* **39**, 249.

Rahn, L. A., Owyoung, A., Coltrin, M. E., and Koszykowski, M. L. (1980a). *C.R.—Conf. Int. Spectrosc. Raman, 7th, 1980,* p. 694.

Rahn, L. A., Farrow, R. L., Koszykowski, M. L., and Mattern, P. L. (1980b). *Phys. Rev. Lett.* **45**, 620.

Rahn, L. A., Mattern, P. L., and Farrow, R. L. (1981). *Symp. (Int.) Combust. [Proc.], 18th, 1980,* p. 1533.

Rahn, L. A., Johnston, S. C., Farrow, R. L., and Mattern, P. L. (1982). *Temperature* **5**, 609.

Rambach, G. D., Dibble, R. W., and Hollenback, R. E. (1979). *West. Sect. Combust. Inst.* *[Pap.]* **WSS/CI-79-51**.

Regnier, P. R., and Taran, J. P. E. (1973). *Appl. Phys. Lett.* **23**, 240.

Regnier, P. R., Moya, F., and Taran, J. P. E. (1973). *AIAA Pap.* **73-702**.

Regnier, P. R., Moya, F., and Taran, J. P. E. (1974). *AIAA J.* **12**, 826.

Robert, D. (1982). *Proc. Int. Conf. Raman Spectrosc., 8th,* p. 269.

Robert, D., and Bonamy, J. (1979). *J. Phys. (Orsay, Fr.)* **10**, 923.

Roh, W. B., and Schreiber, P. W. (1978). *Appl. Opt.* **17**, 1418.

Roh, W. B., Schreiber, P. W., and Taran, J. P. E. (1976). *Appl. Phys. Lett.* **29**, 174.

Roland, C. M., and Steele, W. A. (1980a). *J. Chem. Phys.* **73**, 5919.

Roland, C. M., and Steele, W. A. (1980b). *J. Chem. Phys.* **73**, 5924.

Roquemore, W. M., and Yaney, P. P. (1979). *Proc., NBS Mater. Res. Symp. Charact. High Temp. Vapors Gases, 10th, 1978,* p. 973.

Rosasco, G. S., Lempert, W., Hurst, W. S., and Fine, A. (1983). *Chem. Phys. Lett.* **97**, 435.

Rothman, L. S., and Benedict, W. S. (1978). *Appl. Opt.* **17**, 2605.

St. Peters, R. L. (1979). *Opt. Lett.* **4**, 401.

Schlossberg, H. (1976). *J. Appl. Phys.* **47**, 2044.

Setchell, R. E. (1978). *Sandia Lab. [Tech. Rep.]* **SAND78-1220**.

Shirley, J. A. (1983). Private communication.

Shirley, J. A., Eckbreth, A. C., and Hall, R. J. (1979a). *Proc. 16th JANNAF Combust. Meet.* CPIA Publ. No. 309, p. 487.

Shirley, J. A., Eckbreth, A. C., and Hall, R. J. (1979b). "Investigation of the Feasibility of CARS Measurements in Scramjet Combustion," *Tech. Rep.* R79-954390-8 **NASI-15491**.

Shirley, J. A., Hall, R. J., Verdieck, J. F., and Eckbreth, A. C. (1980a). *AIAA Pap.* **80-1542**.

Shirley, J. A., Hall, R. J., and Eckbreth, A. C. (1980b). *Opt. Lett.* **5**, 380.

Shirley, J. A., Hall, R. J., and Eckbreth, A. C. (1980c). *Proc. Int. Conf. Lasers, 1980.*

Smith, D. C., and Meyerand, R. G., Jr. (1974). In "Principles of Laser Plasmas" (G. Bekefi, ed.), Chap. 11. Wiley (Interscience), New York.

Smith, J. R. (1979). *Sandia Lab. [Tech. Rep.]* **SAND79-8693**.

Song, J. J., Eesley, G. L., and Levenson, M. D. (1976). *Appl. Phys. Lett.* **29**, 567.

Srivastava, R. P., and Zaidi, H. R. (1977). *Can. J. Phys.* **55**, 533.

Stenhouse, I. A., Williams, D. R., Cole, J. B., and Swords, M. D. (1979). *Appl. Opt.* **18**, 3819.

Stephenson, D. A. (1974). *J. Quant. Spectrosc. Radiat. Transfer* **14**, 1291.

Stufflebeam, J. H., Verdieck, J. F., and Hall, R. J. (1983). *AIAA Pap.* **83-1477**.

Switzer, G. L., Goss, L. P., Roquemore, W. M., Bradley, R. P., and Schreiber, P. W. (1980). *AIAA Pap.* **80-0353**.

Taran, J. P. E. (1976). *Proc. Int. Conf. Raman Spectrosc., 5th, 1976* p. 695.

Taran, J. P. E., and Pealat, M. (1982). *Temperature* **5,** 575.

Taylor, D. J. (1983). *Los Alamos Sci. Lab. Rep.* **LA-UR-83-1840.**

Teets, R. E., and Bechtel, J. H. (1981). *Opt. Lett.* **6,** 458.

Tsao, C. J., and Curnutte, B. (1962). *J. Quant. Spectrosc. Radiat. Transfer* **2,** 41.

Verdieck, J. F., Hall, R. J., and Eckbreth, A. C. (1983). *AIAA Pap.* **83-1477.**

Weber, A. (1973). "The Raman Effect" (A. Anderson, ed.), Vol. 2, Dekker, New York.

Williams, W. D., Power, H. M., McGuire, R. L., Jones, J. H., Price, L. L., and Lewis, J. W. L. (1977). *AIAA Pap.* **77-211.**

Wilson-Gordon, A. D., Klimovsky-Barid, R., and Friedmann, H. (1982). *Phys. Rev. A* **25,** 1580.

Wittke, J. P., and Dicke, R. H. (1956). *Phys. Rev.* **103,** 620.

Yacoby, Y., Fitzgibbon, R., and Lax, B. (1980). *J. Appl. Phys.* **51,** 3072.

Yaney, P. P. (1979). "Combustion Diagnostics Using Laser Spontaneous-Raman Scattering." *Air Force Aero Propul. Lab. Rep.* AFAPL-TR-79-2035.

Yee, S. Y., Gustafson, T. K., Druet, S. A. J., and Taran, J. P. E. (1977). *Opt. Commun.* **23,** 1.

Yuratich, M. A. (1979). *Mol. Phys.* **38,** 625.

Yuratich, M. A., and Hanna, D. C. (1976). *Opt. Commun.* **18,** 134.

Zinn, B. T., ed. (1977). "Experimental Diagnostics in Gas Phase Combustion Systems." AIAA, New York.

Zinth, W., Laubereau, A., and Kaiser, W. (1978). *Opt. Commun.* **26,** 457.

INDEX